Concepts in Photobiology

Photosynthesis and Photomorphogenesis

Concepts in Photobiology

Photosynthesis and Photomorphogenesis

Edited by

G.S. SINGHAL
*School of Life Sciences,
Jawaharlal Nehru University, New Delhi, India*

G. RENGER
Technische Universität, Berlin, Germany

S.K. SOPORY
*International Centre for Genetic Engineering and Biotechnology,
New Delhi, India*

K-D. IRRGANG
Technische Universität, Berlin, Germany

GOVINDJEE
*University of Illinois at Urbana Champaign,
Urbana, IL, USA*

Springer-Science+Business Media, B.V.

A C.I.P. catalogue record for the book is available from the Library of Congress

ISBN 978-94-010-6026-4 ISBN 978-94-011-4832-0 (eBook)
DOI 10.1007/978-94-011-4832-0

Originally published by Kluwer Academic Publishers in 1999
Softcover reprint of the hardcover 1st edition 1999

Foreword

Light is an important environment factor that controls plant growth and development. The undoubted satisfaction of studying the various basic and applied aspects of photobiology is that such work can be immediately practical in the field and also immensely rewarding as an area of fundamental research. The advancement in photobiology research in the past several decades has been spectacular. Photosynthesis and photomorphogenesis are currently the two active areas of plant science research.

Photosynthesis by the whole plant is to be seen as a series of balanced but dynamic interactions between individual molecular and physiological mechanisms. On the other hand, the discovery and isolation of phytochrome by Harry A. Borthwick and coworkers and the demonstration of its importance as a pigment that controls photomorphogenic responses is one of the most brilliant and important of all the plant physiological accomplishments. Nonetheless, as a consequence of several exciting discoveries in the photosynthesis and photomorphogenesis during the last two decades, a wealth of genetic and biochemical information about many facets of photobiological research has been appearing. Some of these findings are of great fundamental value and bear promise of contributing significantly towards the improvement of plant productivity. Also, detailed analysis and manipulation of genes encoding photosynthetic components have led to major advances in the area of carbon fixation in higher plants. Fortunately, modern plant photobiology research has been strengthened by recent advances in structural biology along with plant molecular genetics.

The book entitled *Concepts in Photobiology: Photosynthesis and Photomorphogenesis* is unique in bringing together two major basic processes of plant life. The editors are to be congratulated for bringing together the major researchers in the areas of photosynthesis and photomorphogenesis to present recent advances and to review progress and perspectives. This book, with an excellent balance of contributions, both fundamental and applied, presents our current understanding on photobiology research. Topics covered range from such basic concepts as primary photosynthetic processes to more advanced aspects of molecular mechanisms of photomorphogenesis and photoregulation of gene expression in higher plants. The book begins with a summary of basic principles of photobiology to lay the foundations for other exciting major sections that cover different areas of current photobiology research.

As with other fields, the areas of photosynthesis and photomorphogenesis have been growing so rapidly that all researchers involved in these fields are faced with the problem of updating their knowledge. This book enables them to get updated with ease. It gives us the insight with many of the recent findings related to photosynthesis and photomorphogenesis in a single inclusive volume. Various chapters in this book bear testimony to the importance of molecular genetics in the study of photobiology. This comprehensive source of information will certainly attract and stimulate further exploration by the experts involved in this area of research.

Pondicherry University
Pondicherry, India

ARUMUGHAM GNANAM

Preface

Photobiology is an important area of biological research since a very large number of living processes are either dependent upon or governed by solar radiation. Among the various subjects, photosynthesis is one of the most important, and thus popular, topics in both molecular and organismic biology, and it has a considerable impact throughout the world. Almost all life on earth depends upon this process as the source of food, fuel and oxygen. In addition, light is equally essential for growth of plants and recent research on photomorphogenesis has taken off into exciting new developments with the application of newer molecular biological approaches. The present book brings together, and integrates, various aspects of photosynthesis, biology of pigments, light regulation of chloroplast development, nuclear and chloroplast gene expression, light signal transduction, other photomorphogenetic processes and some photoecological aspects under one cover. It was the goal of the editors to present a comprehensive reference text which includes a wide range of "hot topics" that are currently under investigation in the field of photobiology of cyanobacteria, algae and plants. In order to achieve this goal, international authorities, in their specific fields, were selected as authors. The book is aimed primarily at graduate and post-graduate students to be used as a text book. It is, however, also intended to be a resource book for beginning researchers in plant photobiology. Several introductory chapters are designed as suitable reading for undergraduate courses in integrative and molecular biology, biochemistry and biophysics. The book begins with an introductory chapter by the editors; the rest of the chapters and their contents are specifically introduced there. The book is subdivided into three general subject areas: (1) Photosynthesis; (2) Stress response; and (3) Gene regulation and morphogenesis. These topics are divided into sections, each containing three to six chapters. We believe that this organisation enhances the value of the book for a broad audience. Special attention has been given to include introductory chapters with reference lists which provide convenient inroads into the wealth of original papers published in the literature not only for the benefit of the "freshman" students who make their first acquaintance with the fascinating field of different facets of photosynthesis and photomorphogenesis, but also for the participants (both the lecturers and the students) of advanced course on these topics who also need the recent developments and future perspectives. It is the sincere hope of the editors that this book will satisfy these ambitious goals.

We place on record our appreciation of the invaluable help given to us by many colleagues who generously gave of their time and expertise by making suggestions by reading and commenting on one or more chapters. In particular, we thank Drs. S.C. Maheshwari, N. Maheshwari, Sunil Mukherjee, B.C. Tripathy, P. Mohanty and A.K. Tyagi for their help. We are thankful to Dr. Meenakshi Munshi for her help in proof reading some chapters. We are especially thankful to Dr. Sanjay Chauhan and Ms Angela Menke who helped the editors in various ways—proof reading, organisation, checking out the references etc. We take this opportunity to thank all the authors for their timely contribution, their patience in waiting for the publication of this book and, above all, in accepting to modify their chapters as suggested by reviewers and editors.

EDITORS

Noun Shavit (1930–1997)[1]

Dr. Noun Shavit, Professor of Biochemistry at the Ben-Gurion University of the Negev, Beer-Sheva, Israel, one of the leading scientists in the field of bioenergetics of photosynthesis and one of the authors of the chapter on photophosphorylation for this volume, passed away on June 19, 1997. Noun, born in Poland on November 15, 1930 was raised in Argentina and emigrated to Israel as a young man. He graduated in Chemistry at the Hebrew University in Jerusalem and pursued his studies in Bioenergetics from 1961–1964 at the Weizmann Institute of Science in Rehovot as the first Ph.D. student of the late Professor Mordechai Avron. He went on to postdoctoral studies with Professors Paul Boyer (University of California Los Angeles) and Anthony San Pietro (Charles F. Kettering Foundation). From 1964–1968 he worked as a Research Associate in the Weizmann Institute.

Noun started his independent academic career at the Negev Research Institute in Beer-Sheva in 1968. He immediately became active in the pioneering effort to establish the university-level program in biology in Beer-Sheva, which became the nucleus of the future Ben-Gurion University of the Negev (BGU). In addition to his research, Noun dedicated a significant part of his time and energy to the University and contributed greatly to its

[1]Based on an obituary by Aflalo C, Baum H, Chipman DM, McCarty RE and Strotmann H (1997) Noun Shavit (1930–1967). Photosynth. Res. 54: 165–167.

seccessful development. He served as Chairman of the Biology Department, Dean of the Faculty of Natural Sciences, and Director of the Doris and Bertie Black Center for Bioenergetics at BGU, as well as member of many academic steering committees.

Above all Noun will be remembered as a leading biochemist whose scientific contributions were focused on biological energy transduction. His initial studies on photophosphorylation were conducted with lettuce chloroplast thylakoids and photosynthetic bacterial chromatophores, in which the coupling between electron transport and phosphorylation was investigated using partial reactions and specific inhibitors, in the search for energized states or intermediates. He dedicated an important part of his work to the characterization of the activities of the chloroplast ATP synthase and its modulation. His approach involved a broad spectrum of ideas and methodologies, always at the frontiers of knowledge and technology. He initiated fruitful international collaborations and was Visiting Professor in several institutions in Germany, France, Canada and the USA.

With Paul Boyer, Noun studied oxygen and phosphate exchange reactions related with phosphorylation. In Tony San Pietro's laboratory, he discovered the uncoupling potency of the antibiotic nigericin and studied ion translocation in thylakoid vesicles. Later, in his own labratory in Beer Sheva, Noun and his students investigated the mechanism of photophosphorylation and the mode of coupling of ATP formation to the electrochemical proton gradient. They got also involved in the investigation of the ATP synthase, and their contribution was significant to the critical elucidation of its structure and mechanism, a subject which was chosen for the last Nobel Prize in Chemistry (awarded to P. Boyer and J. Walker in 1997). Noun studied the effects of nucleotides and nucleotide analogues on photophosphorylation and coupled electron transport and he established the relation between tight nucleotide binding and regulation of the chloroplast ATP synthase. 1975 he started to cooperate with Heinrich Strotmann; this was the beginning of a long collaboration and friendship. The assessment of the number of nucleotide binding sites, their subunit location and function became the focus of Noun's interest. Using photoreactive nucleotide analogues, he studied the distribution of the label among the α and β subunits. In collaboration with M. Yoshida, Noun extended his studies to the TF_1 of the thermophilic becterium PS3 and used the unique properties of this enzyme to investigate its assembly from isolated subunits, the effect of bound nucleotides and the development of catalytic activity.

Noun Shavit realized early the extraordinary potential of molecular genetics for the understanding of bioenergetic principles. Although many basic molecular studies of the ATP synthase can be done with well characterized standard organisms like *E. coli,* the particular properties of the photosynthetic ATP synthases, e.g. their specific regulatory devices, require work on plants. The object of choice was the green alga *Chlamydomonas reinhardtii,* a model organism for plant photosynthesis. Noun and his associates, in collaboration with H. Strotmann, started a comprehensive program of mutagenesis of CF_0CF_1. Unfortunately, he could only harvest the early fruits of this research.

On the personal level, one cannot refrain from recalling his exuberance, vitality and effervescent sense of humor, his profound esteem and the ease in his direct human relations. Students and research associates found in him a good listener and a source for helpful advice on matters personal or scientific. He received from them in return the same high standards he demanded of himself, as well as the same scientific alertness and joy of creation which characterized him until his last moments. As an open scientist, Noun was ever able and willing to coummunicate his thoughts about and enthusiasm for bioenergetics.

Noun was a loving, attentive family man and received in return the love of his wife Aliza, his two sons and his grandchildren. He was supported bravely by them in his last difficult weeks. Finally he was overwhelmed by cancer after a long, multiple phase struggle, that he conducted till the end with honor, courage and realism.

The bioenergetics community has lost a prominent scientist, an extraordinary personality and good friend.

Selected publications, arranged chronologically

Shavit N and Avron M (1963) The effect of ultraviolet light on photophosphorylation and the Hill reaction. Biochim Biophys Acta 66: 187–195

Avron M and Shavit N (1965) Inhibitors and uncouplers of photophosphorylation. Biochim Biophys Acta 109: 317–331

Shavit N and Boyer PD (1966) Source of oxygen in adenosine triphosphate formed by illumination or by acid-base transition of chloroplast. J Biol Chem 241: 5738–5740

Shavit N and Avron M (1967) The relation of electron transport and photophosphorylation to conformational changes in chloroplasts. Biochim Biophys Acta 131: 516–525

Shavit N, Skye GE and Boyer PD (1967) Occurrence and possible mechanism of ^{32}P and ^{18}O exchange reactions of photophosphorylation. J Biol Chem. 242: 5125–5130

Shavit N, Dilley RA and San Pietro A (1968) Ion translocation in isolated chloroplasts. Uncoupling of photophosphorylation and translocation of K^+ and H^+ ions induced by Nigericin. Biochemistry 7: 2356–2363

Dilley RA and Shavit N (1968) On the relationship of H^+ transport to photophosphorylation in spinach chloroplasts. Biochim Biophys Acta 162: 86–96

Karlish SJ, Shavit N and Avron M (1969) On the mechanism of uncoupling in chloroplasts by ion-permeability inducing agents. Eur J Biochem 9: 291–298

Shoshan V and Shavit N (1973) On the reconstitution of photophosphorylation in chloroplast membranes. Eur J Biochem 37: 355–360

Schmid R, Shavit N and Junge W (1976) The coupling factor of photophosphorylation and the electric properties of the thylakoid membrane. Biochim Biophys Acta 430: 145–53

Shavit N (1977) Bound nucleotides and conformational changes in photophosphorylation. Encyclopedia of Plant Physiol. New Series, Vol. 5: 350–357

Shoshan V and Shavit N (1979) ATP synthesis and hydrolysis in chloroplast membranes. Eur J Biochem 94: 87–92

Shavit N and Strotmann H (1980) Nucleotides tightly bound to chloroplast membranes. Methods Enzymol 69: 321–326

Shavit N (1980) Energy transduction in chloroplasts: Structure and function of the ATPase complex. Annu Rev Biochem. 49: 111–138

Aflalo C and Shavit N (1982) Source of rapidly labeled ATP tightly bound to non-catalytic sites on the chloroplast ATP synthetase. Eur J Biochem 126: 61–68

Tiefert MA and Shavit N (1983) Evaluation by steady-state enzyme kinetics of the role of tightly bound nucleotides during photophosphorylation. J Bioenerg Biomembr 15: 257–276

Aflalo C and Shavit N (1984) A new approach to the mechanism of phosphorylation: modulation of ATP synthetase activity by limited diffusibility of nucleotides near the enzyme. Cur Top Cell Reg 24: 435–445

Bar-Zvi D and Shavit N (1984) Photoaffinity labeling of the soluble chloroplast ATPase with 3'-O-(4-benzoyl)benzoyl ADP. Biochim Biophys Acta 765: 340–346

Bar-Zvi D, Yoshida M and Shavit N (1985) Characterization of nucleotide-binding sites on the chloroplast coupling factor 1 using two photolabile analogs. Biochim Biophys Acta 807: 293–299

Abbott MS, Shavit N, Selman-Reimer S and Selman BR (1986) Photoaffinity labeling of the TF_1-ATPase from the thermophilic bacterium PS3 with 3'-O-(4-benzoyl)benzoyl ADP. FEBS Lett 209: 157–161

Bar-Zvi D, Bar I, Yoshida M and Shavit N (1992) Covalent binding of 3'-O-(4-benzoyl)benzoyl ATP to the isolated α and β subunits and the $\alpha_3\beta_3$ core complex of TF_1. J Biol Chem 267: 11029–11033

Leu S, Schlesinger J, Michaels A and Shavit N (1992) Complete DNA sequence of the *Chlamydomonas reinhardtii atpA* gene. Plant Mol Biol 18: 613–616

Fiedler HR, Schlesinger J, Strotmann H, Shavit N and Leu S (1997) Characterization of *atpA* and *atpB* deletion mutants produced in *C. reinhardtii* cw15: electron transport and photophosphorylation activities of isolated thylakoids. Biochim Biophys Acta 1319: 109–118

Hu D, Fiedler HR, Golan T, Edelmann M, Strotmann H, Shavit N and Leu S (1997) Tentoxin inhibition of ATP synthesis and hydrolysis in *C. reinhardtii* ATP Synthase mutated in codon 83. J Biol Chem 272: 5457–5463

Contents

Section III: Photophosphorylation and CO_2 Metabolism

PART B: STRESS RESPONSE

Section IV: Light Stress

Section V: Other Stress Factors

PART C: GENE REGULATION AND PHOTOMORPHOGENESIS

Part C

Gene Regulation and Photomorphogenesis

Section VI: Chloroplast Molecular Biology and Regulation

Concepts in Photobiology: Photosynthesis and Photomorphogenesis
G.S. Singhal, G. Renger, S.K. Sopory, K-D. Irrgang and Govindjee (Eds)

23. Molecular Biology of Chloroplast Genome

Narendra Tuteja[1] and Krishna K. Tewari[1,2]

[1]International Centre for Genetic Engineering and Biotechnology, Aruna Asaf Ali Marg, New Delhi 11007, India

[2]Department of Molecular Biology and Biochemistry, University of California, Irvine, California, USA

Summary

Chloroplasts are highly polyploid, semi-autonomous organelles containing their own DNA (circular) molecules organized into discrete membrane associated nucleoids. Chloroplast genome structure and overall gene order in higher plants are highly conserved. It contains the genes for ribosomal and transfer RNA, and for a substantial number of proteins for Photosystem I and II, ribosomes and stromal enzyme complexes. Chloroplast genome encodes only some of its proteins and rest are encoded in nuclear genome and which are posttranslationally imported into the chloroplast. Three land plant chloroplast genomes have been completely sequenced: the dicotyledon tobacco (*Nicotiana tabacum*), the monocotyledon rice (*Oryza sativa*) and a liverwort (*Marchantia polymorpha*). In this chapter we have reviewed some of the highlights of research in chloroplast DNA of mainly higher plants which has ultimately led us to the present understanding of the nature of genetic inheritance of plastids. The studies described in this chapter have contributed significantly to the understanding of structure of chloroplast DNA, genes in the chloroplast DNA including light regulated genes and introns, replication and transcription of chloroplast DNA including promoters and transcription terminators.

1. Introduction

Higher plant cells contain a unique class of intracellular organelles, the plastids, which exist in a number of different forms with different functions, but the chloroplast was the first to be discovered and is the best studied of all plastids. Chloroplasts (abbreviated as c-, chl-, cl-, chloro-, cp- or ct-) contain complete photosynthetic machinery of the plants and are found to be involved in the synthesis of starch, lipids, amino acids and nucleotides. Chloroplasts are derived from small proplastids, which are the undifferentiated plastids present in meristematic cells. In nonphotosynthetic plant organs of higher plants the proplastids and chloroplast can also differentiate into special type of plastids which perform other functions, such as chromoplasts in many flowers and fruits or amyloplasts in roots and tubers.

Do chloroplasts contain their own DNA or genetic system like nucleus? This came in light after the discovery of a non-mendelian mutants of the chloroplast phenotype. The original observations of Baur (1909) and Correns

(1909) on variation studies with *Pelargonium zonales* and *Mirablis jalapa* plants respectively, showed that the inheritance of plastids could not be explained by Mendelian inheritance. These observations are now fully understood because of the presence of DNA and ribosome in chloroplast which were first demonstrated by Ris and Plaut (1962) and Lyttleton (1962) respectively. They found DNA-like filaments in low electron density areas within chloroplasts from *Chlamydomonas reinhardii.* Subsequently, DNA was also found in chloroplasts from other organisms and DNA has since been located in proplastids (Edelman et al., 1964), etioplasts (Herrmann and Kowallik, 1970), chromoplasts (Herrmann, 1972) and leucoplasts (Siu et al., 1976).

Chloroplast genome encodes only some of its proteins and rest are encoded in nuclear genome.There are two striking differences between chloroplast and nuclear genome. First, the information content of the chloroplast genome is modest, between 10^{-3} and 10^{-4} fold less than its nuclear counterpart. Second, the chloroplast genome is polyploid; it is present in 100 to 10,000 copies per cell (Bendich 1987); and it represents up to 15% of the total cellular DNA by mass. The chloroplast genomes of tobacco (155,844 bp; Shinozaki et al., 1986), liverwort (121,024 bp; Ohyama et al., 1986), rice (134,525 bp; Hiratsuka et al., 1989) and *Euglena gracilis* (143,170 bp; Hallick et al., 1993) have been completely sequenced. DNA sequence and phylogenetic analyses indicate that chloroplasts are close relatives of free living cyanobacteria and are derived from an endosymbiotic event.

Chloroplast genome is double-stranded, circular which ranges from 120 to 217 kb in size and encodes 120 to 140 genes (Rapp and Mullet, 1991). About half of these are involved in chloroplast protein synthesis and about 30 encode for subunits of the five photosynthetic complexes: Photosystem II (PS II), cytochrome b_6/f complex, Photosystem I (PS I), ATP synthase and ribulose bisphosphate carboxylase/oxygenase. A few chloroplast genes encode functions involved in respiration (Rochaix, 1992). The function of the remaining genes is still unknown. In many cases chloroplast genes encode subunits of large protein complexes such as ribosomes or the photosynthetic electron transport units which also contain proteins encoded by nuclear genes. Therefore, biosynthesis of many chloroplast protein complexes requires coordinated expression of genes located in the nucleus and the chloroplast. In this article we are presenting upto date information of research in chloroplast DNA of higher plants.

2. Structure of Chloroplast DNA

Chloroplast DNA of higher plants is 120 to 160 kb in size and contains about 130 genes. Genetically chloroplasts are not autonomous, unlike their prokaryotic ancestors, they also depend upon nucleus. Usually chloroplast genomes of land plants have a common organisation and gene content. The presence of DNA in chloroplast was first demonstrated by Ris and Plaut (1962) which served as the impetus to study the structure and its expression. They observed DNA-like filaments in low electron density area within chloroplast from *Chlamydomonas.* However, the convincing evidence of the presence of chloroplast DNA (ctDNA) in higher plants was provided by Tewari and Wildman in 1966 by isolating the ctDNA from leaves of tobacco (*Nicotiana tabacum*) and characterizing some of its properties as described (Tewari, 1987).

CtDNA has been isolated from number of organisms and extensive studies have been carried out in *Euglena gracilis, Chlamydomonas reinhardi* and in certain higher plants including pea, tobacco, maize, spinach, rice and *Arabidopsis*. The method for isolating ctDNA from pea, bean, spinach, lettuce, corn and oat plants has been described earlier (Kolodner and Tewari, 1972 and 1975a). Briefly, the leaves were homogenized in Buffer A (4 l/kg leaves) containing 0.3M mannitol/0.05M Tris/0.003 M ethylenediaminetetraacetic acid (EDTA)/0.001 M mercaptoethanol/0.1% bovine serum albumin, pH 8.0, with two 5 s bursts in a Waring blender at medium power. The homogenates are filtered through two layers of cheesecloth and four layers of Miracloth (Calbiochem) and are centrifuged for 10 min at 40 g at 4°C to remove nuclei. The supernatant is centrifuged at 1020 g for 15 min at 4°C and the resulting crude chloroplast pellet is suspended in 200 ml of Buffer A. $MgCl_2$ (0.01M) and DNAse I (50 μg/ml) are added and the suspension is incubated for 1 hr at 4°C. At the end of the incubation, 600 ml of Buffer B containing 0.3 M sucrose/0.05 M Tris/0.02 M EDTA, pH 8.0, is added and the suspension is centrifuged at 1500 g for 15 min at 4°C. The pellet is washed twice by suspending it in 600 ml of Buffer B and centrifuging for 15 min at 1500 g. The final pellet is suspended in 48 ml of Buffer C containing 0.05 M Tris/0.02 M EDTA, pH 8.0. Pronase (200 μg/ml) and 12 ml of Buffer C containing 10% sodium sarkosyl is added and the suspension is incubated for 1/2 hr at 37°C. The ctDNA is isolated from this lysate by extracting with an equal volume of phenol buffered with 0.1 M Tris, pH 12.0. The aqueous phase is reextracted with an equal volume of phenol. Two volumes of 95% alcohol are added to the aqueous phase, the mixture is kept in a freezer overnight, and the precipitate is collected by centrifugation and dissolved in 5 ml of Buffer C. The solution is incubated for 2 hr at 37°C with 50 μg/ml of RNAse and 50 units/ml of RNase T_1 followed with 200 μg/ml of pronase and a further incubation of 2 hr. At the end of incubation, two phenol extractions are carried out and the aqueous phase is dialyzed against a total volume of 10 l of SSC with five changes in 24 hr.

The method described above yields pure ctDNA from higher plants uncontaminated with nuclear and mitochondrial DNA. The success of the method depends upon the fact that higher plant chloroplasts can be obtained in relative intact condition, therefore, it is possible to remove contaminating nuclear DNA by treating with DNase. The contaminating mitochondria are quantitatively removed during washing with Buffer B. However, the ctDNA from *Euglena* and *Chlamydomonas* cannot be obtained by using DNase because the physical methods used to obtain a cell-free homogenate from these organisms invariably result in broken chloroplasts. The isolation of ctDNA from these organisms is, therefore, carried out by isolating the DNA from cell fractions enriched for chloroplasts by differential or gradient centrifugations (without the DNase treatment) and separating the ctDNA in CsCl density gradients.

The density profile of the DNA isolated from 12000 g pellet of a cell free homogenate from a pea plant (after removing the nuclear fraction by centrifuging at 100 g) without DNase treatment is shown in Fig. 1. There are three DNA bands in such a fraction. The major band at a denisty of 1.684 g/cm^{-3} represents the DNA of nuclei (nDNA). The shoulder of the nDNA at a density of about 1.687 g/cm^{-3} represents the ctDNA. The band at a density of 1.705 g/cm^{-3} arises from mitochondria. Since the buoyant density of ctDNA was found to be very close to that of nDNA, a conclusive evidence for the presence of organelle DNA was provided by denaturation and renaturation studies. On denaturation, the buoyant density of ctDNA increased

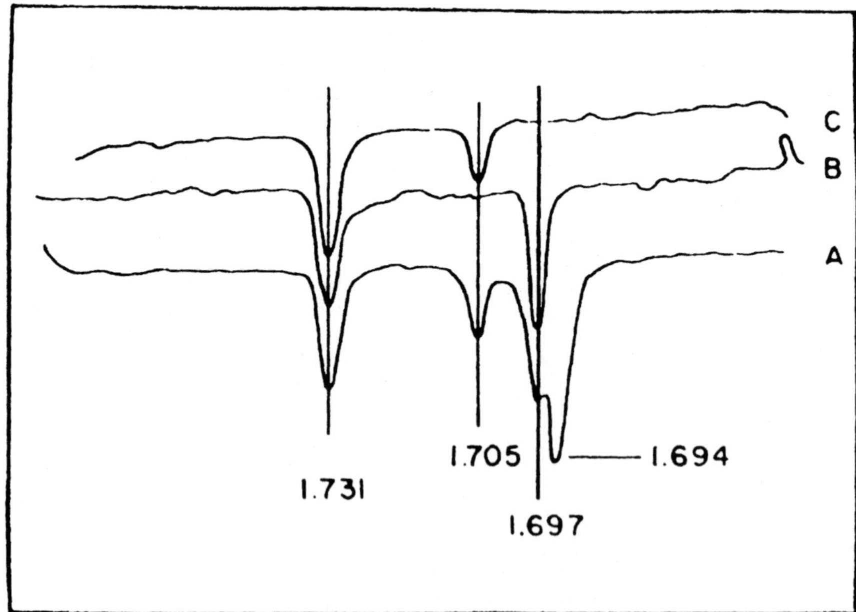

Fig. 1 Photoelectric scans of DNA centrifuged in CsCl of density 1.710 g/cm^{-3} for 18 hrs at 28° using a Beckman model E analytical ultracentrifuge equipped with photoelectric scanner. *Micrococcus lysodeikticus* DNA of density 1.731 g/cm^{-3} was used as a marker. (A) DNA isolated from 12,000 g pellet containing chloroplasts and mitochondria. The picture shows a significant amount of nuclear contamination (DNA of density 1.694 g/cm^{-3}). Although the difference in densities between nuclear and chloroplast DNA is only 0.003 g/cm^{-3}, the banding pattern clearly shows two components. (B) DNA from chloroplast fraction after DNAse treatment. (C) DNA from mitochondrial fraction after DNAse treatment.

by 0.013 g/cm^{-3}. Incubation of the denatured DNA at 60°C for 4 hr at a concentration of 20 μg/ml in 0.15 M NaCl/0.015 M sodium citrate, pH 7.0, resulted in about 90% renaturation as evidenced by its banding at a buoyant density of 1.699 g/cm^{-3} (Fig. 2). The nDNA, on the other hand, showed an increase of buoyant density by 0.013 g/cm^{-3} on denaturation but renatured to only 20%. These differences clearly identified the ctDNA to be quite different from the nDNA. In later studies Kolodner and Tewari (1975 b) found that ctDNA from different higher plants had practically the same buoyant densities and behaved in the same manner as the original observations with tobacco ctDNA.

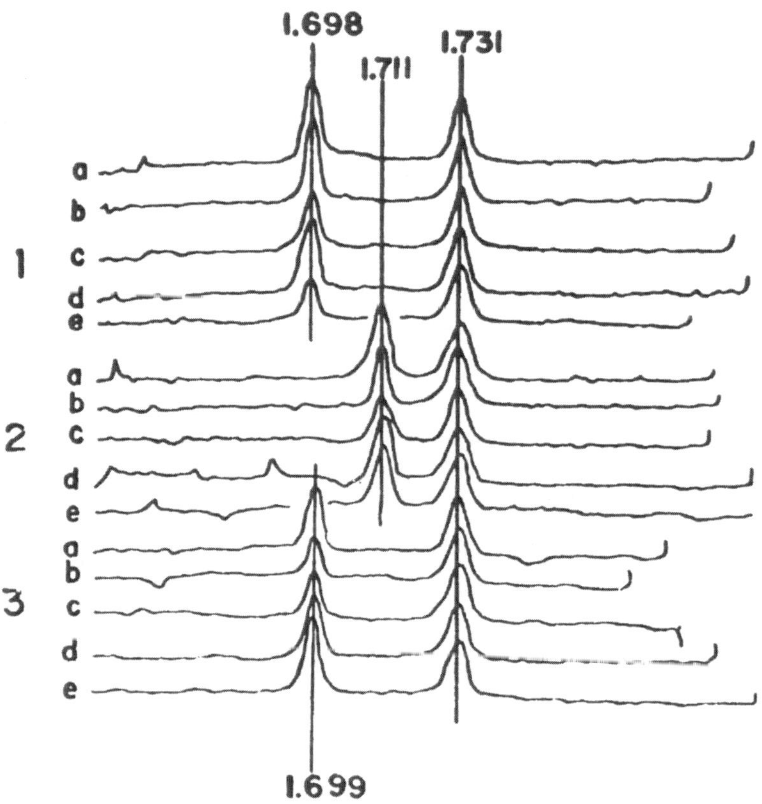

Fig. 2 Photoelectric scans of DNA centrifuged in CsCl of density 1.700 g/cm^{-3} for 18 hr at 18°C, 149,000 g, in a Beckman model E analytical ultracentrifuge. *Micrococcus leuteus* DNA of density 1.731 g/cm^{-3} has been used as a marker. (1) native ctDNA; (2) denatured ctDNAs; (3) renatured ctDNAs. (a) Pea; (b) Lettuce; (c) Spinach; (d) Bean; (e) Corn.

The structure of ctDNA from higher plants was extensively analyzed by Kolodner and Tewari (1972). The DNase treated chloroplasts from pea leaves was centrifuged in CsCl/Ethidium bromide (EtBr) density gradients. Two bands of DNA were clearly seen. When the lower band ctDNA was examined in the electron microscope, more than 90% of the molecules were present in supertwisted forms (Fig. 3). Most of the remaining molecules were relaxed circles. The ctDNA molecules that were found in the region of the gradient between the upper and lower bands mostly consisted of supertwisted circular DNA molecules. After nicking the supertwisted ctDNA molecules, there were three species of circular molecules, double length circular molecules and catenated dimers that appeared to consist of two topologically interlocked single length circular molecules. The ctDNA, similarly, from all other plants like spinach, lettuce, oats and corn were found to contain circular catenated dimers.

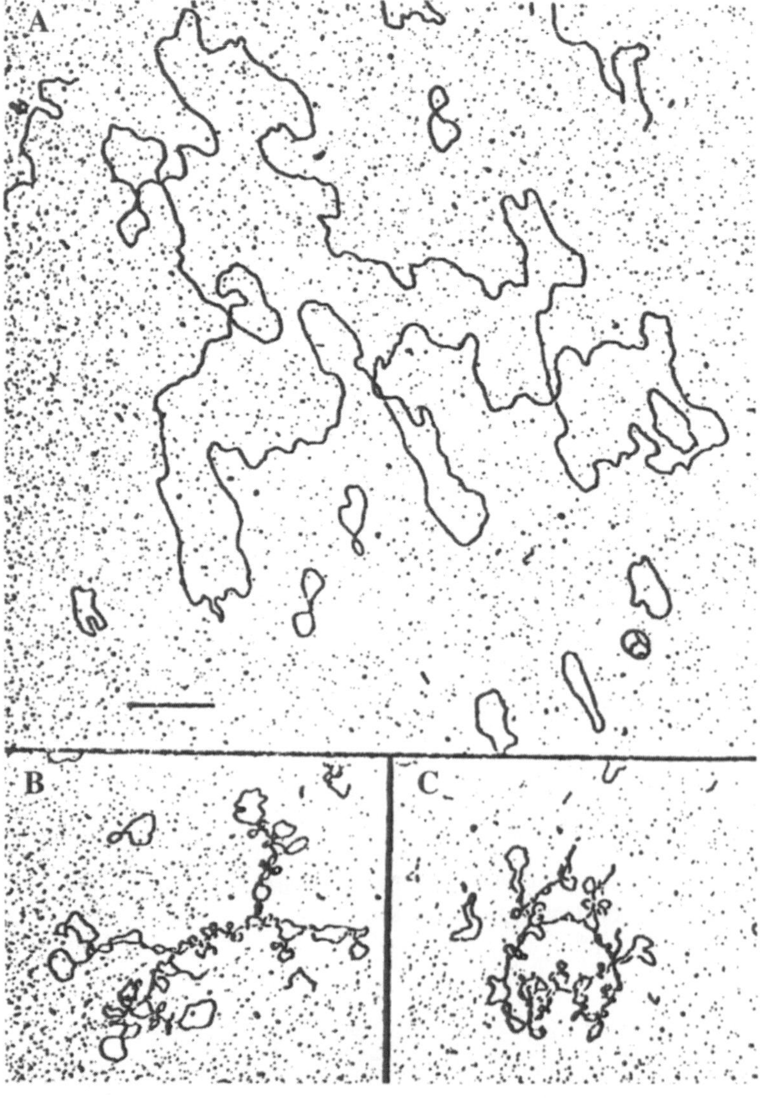

Fig. 3 Relaxed circular and supertwisted ctDNA molecules. (A) Relaxed circular lettuce ctDNA molecules. (B, C) supertwisted pea and oat ctDNA molecules, respectively.

All of the circular ctDNA molecules were found to contain the same genetic information. This was established by the analysis of molecular size of ctDNA by electron microscopy, renaturation kinetics and denaturation mapping of the individual circular ctDNA molecules (Tewari, 1979). One of the most intriguing observation was the presence of covalently linked ribonucleotides in the closed circular ctDNA (Kolodner et al, 1975). Pea and spinach ctDNAs were found to contain a maximum of 18 ± 2 ribonucleotides. Similarly, lettuce ctDNA was found to contain 12 ± 2 ribonucleotides. The ribonucleotides have been even mapped in pea ctDNA but the significance of the individually inserted ribonucleotides in supercoiled DNA is still not understood. The ribonucleotides could arise from non-stringent DNA polymerase, but in that case, they would not be located in specific positions in the ctDNA molecules. It is possible that ribonucleotides are remnants of DNA primers which are generally believed to initiate DNA replication. It should be pointed out that all systems in which RNA primers for DNA replication have been studied, the primer is completely excised in the mature DNA except under abnormal conditions. In the absence of any convincing arguments on the role of ribonucleotides in ctDNA, all one could say that they may yet be involved as recognition sites in the replication, transcription or recombination.

The structure of ctDNA in all higher plants has been found to be similar except for the fact that ctDNA from all higher plants contain a large inverted repeat except from those plants that belong to the family of *Leguminosae* (e.g. pea, broad bean, alfalfa etc.) (Kolodner and Tewari, 1975c) and pine ctDNA (White, 1990). When preparations of nicked circular lettuce, spinach and corn ctDNA molecules were denatured, DNA molecules that contained one duplex region of 22–24 kbp was observed (Fig. 4). The ctDNA from pea, on the other hand, did not show any intramolecular renaturation. Similar studies with circular dimers showed that these molecules were in head to tail configuration.

The studies on the structure of the ctDNA were augmented by the report of the first restriction maps of maize ctDNA presented by Bedbrook and Bogorad (1976). The closed circular molecules of 85×10^6 dalton from *Zea mays* were isolated, digested with the restriction endonucleases SalI, BamHI, and EcoRI and the resultant fragment sized by agarose gel electrophoresis. A map of maize ctDNA showing the relative location of all the SalI recognition sites and many of the BamHI and EcoRI sites was determined. The restriction map also confirmed the presence of an inverted repeat. The two copies of about 15% of the genome was found to be in an inverted orientation with respect to one another and were separated by a nonhomologous sequence representing approximately 10% of the genome length.

The physical characterization of the chloroplast genome and the successful completion of the restriction map paved the way for a complete sequencing of the chloroplast genome from *Marchantia polymorpha* by Ohyama et al., in 1988. Using a clone bank of ctDNA fragments, the circular genome of *Marchantia* was found to be of 121,024 bp including the two inverted repeats of 10,058 bp, a large single copy region of 81,095 bp and a single copy region of 19,813 bp.

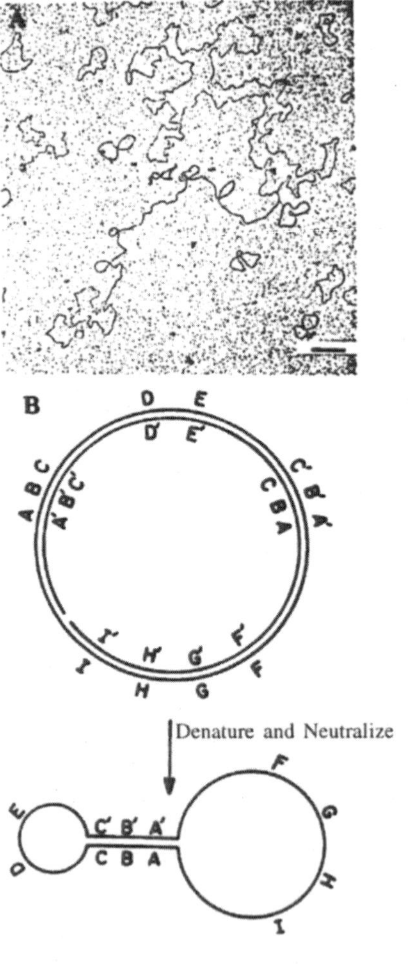

Fig. 4 The self-renatured circular ctDNA molecule. (A) Electron micrograph of a self-renatured spinach ctDNA molecule. The small circles are single-stranded and double-stranded ϕX DNA. The bar indicates 1.0 μm. (B) Illustration of the formation of a self-renatured molecule from a circular molecule containing an inverted repeat.

The inverted repeat (IR), found in most ctDNA, ranges from 6 to 76 kb in size (Palmer, 1985). These IR segments are separated by one large single-copy region and one small single-copy region. They are found to contain genes encoding the ct rRNAs, certain tRNAs and often one or more genes specifying proteins (Sugiura and Wakasugi, 1989). Within the IR, the tRNA operon is usually oriented with the 23S rRNA gene closer to the small single-copy region and the 16S rRNA gene closer to the large single-copy region. The two repeats are identical in sequence of an active copy correction system. About two-thirds of the variation in size among the land plant ct genomes is accounted for the expansion or contraction of the IR (Palmer, 1991). The smallest ct genomes among land plants are found in conifers

(Strauss, et al. 1988) and in six tripes of the legume family Fabaceae (Palmer 1991), which have lost the inverted repeat during evolution and thus contain only a single copy of each of the rRNA genes. The ctDNA of black pine (*Pinus thunbergii*) has short inverted repeat sequence which contains a tRNA gene and part of the 3′ portion of the *psbA* gene, but does not contain the rRNA genes. On the other hand, species with largest ctDNA often have expanded IR (e.g. *Pelargonium hortorum* has a 76 kb IR encompassing nearly half of the 216 kb ctDNA, in which many genes normally in the single-copy region have been duplicated [Palmer, 1991]).

Restriction sites on the ctDNA of western white pine (*Pinus monticola*) have been mapped (White, 1990) and some of the genes have been located. The ctDNA of white pine also lacks the IR characteristic of most angiosperms. IR containing rRNA genes are also found in the ctDNA of brown alga *Dictyota dichotoma* and the golden-brown algae *Olisthodiscus luteus* and *Ochromonas danica.* The ctDNA of the brown alga *Phylaiella littoralis* contains two different circular DNA molecules. The larger (130 kb) molecule resembles a typical land plant ct genome, with two rRNA operons in an IR. The smaller (58 kb) molecule contains a 16S pseudogene sequence (Harris et al., 1994).

On the basis of IR, Sugiura (1992) has classified ctDNA into three groups. Group I: ctDNA lacking IRs, e.g. pea; Group II: ctDNA containing IRs, e.g. Tobacco and *Chlamydomonas;* Group III: ctDNA with tandem repeats, e.g. *Euglena.*

Like ctDNA of most higher plants, *Chlamydomonas* species are also divided into two single-copy regions by a large inverted repeat sequence, part of which encodes the ct rRNA genes. Recombination of *Chlamydomonas* ctDNA occurs more frequently in the large inverted repeat sequence than in the single copy regions (Lemieux et al., 1990).

In angiosperms, the ctDNA is quite conserved in structure and generally follows a uniparental-maternal mode of inheritance. Although there is evidence for the inheritance of paternal plastids in many species, including several of the Leguminosae (Harris and Ingram, 1991). Analysis of restriction fragment length polymorphisms (RFLPs) now use to observe the inheritance of ctDNA. RFLP analysis has shown only maternal ctDNA inheritance in *Glycine* (Hatfield et al., 1985) and *Pisum sativum* (Polans et al., 1990). However, biparental or even predominantly paternal ctDNA inheritance was also reported in *Medicago sativa* (Masoud et al., 1990). The occurrence of novel non-parental ctDNA variants in natural sexual interspecific hybrids between *Populus deltoids* var *deltoides* and *P. nigra, P. canadensis* and also intraspecific ctDNA variants have been reported (Rajora and Dancik, 1995).

The extrachromosomal elements are normally absent in ct of higher plants. Staub and Malgia (1994) reported the presence of 868-bp ctDNA minicircle, NICE1 that formed in *Nicotiana tabacum* chloroplast during transformation, as an unexpected product of homologous recombination.

3. Genes in Chloroplast DNA

Chloroplast genome is not large enough to contain all the regulatory genes. Chloroplast gene expression is coordinated with expression of nuclear genes encoding chloroplast proteins during chloroplast development. All the reported chloroplast genes encode either components of ct transcription or translation, or proteins involved in photosynthesis. The apparent prokaryotic nature of the ctDNA also makes it unlikely that it contains genes that can interpret various eukaryotic types of developmental information such as timing during morphogenesis and cellular position within an organ. The chloroplast differentiation involves organ-specific, developmental stage specific, and cell specific gene regulation.

Restriction mapping and cloning confirmed the circular nature of ctDNA. The development of the molecular biology of ctDNA has closely paralleled the development of technology in analyzing the structure and information content of DNA. The knowledge that DNA can be identified by analytical ultracentrifugation because the buoyant density of DNA depended upon the base composition and conformation of DNA a search for organelle DNA started in earnest. Similarly, the development of hybridization technique between single-stranded (ss) DNAs and ss DNA and RNA led to the search of genes in ctDNA. In early experiments, it was clearly shown that ctDNA hybridized with rRNA from 70S ribosomes (Tewari and Wildman, 1968), tRNA genes and messenger RNAs (Tewari, 1979).

CtDNA from higher plants contains relatively constant set of components for the protein-synthesizing apparatus of the organelle (4 rRNAs, 30 to 31 tRNAs, about 21 ribosomal proteins and 4 RNA polymerase subunits) and for photosynthesis (about 28 thylakoid proteins and 1 soluble protein, the ribulose-1, 5-bisphosphate carboxylase/oxygenase [rubisco] large subunit). It is also now found that 11 subunits of mammalian mitochondrial complex I (the *ndh* genes) are encoded by ctDNA in flowering plants and *Marchantia* species (Harris et al., 1994). All the known ct genes are presented in Table 1. The gene nomenclature follows the proposal of Hallick (1989).

The detailed studies on the ribosomal genes in ctDNA was taken up by Thomas and Tewari (1974 a, b). Their data showed that there were two rRNA genes in ctDNA from higher plants. The competition hybridization studied showed that base sequences of ct rRNA from the various plants were similar. All these studies were confirmed by other workers but it became clear that pea ctDNA could not contain two rRNA genes. Bedbrook and Bogorad (1976) had made the restriction map of maize ctDNA and showed that rRNA gene was present in the reverse repeat of ctDNA. Thus, the maize ctDNA contained two rRNA genes. The pea ctDNA did not contain the reverse repeat in structure and finally it was found that it contained only one rRNA gene (Chu et al., 1981).

The 70S ribosome are found in chloroplasts which is different from their larger counterparts (80S) found in cytoplasm. The 16S rRNA is found to be

Tabel 1 Chloroplast Genes

Genes for the genetic apparatus

Genes	Products	Remarks
23S rDNA	23S rRNA	In tobacco, rice, pea, *Chlamydomonas, Euglena,* alder
16S rDNA	16S rRNA	In tobacco, rice, pea, *Chlamydomonas, Euglena*
7S rDNA	7S rRNA	In tobacco, rice, pea, *Chlamydomonas, Euglena*
5S rDNA	5S rRNA	In tobacco, rice, pea, *Chlamydomonas, Euglena*
4.5S rDNA	4.5S rRNA	In tobacco, rice, pea
3S rDNA	3S rRNA	In *Chlamydomonas*
*trn*A-UGC	Ala-tRNA (UGC)	In tobacco, rice, pea, *Chlamydomonas, Euglena*
*trn*R-ACG	Arg-tRNA (ACG)	In tobacco, rice, pea, *Euglena*
*trn*R-UCU	Arg-tRNA (UCU)	In tobacco, rice, pea,
*trn*R-CCGΔ	Arg-tRNA (CCG)	In tobacco, rice, liverwort
*trn*N-GUU	Ans-tRNA (GUU)	In tobacco, rice, pea, *Chlamydomonas, Euglena*
*trn*D-GUC	Asp-tRNA (GUC)	In tobacco, rice, pea, *Euglena*
*trn*C-GCA	Cys-tRNA (GCA)	In tobacco, rice, *Euglena*
*trn*Q-UUG	Gln-tRNA (UUG)	In tobacco, rice, *Euglena*
*trn*E-UUC	Glu-tRNA (UUC)	In tobacco, rice, pea, *Euglena*
*trn*G-GCC	Gly-tRNA (GCC)	In tobacco, rice, pea, *Euglena*
*trn*G-UCC	Gly-tRNA (UCC)	In tobacco, rice, pea, *Euglena*
*trn*H-GUG	His-tRNA (GUG)	In tobacco, rice, pea, *Chlamydomonas, Euglena*
*trn*I-GAU	Ile-tRNA (GAU)	In tobacco, rice, *Chlamydomonas, Euglena*
*trn*I-GAU	Ile-tRNA (CAU)	In tobacco, rice,
*trn*L-UAA	Leu-tRNA (UAA)	In tobacco, rice, *Chlamydomonas, Euglena*
*trn*L-CAA	Leu-tRNA (CAA)	In tobacco, rice, pea
*trn*L-UAG	Leu-tRNA (UAG)	In tobacco, rice,
*trn*K-UUU	Lys-tRNA (UUU)	In tobacco, rice, pea, *Euglena*
*trn*fM-CAU	fMet-tRNA (CAU)	In tobacco, rice, pea, *Euglena*
*trn*M-CAU	Met-tRNA (CAU)	In tobacco, rice, *Euglena*
*trn*F-GAA	Phe-tRNA (GAA)	In tobacco, rice, *Chlamydomonas, Euglena*
*trn*P-UGG	Pro-tRNA (UGG)	In tobacco, rice, pea, *Chlamydomonas*
*trn*S-GGA	Ser-tRNA (GGA)	In tobacco, rice,
*trn*S-UGA	Ser-tRNA (UGA)	In tobacco, rice,
*trn*S-GCU	Ser-tRNA (GCU)	In tobacco, rice, *Euglena*
*trn*T-GGU	Thr-tRNA (GGU)	In tobacco, rice, pea,
*trn*T-UGU	Thr-tRNA (UGU)	In tobacco, rice, *Euglena*
*trn*W-CCA	Trp-tRNA (CCA)	In tobacco, rice, pea, *Euglena*
*trn*Y-GUA	Tyr-tRNA (UGA)	In tobacco, rice, pea, *Chlamydomonas, Euglena*

(Contd.)

Genes	Products	Remarks
*trn*V-GAC	Val-tRNA (GAC)	In tobacco, rice, pea
*trn*V-UAC	Val-tRNA (UAC)	In tobacco, rice, *Euglena*
*rps*2	30S ribosomal protein CS2	In tobacco, rice, pea,
*rps*3	30S ribosomal protein CS3	In tobacco, rice, maize
*rps*4	30S ribosomal protein CS4	In tobacco, rice,
*rps*7	30S ribosomal protein CS7	In tobacco, rice, *Chlamydomonas, Euglena, maize*
*rps*8	30S ribosomal protein CS8	In tobacco, rice,
*rps*9	30S ribosomal protein CS9	In cryptomonas
*rps*10	30S ribosomal protein CS10	In cryptomonas
*rps*11	30S ribosomal protein CS11	In tobacco, rice, pea,
*rps*12	30S ribosomal protein CS12	In tobacco, rice, pea, *Chlamydomonas, Euglena,* maize
*rps*13	30S ribosomal protein CS13	In tobacco, rice,
*rps*15	30S ribosomal protein CS15	In tobacco, rice
*rps*16	30S ribosomal protein CS16	In tobacco, rice
*rps*18	30S ribosomal protein CS18	In tobacco, rice
*rps*19	30S ribosomal protein CS19	In tobacco, rice
*rpl*2	50S ribosomal protein CL2	In tobacco, pea, *Chlamydomonas,*
*rpl*5	50S ribosomal protein CL5	In *Euglena*
*rpl*13	50S ribosomal protein CL13	In cryptomonas
*rpl*14	50S ribosomal protein CL14	In tobacco, rice, pea
*rpl*16	50S ribosomal protein CL16	In tobacco, rice, maize
*rpl*20	50S ribosomal protein CL20	In tobacco, rice
*rpl*21*Δ*	50S ribosomal protein CL21	In liverwort
*rpl*22	50S ribosomal protein CL22	In tobacco, rice,
*rpl*23	50S ribosomal protein CL23	In tobacco, rice, pseudo gene in spinach
*rpl*27	50S ribosomal protein CL27	In algae
*rpl*31	50S ribosomal protein CL31	In *Porphyra*
*rpl*32	50S ribosomal protein CL31	In maize
*rpl*33	50S ribosomal protein CL33	In tobacco, rice
*rpl*36	50S ribosomal protein CL36	In tobacco, rice, pea
*rpo*A	RNA polymerase subunit α	In tobacco, rice, pea, *Chlamydomonas, Euglena*
*rpo*B	RNA polymerase subunit β	In tobacco, rice, pea, *Chlamydomonas*
*rpo*C1	RNA polymerase subunit β'	In tobacco, rice, pea, *Chlamydomonas*
*rpo*C2	RNA polymerase subunit β''	In tobacco, rice
*tuf*A	Elongation factor Tu	In tobacco, *Chlamydomonas, Euglena,* algae
*inf*A	initiation factor 1	In pseudogene
IF-2	initiation factor 2	*In Euglena*
IF-3	initiation factor 3	*In Euglena*
mat K	maturase-like protein	In potato
spr A	Small Plastid RNA	
irf 168 (*ycf*3)	Intron containing reading frame (168 codons)	In tobacco (*gene for other system*)
clp P	ATP dependent protease, (proteolytic subunit)	In wheat, conifer (*gene for other system*)

Genes for the Photosynthetic apparatus

Genes	Products	Remarks
*rbc*L	RuBisCo, large subunit	In tobacco, rice, pea, *Chlamydomonas, Euglena*
*rbc*S	RuBisCo, small subunit	In red and brown algae
*psa*A	PS I, p700 apoprotein A1, 83 kDa	In tobacco, rice, pea, *Chlamydomonas, Euglena*
*psa*B	PS I, P700 apoprotein A2, 83 kDa	In tobacco, rice, pea, *Chlamydomonas, Euglena*
*psa*C	PS I, 9-kDa polypeptide	In tobacco, rice, pea
*psa*I	I-protein (4.6 kDa)	
*psa*J	J-protein (4.8 kDa)	
*psa*M	M-protein 3.5 kDa	
*psb*A	PS II D1 protein, 32 kDa	In tobacco, rice, pea, *Chlamydomonas, Euglena*
*psb*B	PS II, 47 kDa polypeptide	In tobacco, rice, pea, *Chlamydomonas, Euglena*
*psb*C	PS II, 43 kDa polypeptide	In tobacco, rice, pea, *Chlamydomonas, Euglena*
*psb*D	PS II, D2 protein, 32 kDa	In tobacco, rice, pea, *Chlamydomonas*
*psb*E	PS II, cytochrome b_{559} (9 kDa)	In tobacco, rice, pea
*psb*F	PS II cytochrome b_{559} (5 kDa)	In tobacco, rice, pea
*psb*G	PS II G protein	In tobacco, rice, pea
*psb*H	PS II 10 kDa, phosphoprotein	In tobacco, rice, pea
*psb*I	PS II PsbI, 4.5 kDa	
*psb*J	PS II PsbJ, 4 kDa	
*psb*K	PS II PsbK, 3.5 kDa	
*psb*L	PS II PsbL, 4 kDa	
*psb*M	PS II PsbM, 4 kDa	
*psb*N	PS II PsbN, 4 kDa	
*psb*T	PS II PsbT, 4 kDa	
*pet*A	b/f complex, cytochrome f	In tobacco, rice, pea,
*pet*B	b/f complex, cytochrome b_6	In tobacco, rice, pea,
*pet*D	b/f complex, subunit IV	In tobacco, rice, pea,
*pet*G	b/f complex, subunit V	
*atp*A	H^+-ATPase, subunit $CF_1\alpha$	In tobacco, rice, pea, *Chlamydomonas, Euglena*
*atp*B	H^+-ATPase, subunit $CF_1\beta$	In tobacco, rice, pea, *Chlamydomonas, Euglena*, brown algae
*atp*D	H^+-ATPase, subunit CF	In *Odontella sinensis*
*atp*E	H^+-ATPase, subunit CF_1, ε	In tobacco, rice, pea; *Chlamydomonas*, brown alga
*atp*F	H^+-ATPase, subunit CF I, 18–19 kD	In tobacco, rice, pea, *Chlamydomonas*
*atp*G	H^+-ATPase, subunit, CF_0 II	
*atp*H	H^+-ATPase, subunit CF_0 III 8 kDa	In tobacco, rice, pea
*atp*I	H^+-ATPase, subunit CF_0 IV, 27 kDa	In tobacco, rice, pea

(Contd.)

Genes	Products	Remarks
*ndh*A	NADH dehydrogenase, ND1	In tobacco
*ndh*B	NADH dehydrogenase, ND2	In tobacco
*ndh*C	NADH dehydrogenase, ND3	In tobacco
*ndh*D	NADH dehydrogenase, ND4	In tobacco, pea
*ndh*E	NADH dehydrogenase, ND4L	In tobacco
*ndh*F	NADH dehydrogenase, ND5	In tobacco
*ndh*G	NADH dehydrogenase, ND6	
*ndh*H	NADH dehydrogenase, 49 kDa protein	
*ndh*I (frxB)	NADH dehydrogenase, 18 kDa protein	
*ndh*J		ORF 158-169 in LSC
*ndh*K (psbG)	NADH dehydrogenase, 27 kDa protein	
frxA		In Liverwort
frxB		In Liverwort
frxC	31 kDa protein	
Tpi PI	single copy gene	In spinach
SCE 70		In spinach
Omp 24		In spinach
ChlN		In *Chlamydomonas,* liverwort
ChlB		In *Chlamydomonas*
FBPases	39 kDa	In pea, *Arabidosis*, potato, spinach, wheat
rpl 31	50S ribosomal protein CL-31	In *Porphyra*

associated with 30S subunit while 23S, 5S and 4.5S rRNAs are associated with 50S subunit. The ct-RNAs are highly conserved at the sequence level and are closely related to eubacterial sequences, which include cyanobacteria (Harris et al., 1994). For example 4.5S rRNA is present in chloroplast of higher plants and it is homologous to the 3′ end of the 23S rRNA of prokaryotes (Edwards et al., 1981). In 1988 Gray reported eight noncontiguous conserved primary sequences in 16S rRNA interspersed among nonconserved sequences. The predicted secondary structures of these molecules are conserved, and all the 45 helices reported in *E. coli* 16S rRNA, are also found in 16S rRNA of *Euglena gracilis, Chlamydomonas* species, maize and tobacco (Harris et al., 1994). Comparative studies of 16S and 5S rRNA sequences also supported both the probable origin of chloroplasts from endosymbiotic cyanobacteria and the hypothesis that land plants derive from one branch of chlorophyte algae.

Chloroplast rRNA genes are normally arranged in an operon transcribed in the order 16S-23S-4.5S-5S (Fig. 5). In higher plants about 95 nucleotides homologous to the 3′ terminus of the *E. coli* 23S molecule constitutes a 4.5S rRNA molecule, separated from the remainder of the 23S gene by a transcribed spacer (Harris et al., 1994). In higher plants the tRNA genes (tRNA Ile and tRNA Ala) are split by introns but within the spacer between the 16S and 23S genes (Fig. 5). However, in algae these tRNA genes are uninterrupted.

The 16S rRNA sequences in tobacco can be divided into three functional domains: 5′, central, and 3′, which correspond to residues 26 to 557, 564 to 912 and 926 to 1391 of *E. coli*, respectively. Each of these domains comprises helices and loops whose secondary structure is phylogenetically conserved (Gray, 1988). The secondary-structure model of 23S rRNA in tobacco shows presence of six domains with 95 helices (Gutell et al., 1994). Domain V is the site for binding tRNA to 50S subunit. In higher plants the 5S rRNA gene is followed by a tRNA Arg gene in the same orientation and by a tRNA Asn gene on the opposite strand (Kato et al., 1985; Shapiro and Tewari, 1986). Through primer extension experiments in maize, it is shown that tRNA Arg gene, which is separated from the 5S gene by a 252-bp spacer, is cotranscribed with the rRNA operon (Dormann-Przybyl et al., 1986).

Fig. 5 Arrangement of the rRNA operons in higher plants showing the tRNA genes (tRNA^{Ile} and tRNA^{Ala}) are split by introns within the spacer between the 16S and 23S genes.

The hybridization studies between ctDNA and radioactively labeled rRNAs has shown that there are about 30–31 tRNA genes in ctDNA (Harris et al., 1994). Meeker and Tewari (1980) carried out an elaborate and cumbersome approach to identify the presence of each of the amino acid specific tRNA in ctDNA by first acylating the tRNAs with radioactively labeled amino acid and then hybridizing the labeled aminoacyl-tRNAs with ctDNA. From the hybridization of labeled aminoacyl-tRNA to ctDNA, it was not possible to calculate the number of genes for each of the tRNAs because the amounts of specific aminoacyl tRNA in total ct-RNAs were not known and it was not possible to know whether all of the specific aminoacyl tRNAs have been acylated. However, 17 aminoacyl tRNA genes were identified with this technique and one could even calculate that some of the genes were present in multiple copies.

The presence of rRNA and tRNA genes can account for not more than 4–5% of the chloroplast genome. The question whether all of the base sequence of ctDNA are transcribed in chloroplasts was investigated by Oishi et al. (1981). Poly U sepharose separated total RNA of chloroplast was hybridized to the pea ctDNA. About 45–50% of the ctDNA was found to hybridize with the ctDNA. These data showed that the entire information content of the ctDNA is expressed in chloroplasts.

Chloroplast genomes encode all the tRNAs involved in chloroplast protein synthesis. In higher plants, the tRNA genes (*trn*) are scattered over the ctDNA. No ct-genes code for their own 3′-CCA end; hence these must be added postransriptionally. All the 61 possible codons are used in the sequences coding for polypeptides in chloroplasts.

The search for genes in ctDNA that code for specific protein was carried out by various methods. The advantage was taken of the fact that cycloheximide inhibis the protein synthesis of cytoplasmic ribosomes and chloroamphenicol inhibits the synthesis of chloroplast protein. Purified chloroplasts were incubated in the presence or absence of antibiotics with labeled amino acids and the synthesized protein analyzed in SDS/PAGE gels along with the purified enzymes as markers. The complete nucleotide sequence of ctDNA from *M. polymorpha* by Ohyama et al. (1986) from tobacco by Shinozaki et al. (1986), and rice by Hiratsuka et al. (1989) was the hallmark of achievement. In 1966, there were doubts whether the ctDNA was real entity. Twenty years later, the complete sequences of ctDNA were available. The nucleotide sequences confirmed the presence of rRNA, tRNA and mRNA genes that had been individually established before and identified some new ones. The details of the genes present in ctDNA are also well described by Sugiura (1992).

The genes for chloroplast ribosomal proteins (*rpl* represent large subunit proteins and *rps* for small subunit proteins) found through homology with their corresponding genes in *E. coli.* Sugita and Sugiura in 1983 isolated first ribosomal protein gene, *rps*19, from tobacco. For example, seven of the ribosomal proteins from ct ribosomes of spinach have been purified (Schmidt et al., 1992) and shown by amino-terminal sequencing to be equivalent to seven *E. coli* proteins (S12, S16, S19, L20, L32, L33 and L36), whose homologs are encoded by ctDNA. The same subset of ribosomal proteins is encoded in the ctDNA of each higher plant with only few exceptions, while marked variations occur in some algal groups.

In general, the ct-encoded ribosomal proteins show more immunological cross-reactivity with bacterial ribosomal proteins than those encoded in nuclear genes (Randolph-Anderson et al., 1989). However biogenesis of ct-ribosomes requires expression of both chloroplast and nuclear genes encoding different ribosomal proteins, as well as ct-genes encoding the component rRNAs.

The ct-ribosome of higher plants also contains at least five novel proteins that are not homologous to any of the *E. coli* ribosomal proteins (Subramanian, 1993). But these novel proteins are encoded in the nucleus. Probably unidentified open reading frames in the ctDNA encode some more novel ct-ribosomal proteins which still has to be seen. It is known that all the mRNAs translated by the ct-ribosome are also transcribed within the chloroplast (Subramanian, 1993).

By expressing maize ct-*rpl*23 gene in *E. coli* and raising antibodies against the expressed protein, Bubunenko et al. (1994) showed that chloroplast L23-like protein is absent in spinach chloroplast ribosomes. They also reported the presence of a new eukaryotic cytosolic-type protein from spinach chloroplast ribosome which has taken over the function of the missing prokaryotic-type L23 protein. Many nuclear genes have also been identified which code for components of the chloroplast translational system. These include the genes for over 15 ribosomal proteins of the chloroplast ribosome and the genes for translational factors EF-Tu and EF-G. The gene for L35 is nuclear in spinach but is organellar in *Cyanophora paradoxa* (Bryant and Stirewalt, 1990). Similarly the light-regulated *tufA* gene which encodes translational elongation factor EF-Tu is organellar in *Euglena gracilis* but nuclear in all land plants. A gene for EF-Tu from tobacco chloroplast has also been reported (Ursin et al., 1993). In the case of L21 the gene is nuclear in flowering plants but is chloroplastic in the lower land plant *M. polymorpha.* The *rpl*22 gene is chloroplastic in most land plants and *Euglena* but is nuclear in the legume family (Bubunenko et al., 1994). The gene for *rpl*27 is present in ctDNA of *Pleurochrysis carterae* but not found in ctDNA of land plants (Fujiwara et al., 1994). A putative gene for the initiation factor IF-1 (*infA*) was found in the *rpl*23 gene cluster in spinach. A putative wheat chloroplast gene (*clpP*) which encodes the proteolytic subunit of an ATP-dependent protease has also been reported (Gray et al., 1990). The *ClpP* gene from ct-genomic library of the conifer *Pinus contorta* has also been isolated by Clarke et al. (1994) and these authors also suggested that the *ClpP* gene in conifers is part of an operon which includes the first exon of the *rps12* and the entire *rpl20* gene and is expressed in a light-independent manner as a polycistronic precursor which later undergoes post-transcriptional processing to give the mature monocistronic *clpP* mRNA.

A cDNA clone for chloroplast translational initiation factor-3 (IF-3_{chl}) from *Euglena gracilis* has been reported (Lin et al., 1994). Sequence analysis showed that IF-3_{chl} mRNA contains the spliced leader found at the 5′ end of nuclear encoded mRNAs in *E. gracilis*. The activity of IF-3 is inducible by light. Lin et al. (1994) also reported that the IF-3_{chl} mRNA is present in approximately equal amounts in both dark and light grown cells which suggested that the light-dependent induction of IF-3_{chl} activity could be post-transcriptional. IF-3 is also known for its proof reading role in initiation by promoting the selection of the initiator tRNA as opposed to elongator tRNA in the 30S initiation complex (Hartz et al., 1990). Analog of IF-3 is not found in the eukaryotic cytoplasmic protein biosynthetic system IF-3_{chl} can also replace *E. coli* IF-3 in promoting 70S initiation complex formation on *E. coli* ribosomes. This observation suggests that the chloroplast factor has retained many of the features and functions of the corresponding prokaryotic factors. A cDNA clone encoding a cognate 70 kDa heat shock protein of the spinach chloroplast envelope (SCE 70) has also been reported (Ko et al., 1992). A gene product of eukaryotic translation initiation factor 4A (eIF-4A) is also present in tobacco chloroplast where it is ribosome-associated (Owttrim et al., 1994). The genes for RNA polymerase subunits will be described under the "transcription of chloroplast DNA" section.

Rubisco or fraction 1 protein (E.C. 4.1.1.39) is the major stromal protein of chloroplast and consists of eight similar large subunits (LS) of 55 kDa encoded by ctDNA and eight similar small subunits (SS) of 12 kDa encoded by nDNA. However in sea alga, *Olisthodiscus luteus,* the SS gene (*rbcS*) is present in ctDNA (Sugiura, 1992). The first chloroplast protein gene cloned and sequenced was the LS gene (*rbcL*) from maize.

However, it was Kawashima and Wildman (1972) who established that the two polypeptides of fraction I protein (Ribulose 1, 5 diphosphate carboxylase) were coded by nuclear and ctDNA. The paper was a clever combination of biochemical techniques and classical breeding. Tryptic peptides were resolved from the small subunit of highly purified Fraction I protein obtained from *Nicotiana tabacum, Nicotiana glutinosa, Nicotiana glauca* and four reciprocal F1 hybrids: *N. tabacum* × *N. glutinosa; N. tabacum* × *N. glauca.* Information for the synthesis of an extra *N. tobacum* peptide was transferred by pollen to *N. glutinosa* egg cells and, therefore, the Mendelian mode of inheritance signified nuclear DNA as containing the code for the primary structure of the small subunit. Two differences in peptides between *N. tabacum* and *N. glauca* were also inherited in a Mendelian manner. Transfer of the new *N. tabacum* information to *N. glauca* egg cells also suppressed the synthesis of the *N. glauca* type of Fraction I protein.

The thylakoid membranes system contains the ct-genes encoding proteins of Photosystem I (PS I), Photosystem II (PS II), the cytochrome b/f complex, ATP synthase and the nuclear-coded light-harvesting chlorophyll protein complex. PS I and PS II are two multisubunit pigment-protein complexes in the thylakoid membranes in the chloroplast of plants, algae and in cyanobacteria. Both the systems contain specialized chlorophyll molecules, the "reaction centers", in which absorption of light energy initiates a series of redox reactions that result in the production of high-energy molecules such as NADPH and ATP (Pakrasi, 1995).

The PS I complex binds, P700 and can photoreduce ferredoxin (the iron sulfur protein) and oxidize a copper protein, plastocyanin. The complex is composed of two moieties: (1) a chlorophyll a binding core complex and (2) a chlorophyll a/b binding peripheral antenna called light-harvesting complex of PS I (LHCI). It comprises of at least 11 different subunits and about 100 chlorophyll molecules (Pakrasi, 1995). Out of 11 subunits 6 are encoded by the genes of ctDNA (*psaA, psaB, psaC, psaI, psaJ* and *psaM*) and 5 are encoded by genes of nDNA (*psaD, psaE, psaF, psaK* and *psaL*). The products of the *psaA* and *psaB* ct-genes are major 60 kDa subunits of the core complex and bind all of the about 90 chlorophyll a and 14 β-carotene molecules besides P700 as well as the primary acceptor A_0 and A_1 at the interface between the 2 homologous subunits. A leucine zipper motif was also reported to be present in both the *psaA* and *psaB* gene products and its role is speculated in mediating interactions of the PS I heterodimer (Webber and Malkin, 1990). The *psaC* gene product is a ferredoxin-type protein which contains nine conserved Cys residues. *psaI* gene product shows sequence homology to a membrane-spanning helix of D2 (in the reaction center of PS II) and it is suggested to play a role in the binding of the phylloquinone cofactor in the

PS I core complex. The functional roles of gene products of *psaI* and *psaM* are not known (Pakrasi, 1995).

The function of PS II complex is to use light energy for the reduction of plastoquinone by water, which can donate electrons to the PS I reaction center through a water-soluble electron carrier (plastocyanin for plants) located in the thylakoid lumen. Till to date at least 14 genes have been reported from ct-genome of PS II system. These are *psbA, psbB, psbC, psbD, psbE, psbF, psbH, psbI, psbJ, psbK, psbL, psbM, psbN* and *psbT* (Pakrasi, 1995). At least 6 nuclear encoded genes in PS II were also reported which include *psbO, psbP, psbQ, psbR, psbS and psbW* (Pakrasi, 1995). The first isolated gene of PS II was *psbA* (for 32 kDa D1 protein) from spinach and *Nicotiana debneyi* (Sugiura, 1992). In higher plants the genes for PS II are continuous while in case of some of the algae (e.g. *Euglena gracilis*) the *psb* genes are split by introns (Sugiura, 1992). The *psbB* gene product can be detected only after illumination and also a nuclear factor has been found to influence translation of *psbB*, along with *psbA* during light-induced development (Gamble and Mullet, 1989). In green plants and cyanobacteria *psbE* and *psbF* genes encode α and β subunits of cytochrome b_{559} which are small membrane-spanning proteins. The *psbL* gene encodes the polypeptide (4 kDa) which is a small integral membrane protein essential for function of PS II (Pakrasi, 1995).

There are at least 7 genes reported for cytochrome b/f complex, of which 4 are present in ctDNA and 3 in nDNA. The ct-genes *petA, petB, petD* and *petG* encode cytochrome f, cytochrome b_6, subunit IV and subunit V components of the complex respectively. The *petC, petE* and *petF* are nDNA genes which encode Rieske Fe-S protein, plastocyanin and ferredoxin-NADP oxidoreductase respectively. The genes *petA, petB* and *petD* are first sequenced in pea and spinach. In higher plants the *petB* and *petD* genes are found to be clustered with *psbB* and *psbH* and also contain single introns with only 6–8 bp exons (Sugiura, 1992).

Recently, a new ct-gene for small subunit of cytochrome b_6/f complex, the *petL*, has been reported from *Chlamydomonas reinhardtii* (Takahashi et al., 1996). The small subunit is encoded by the hypothetical chloroplast open reading frame 7, *ycf7* (ORF 43) in the *psaC* operon of *C. reinhardtii*. Generation and analysis of chloroplast transformants with disrupted *ycf7* (*petL*) show that the encoded protein *ycf7* is important for photoautotrophic growth as well as for electron transfer efficiency and stability of the cytochrome b_6/f complex (Takahashi et al., 1996).

The ATP synthase (F_0F_1-type ATPases) is composed of a membrane-integrated (F_0 or proton pore CF_0) and peripheral sector (F_1 or CF_1), the extrinsic enzyme that catalyzes ATP synthesis-hydrolysis. CF_0 is responsible for transmembrane proton translocation and consists of four subunits (CF_0I to IV) and CF_1 comprising the catalytic sites for reversible ATP synthesis, contains five different subunits, designated α, β, γ, δ and ε, present in the ratio of 3:3:1:1:1. The genes for at least eight subunits are present in the chloroplast genome. The genes for the β and ε subunits (*atpB and atpE*) were isolated and sequenced from maize and spinach. The genes for the three CF_0 subunits (*atpI, atpH, atpF*) were found to be clustered just before *atpA*. The genes *atpG* and *atpD* which are nucleus encoded in chlorophyll a + b plants, are present in the *Odontella sinensis* chloroplast gene cluster (Pancic et al., 1992).

In cyanobacteria, the genes are arranged in two clusters, the *atpB* gene cluster containing the genes for β subunit (*atpB*) and ε (*atpE*), and the *atpA* gene cluster harbouring all CF_0 (*atpG, atpH, atpI* and *atpF,* encoding the subunits b', c, a and b which are homologous with subunits II, III, IV and I respectively of higher plants) and the CF_1 genes *atpD* and *atpA* coding for subunits δ and α.

The existence of respiratory chain NADH dehydrogenase in the chloroplast of higher plants are also reported. There are at least 11 putative ct-DNA genes (*ndhA* to *ndhK*) in higher plants whose predicted amino acid sequences are similar to those of components (ND 1, 2, 4, 4L, 5, 6, 49 kDa, 18 kDa, ORF 158–169 in LSC and 27 kDa) of the respiratory chain NADH dehydrogenase from human mitochondria. The tobacco *ndhA* and *ndhB* genes contain single introns (Sugiura, 1992).

The actively-transcribed gene for a subunit of tryptophan synthase, *trpA*, has been reported from ct-genome of the "primitive" unicellular red alga *Cyanidium caldarium* strain RK-1 (Ohta et al., 1994).

The chloroplast triophosphate isomerase (EC 5.3.1.1) from spinach is encoded by a single-copy gene, *tpiP1,* which arose during plant evolution through duplication of a pre-existing nuclear gene for the cytosolic enzyme (Henze et al., 1994). A cDNA clone of pea chloroplast fructose-1, 6-bisphosphatase (FBPase; EC 3.1.3.11) has been reported by Carrasco et al. (1994).

3.1 Light-regulated Chloroplast Genes

Chloroplast gene expression in higher plants is regulated at different levels to make sure the coordinated expression of chloroplast and nuclear genes. Light not only provides energy for plant growth and development but also acts as a stimulus for regulation of many metabolic processes. Usually light induction of chloroplast biogenesis is accompanied by increased synthesis of many proteins which were not detectable in the dark. Many ct-genes are light-regulated at the level of transcription. In many cases increase in the transcript levels from these genes occurs in etiolated seedlings and dark-adopted plants in response to light. This increase is mediated by the photoreceptor phytochrome and is regulated at the transcriptional level. Usually the DNA elements responsible for light-responsive expression are located within 5′ upstream sequences (Gilmartin et al., 1990). However, nuclear run-on experiments with petunia *rbcS* show that both upstream and downstream sequences play a role in the transcriptional regulation of these genes (Dean et al., 1989).

Light-responsive gene expression of pea *rbc-3A* revealed that presence of multiple regulatory elements at upstream region are involved in maintaining the *rbcS* expression at various light influences (Gilmartin et al., 1990). The chloroplast *rbcS* gene is the most extensively studied light-responsive gene (Tobin and Silverthorne, 1985). In barley, the initial activation of ct-genes expression and accumulation of most ct-proteins is light independent but chlorophyll and chlorophyll-apoprotein accumulation requires light. The transcription of *psbA,* the gene encoding 32 kDa D1 protein of PS II, *rbcL* and 16S rDNA were shown to be differentially enhanced in response to light (Klein and Mullet, 1990). In tobacco chloroplast the translation of *psbA* mRNA was also regulated by light via the 5′-untranslated region (Staub and Maliga, 1994). The chlorophyll and chlorophyll-apoprotein accumulation requires light (Apel, 1979; Klein et al., 1988). When the plants are illuminated, only then photosynthetically active matured chloroplasts are formed. Sexton et al. (1990) showed that the light-induced *psbD-psbC* RNAs arise from an unusual promoter which mediates transcription initiation from a 23 nucleotide region. They also proposed a transcriptional regulatory mechanism for a chloroplast operon, in which a light-induced switch in promoter utilization resulted in maintenance of *psbD-psbC* gene expression in chloroplasts of illuminated barley. A 47 kDa protein was isolated which binds specifically 36 bases of 5′ leader of the chloroplastic *psbA* mRNA and shown to be correlated with the level of translation of *psbA* mRNA observed in light- and dark-grown cells (Danon and Mayfield, 1991).

The chloroplast protein synthesis factors EF-G, EF-Ts, EF-Tu, IF-2 and IF-3 are also light induced (Breitenberger et al., 1979; Fox et al., 1980; Sreedharn et al., 1985; Gold and Spremulli, 1985; Kraus and Spremulli, 1986; Akkaya and Breitenberger, 1992). Leaf chloroplast of the *Arabidopsis* pale cress (*pac*) gene is regulated by light and appears to be a novel component of a light-induced regulatory network that controls the development of leaves and chloroplasts (Reiter et al., 1994). The immunosuppressive drug receptor, immunophilins, is shown to be present in the chloroplast of fava bean and also regulated by light (Luan et al., 1994).

3.2 Introns

Introns in chloroplast genes can be classified into group I or group II on the basis of several conserved sequences they contain and it is suggested that they evolved from or into mobile elements (Michel and Dujon, 1983). The intron DNA sequence is suggested to be transferred by double-strand breakage-repair mechanism and consistent with this process, flanking genetic markers are frequently co-converted (Dujon, 1989). Many group I introns are capable of self splicing *in vitro* and shared with group II introns (Garriga and Lambowitz, 1984; Gott et al., 1986; Partono and Lewin, 1988). Group I introns have been found in many different locations in chloroplast. Many Group I introns contain either an internal open reading frame (ORF) or an ORF which is in phase with the preceding exon. These intron ORFs can be divided into at least two classes according to their function. The first class of intron ORFs encodes RNA maturases which are essential for *in vivo* splicing (Burke, 1988). The second class encodes DNA endonucleases which confer genetic mobility to the intron in which they are encoded.

In *C. reinhardtii* ct-genome the genes for the 23S ribosomal RNA are interrupted by a 888 bp group I intron which is located close to the 3′ end of the gene. This was the first report of introns in chloroplast genes (Rochaix and Malnoe, 1978; Rochaix 1978; Allet and Rochaix, 1979). This intron has an internal ORF that encodes a predicted protein of 163 amino acid residues (Rochaix et al., 1985). Durrenberger and Rochaix (1991) reported that the protein encoded by self splicing LSU (large subunit) intron in *C. reinhardtii* chloroplast contained double-strand (ds) DNA endonuclease activity specific for a sequence close to the homing site of the LSU introns. They also showed that this LSU intron is mobile *in vivo*.

Group I intron is also found in the gene for a leucine tRNA with a UAA anticodon [tRNALeu (UAA)] in chloroplast of many higher plants (Cech, 1988). Self-splicing group I introns can be identified by incubating unspliced RNA with [^{32}P] GTP *in vitro* (Yuan et al., 1990). The first step in splicing is attacked by the 3′ hydroxyl group of guanosine at the 5′ splice site, with the guanosine derivative remaining covalently attached to the 5′ end of the linear excised intron (Kruger et al., 1982). The chloroplast tRNA Leu introns cannot splice *in vitro*, even in the presence of extracts made from chloroplast (Xu et al., 1990).

More than 16 chloroplast genes from higher plants are known to contain introns. These genes are: *trnL*-UAA, *trnI*-GAU, *trnA*-UGC, *trn*V-UAC, *trn*G-UCC, *trn*K-UUU, *rps*16, *rpl2*, *rpl*16, *rpo*C1, *clpP, petB, petD, atpF, ndhA, rdhB* (Sugiura, 1992). Most of these higher plant genes contain single introns except *clpP* gene which contains 2 introns. However *Euglena* and *Chlamydomonas* contains multiple introns (Plant et al., 1988). The *rps16* introns of tobacco, mustard and rice are 860 bp, 887 bp and 809 bp long respectively. The *rps*16 intron of maize could be classified as a group II intron with six characteristic stem-loop structure (Kanakari et al. 1992). The *Euglena gracilis* chloroplast genome contains introns-within-introns, termed twintrons. Twintrons are excised from pre-mRNAs by sequential splicing of the individual introns (Drager and Hallick, 1993). A mustard chloroplast gene *trnG*-UCC is reported to split by a 717-bp group-II introns (Liere and Link, 1994).

In the tobacco chloroplast genome, 18 genes contain introns of 503–2526 bp in length, six for tRNA genes and 12 for protein-encoding genes (Sugita and Sugiura, 1996). Most introns can be folded into a characteristic, evolutionarily conserved secondary structure. Chloroplast group II-type introns have conserved boundary sequences (5′-GTGYGRY ... RYCNAYYyYRAY-3′) which are similar in part to those of nuclear pre-mRNA introns (Sugita and Sugiura, 1996). No self-splicing has been reported for chloroplast group II-type introns.

4. Replication of Chloroplast DNA

Higher plant chloroplast DNA can be isolated as a covalently, closed circular molecule with a molecular mass of about 85–95 × 10^3 kDa (Kolodner and Tewari, 1975a) but so far, little is known about the DNA replication mechanism in the higher plant systems. Although the mechanism of DNA replication has been well defined in plasmids, bacteriophages, bacteria, viruses and to a lesser extent, in yeast (Tomizawa, 1986; Mecsas and Sugden, 1987; Bramhill and Kornberg, 1988; Alfano and McMacken, 1989; Thommes and Hubscher, 1990; MacAllister et al., 1991; Kornberg and Baker, 1991; Yang et al., 1991; Hubscher and Spadari, 1994). Most studies of the replication of DNA in plants have investigated chloroplast DNA (Tewari, 1987). Denaturation mapping (Kolodner and Tewari, 1975a) and restriction endonuclease analysis (Bedbrook and Bogorad, 1976) have reported that majority of the circular molecules in the chloroplasts of a given species are identical in sequence. Chloroplast DNA replication in higher plants is not coupled to the synthesis of nuclear DNA. Timing of ctDNA amplification is under developmental control (Lammpa and Bendich, 1979; Lawrence and Possingham, 1986). The studies with mutant chloroplasts completely lacking ribosomes and inhibitors of protein synthesis suggested that almost all proteins essential for ctDNA replication and chloroplast multiplication are nuclear-encoded (Walbot and Coe, 1979; Heinhorst et al., 1985).

The replication of ctDNA was studied by analyzing the structure of replicative intermediates in electron microscope. Chloroplast DNA was centrifuged in CsCl-EtBr gradients and the fractions starting from the supercoiled region to the circular DNA molecules were analyzed by formamide spreading technique (Kolodner and Tewari, 1975 b,c). Pea ctDNA was found to contain two D-loops (Fig. 6) which were located at two adjacent sites about 7 kb apart. The inner distance between the two D-loops which were of about 800 bp was highly variable, indicating that the two D-loops expand towards each other. Small Cairns type of replicative forked structures ranging in size from 7.2 to 10.6 kb were observed in closed circular pea ctDNA. This Cairns structure mapped at the position of D-loops and probably result when the two displacing strands expand towards each other. The corn ctDNA was also found to contain the D-loops that were about 860 bp long and were separated by 7.0 kb. Replicative forked structures of the Cairns type were found in the DNA from lower, middle and upper bands. The extent of the replication of ctDNA ranged from 38–88%. The finding of Cairns replicative intermediates in the lower and middle bands of the CsCl-EtBr density gradients, and the correlation between higher banding position and larger amount of

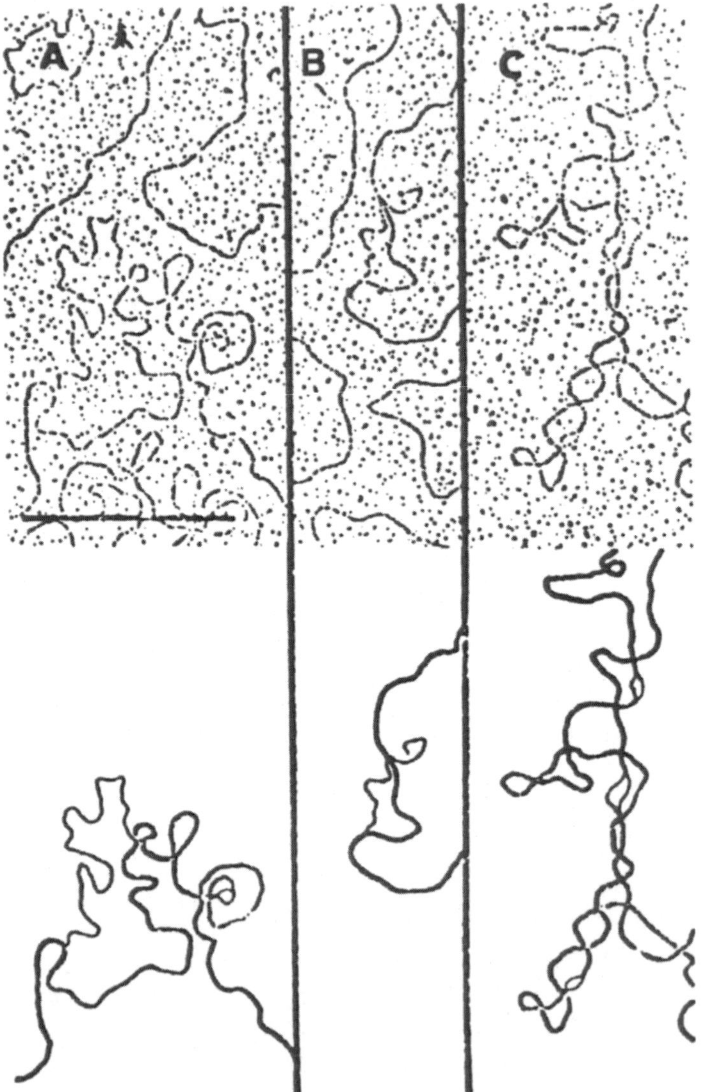

Fig. 6 **High magnification electron micrographs of ctDNA molecules. (A) molecule containing one D-loop and one den-loop from the pea ctDNA. (B) A branch migrating D-loop from corn ctDNA. (C) A pea ct-DNA molecule containing three den-loops. The thin and thick lines in the drawings represent the single-stranded regions, respectively. The bar indicates 1 μm.**

replication suggested that the replication takes place on a covalently closed circular template and is accompanied by nicking and closing cycles. In the pea ctDNA preparations, circular molecules with an attached double stranded tail were also observed. The lengths of tails in the pea ctDNA ranged from 1.5 to 124% of the length of the attached monomer length circular molecule.

Based upon the above results, a model for the replication of pea and corn ctDNA is presented in Figs. 7 and 8. The ctDNA replication is initiated by the formation of two D-loops, whose displacing strands are complementary to the opposite parental strands of ctDNA. The two displacing strands expand towards each other and initiate the formation of Cairns replicative forked structures. The small Cairns forked structure expand bidirectionally until termination takes place at a site that is 180° around the circular ctDNA molecule from the initiation site. Separation of the daughter molecules takes place yielding two circular molecules that each have a single strand break of small gap at the same site located in opposite daughter strands (Fig. 7). In

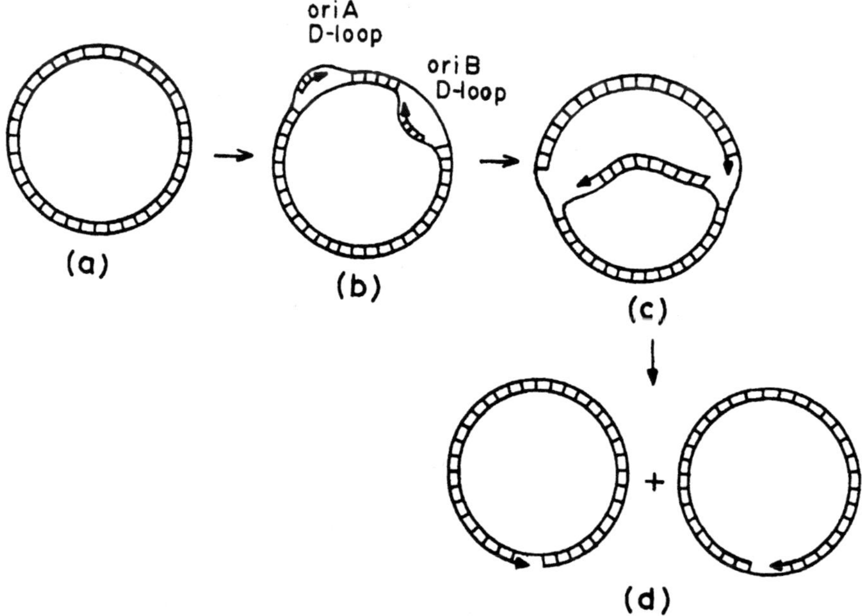

Fig. 7 **A model for the replication of ctDNA. (a) closed circular parental molecule; (b) for oriA D-loop and oriB D-loop containing molecule; (c) Cairns type of replicative intermediate of expanded D-loops containing molecules; (d) nicked progeny molecules.**

ctDNA, the nicked circles could be sealed to close the circle on 3′-OH group or each nicked progeny molecule could be extended by a DNA polymerase molecule. This would displace a single stranded tail from the molecule and this tail could be filled by discontinuous duplex synthesis to yield a molecule with a double stranded tail. The tails might than be converted to circular molecules by intrastrand recombination event. If both progeny forms a Cairns round of replication initiated rolling circle synthesis (Fig. 8), two types of rolling circles would be formed. In each case, the tip of the tail would map at the same site, but the sequence of the two types of tails would extend in the opposite direction from this site. The denaturation maps with pea ctDNA rolling circle confirmed that this was the case.

The locations of the two replication origins in pea ctDNA was mapped by electron microscopic analysis of restriction digests of supercoiled ctDNA cross linked with trioxalen (Meeker et al., 1988). Both origins of replication identified as D-loops were present in the 44 kb Sal I fragment. The first D-loop (OriA) was located at 9.0 kbp from the closest Sal I restriction site.

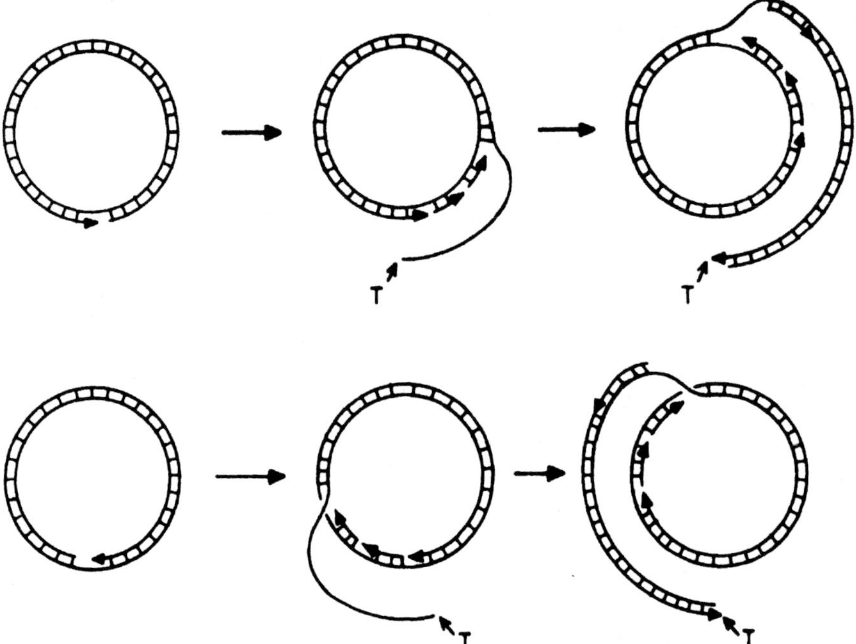

Fig. 8 Rolling circles model for the replication of ctDNA. The "T" indicates the terminus of the Cairns round of replication which is 180° opposed to the origins of D-loop synthesis.

The average size of this D-loop was 0.7 kb. The second D-loop (OriB) started 14.2 kb in from the same restriction site and ended at about 15.5 kb giving it a size of about 1.3 kb. The orientation of these two D-loops on the restriction map of pea ctDNA was determined by analyzing Sma I, Pst I and Sal I/Sma I restriction digests of pea ctDNA. One D-loop was mapped in the spacer region between the 16 and 23 S rRNA gene. Denaturation mapping of recombinants pCT 12-7 and pCB 1-12, which contain both D-loops, confirmed the location of the D-loops in the restriction map of pea ctDNA. Denaturation mapping also revealed that the two D-loops had different base compositions.

An *in vitro* DNA replication was developed by Meeker et al. (1988), in which the recombinants pCT 12-7 and pCB 1-12 were found to be highly active in DNA synthesis when used as a template. Analysis of the *in vitro* synthesized DNA with either of these recombinants showed that full length template DNA was synthesized. Recombinants from other regions of the pea chloroplast genome showed no significant DNA synthesis *in vitro.* These experiments confirmed that the observed D-loops region were truly origin of replications sequences. A detailed analysis of the OriA region identifying to about 300 bp has also been reported by Nielsen et al. (1993).

A partially purified replicative system from pea chloroplasts was developed by Reddy et al. (1994) which entirely mimics the replicative intermediate of pea ctDNA found *in vivo*. When sequences of either of the two chloroplast origins of replication (OriA and OriB) were used as templates, the replicative intermediates were found to have Sigma structures (Fig. 9). Electron microscopic analysis of the Sigma structures revealed that the initiation site

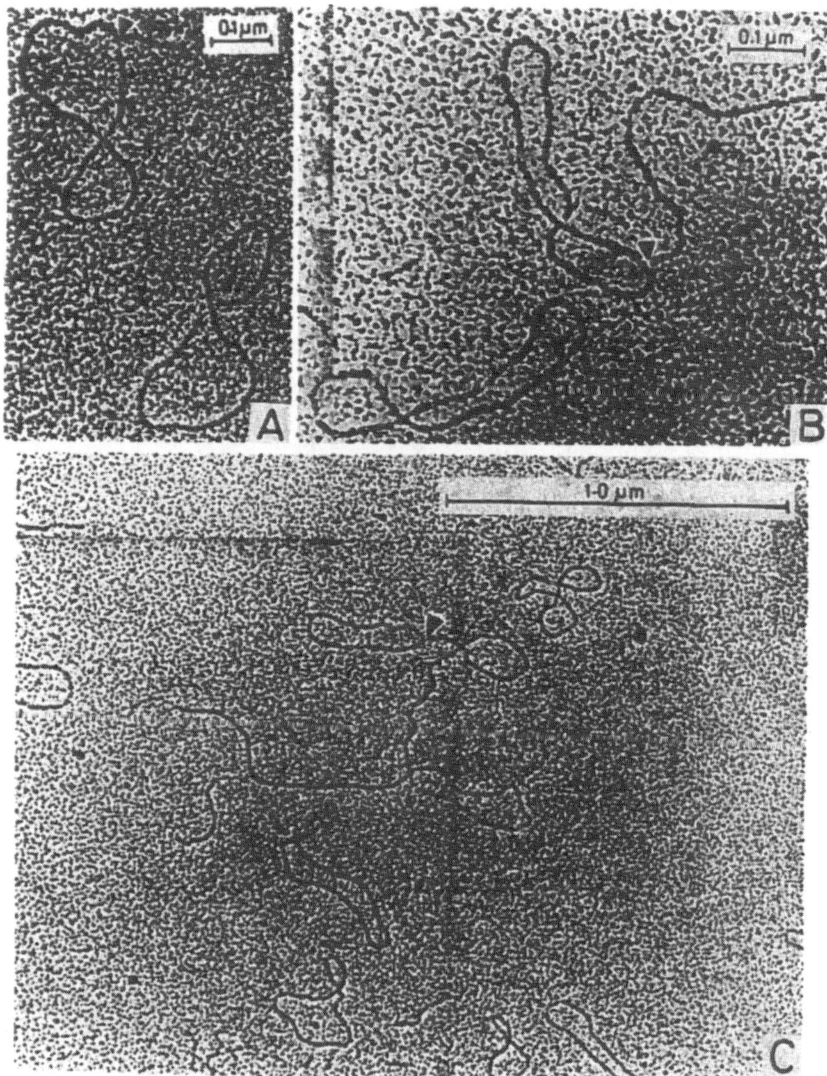

Fig. 9 Structure of the replicated molecules. Tails of various sizes of the lariats are shown with the forkpoints marked by single arrows. Corresponding unreplicated molecules are also shown for dimensional comparisons. Scales are at the top of each figure.

of *in vitro* replication maps near the displacement-loop regions where replication initiates also *in vivo*. Although the observed replication initiation in the OriA recombinant template was chloroplast specific, the mode of replication was different from that observed *in vivo* with intact ctDNA. However, when the template contained both the OriA and OriB sequences, the *in vivo* replication proceeded in the theta mode, the mode of replication usually observed *in vivo*.

The understanding of the replicative intermediates and the mapping of the Ori regions of ctDNA gene gave impetus to use recombinant containing Ori region as vectors for transformation (Daniell et al., 1990). Cultured NT1 tobacco cells were bombarded with tungsten particles coated with pUC18 (negative control), 35S CAT (nuclear expression vector), pHD407 (chloroplast expression vector with replicon). Sonic extracts of cells bombarded with pUC 18 showed no detectable *cat* activity. Nuclear expression of *cat* reached two thirds of the maximal activity 48 hrs after bombardment and the maximal at 72 hrs. Cells bombarded with chloroplast expression vectors showed a low level of expression until 48 hrs of incubation. A dramatic increase in the expression of *cat* was observed 24 hrs after the addition of fresh medium to cultured cells in samples bombarded with pHD407; the replicon less vector pHD312 showed about 50% of the maximal activity. The expression of nuclear *cat* and the repliconless chloroplast vector decreased after 72 hrs, but a high level of chloroplast expression maintained in cells bombarded with pHD407. Organelle specific expression of *cat* in appropriate compartments was checked by introducing various plasmids into tobacco protoplasts by electroporation. Although the nuclear vector 35S-CAT showed expression of *cat*, no activity was observed with any chloroplast vectors.

The stable formation of plastids in higher plants was, however, achieved by the group of Maliga (Svab et al., 1990). Plastid transformation was obtained by bombarding the leaves of *N. tabacum* with P2S148 plasmid. Plasmid P2S148 contained 3.7 kbp plasmid DNA fragment encoding the 16S rRNA. In the 16S rRNA encoding DNA, a spectinomycin resistance mutation as flanked on the 5′ side and on the 3′ side was a Pst I site generated by ligating an oligonucleotide in the intergenic region. Transgenic lines were selected by spectinomycin resistance and distinguished from spontaneous mutants by the flanking, cotransformed streptomycin resistance and Pst I markers. Regenerated plants were homoplasmic for the spectinomycin resistance and the Pst I markers, heteroplasmic for the unselected streptomycin resistance trait. Transgenic plastid traits were transmitted to the seed progeny. The transgenic plastid genomes were products of multistep processes, involving DNA recombination, copy correction and sorting out of plastid DNA copies.

Various enzymes which are involved in ctDNA replication have been isolated. DNA polymerase which is responsible for synthesis of new daughter strands have been purified to homogeneity from pea chloroplasts (Mckown and Tewari, 1984; Tewari, 1986). A γ-like DNA polymerase, which is suggested to be involved in ctDNA replication, have also been isolated from chloroplast of spinach (Sala et al., 1980), cultured soybean cells (Heinhorst et al. 1990) and *Chlamydomonas reinhardtii* (Wange et al., 1991). The gene for ctDNA polymerase has not been found yet.

The DNA topoisomerases which are required for interconversion of different topological constraints have also been isolated from higher plants. These enzymes are essentially of two types (Gellert, 1981). Type I topoisomerases introduce transient single-stranded breaks into DNA, pass an intact single strand of DNA through the broken strand, and then re-ligate the break. In contrast, type II topoisomerases make transient double-stranded breaks into one segment of DNA and pass an intact duplex through the broken DNA before resealing the break (Wang, 1985; Watt and Hickson, 1994). From higher plants both topoisomerase I (Siedlecki et al., 1983; Nielson and Tewari, 1988; Fukata et al., 1991) and topoisomerase II (Lam and Chua, 1987) specific to chloroplasts have been isolated. In contrast to previously reported pea chloroplast enzyme (112 kDa) which was prokaryotic type 1 (Nielsen and Tewari, 1988), a eukaryotic type 1 topoisomerase of 69 kDa was also reported (Mukherjee et al., 1994) which relaxes both positive and negative supercoils in topological steps of unity without requiring magnesium ions. Although a gene for topoisomerase I (topo 1) from *Arabidopsis* has been reported (Kieber et al., 1992) a gene specific for ctDNA has still not been found.

DNA ligase, which seals the single-strand breaks in DNA by forming a phosphodiester linkage between the 3′ OH group on one side of the break and the 5′ phosphate group on the other side of the break, has been reported in pea but the genes for DNA ligases from chloroplast genome have not been discovered yet. Similarly, the gene for chloroplast DNA primase has not been reported till to date. However, the pea chloroplast DNA primase has been identified and characterised (Nielsen et al., 1991).

A DNA helicase, which unwinds the duplex DNA by interacting with ssDNA or ss-/ds-junction and translocating unidirectionally along the bound strand into the duplex region disrupting the hydrogen bonds between the two strands, has also been purified to homogeneity and characterized (Tuteja et al., 1996) from pea chloroplasts. However, the gene for chloroplast DNA helicase has also not been found. The unwinding by DNA helicases are ATP dependent and they are also known for their analog to motor proteins such as myosin, kinesin and dynein which also uses ATP for their action. The helicases are known to play essential role in DNA replication as well as other DNA transactions (Tuteja and Tuteja, 1996; Tuteja, 1997). The effect of different DNA-interacting ligands on chloroplast DNA helicase activity has also been studied (Tuteja and Phan, 1998). This study showed that the DNA-interacting ligands actinomycin C_1, ethidium bromide, daunorubicin and nogalamycin were inhibiting the DNA unwinding activity of pea chloroplast DNA helicase with an apparent Ki of 2.9 μM, 3.0 μM, 1.4 μM and 1.0 μM, respectively. These inhibitors also inhibited the ATPase activity of pea chloroplast DNA helicase. Similar results were also reported for human DNA helicase II (Tuteja et al., 1997). These results indicated that the intercalation of the inhibitors into DNA generates a complex that impedes the translocation of chloroplast DNA helicase, resulting in both inhibition of unwinding activity and ATP hydrolysis (Tuteja and Phan, 1998). This kind of study could be useful for understanding the mechanism of chloroplast DNA unwinding. Recently, second helicase from pea, called chloroplast DNA helicase II, has been purified to homogeneity and characterized (Tuteja et al.; unpublished data). This enzyme is stimulated by replication fork-like structure of the DNA substrate. Since the critical role of the fork structure in DNA replication has been well documented, it seems likely that pea DNA helicase II is involved in chloroplast DNA replication.

Mostly, these replicative enzymes are present in extremely low abundance in chloroplast which give problems in obtaining enough amount of these enzymes for raising antibodies or for microsequencing of the protein. That could be a probable reason for delay in getting the genes for these enzymes.

5. Transcription of the Chloroplast DNA

Gene transcription is responsible for controlling the plant development, differentiation and also environmental responses. The *trans*-factors which select genes to be transcribed and cis elements on the ctRNA play the central roles in transcription process. The biogenesis of chloroplast requires the coordinate expression of both specific nuclear genes and chloroplast genes. In general, the chloroplast gene expression is largely regulated post-transcriptionally, wherase nuclear gene expression in plant is regulated at the level of transcription (Harris et al., 1994). The transcription of chloroplast genes is catalyzed by a DNA-dependent RNA polymerase which is distinct from nuclear RNA polymerases (Bottomley et al., 1971; Briat and Mache, 1980; Tewari and Goel, 1983).

Actually, the convincing evidence that chloroplasts did contain their own DNA led to the search of DNA dependent RNA polymerase in chloroplasts. The identification of chloroplast specific enzymes for DNA replication and transcription has always been diffcult because similar enzymes are present in the nucleus. Unlike ctDNA, these enzymes cannot be identified conclusively. This problem has been most evident in the case of transcription. Even though Tewari and Wildman (1969) unequivocally demonstrated the presence of DNA dependent RNA polymerase in tobacco leaves, the nature of the RNA polymerase and mechanism of transcription is far from understood.

Previously, it was suggested that chloroplasts contain two RNA polymerases (Greenberg et al., 1984; Gruissem et al., 1986). One of these activities, referred to as the "soluble activity", could be readily dissociated from ctDNA by high salt extraction and accepted exogeneous DNA as template. It can transcribe, in a template-dependent manner, only the genes for mRNA and tRNA (Greenberg et al., 1984; Orozco et al., 1985; Gruissem et al., 1986). Another RNA polymerase was found to be associated with the transcriptionally active chromosome (TAC) fraction, which did not accept exogeneous DNA as template and was shown to transcribe all three classes of ct-genes with preference to rRNA genes (Briat and Mache, 1980; Rushlow et al., 1980; Blanc et al., 1981; Narita et al., 1985; Reiss and Link, 1985; Lebrum et al., 1986). TAC was first isolated from *Euglena gracilis* and later from many higher plants (Hallick et al., 1976). On the basis of general transcription activity assayed by nucleotide incorporation into acid-insoluble material, attempts at purification of RNA polymerase from chloroplast of pea, spinach, maize and tobacco have resulted in preparations containing 7–14 polypeptides with molecular weight ranging from 22 to 180 kDa (Smith and Bogorad, 1974; Lerbs et al., 1983; Greenberg et al., 1985; Lerbs et al., 1988).

Table 2 shows the different polypeptides present in the highly purified RNA polymerase from maize, pea and spinach (Hu and Bogorad, 1990; Rajshekhar et al., 1991; Igloi and Kossel, 1992). Even though Table 2 shows very few polypeptides for maize chloroplast, the preparation really contained many more polypeptides which were not addressed to. In case of pea, the

Table 2 Components and Properties of Highly Purified Chloroplast RNA Polymerase (Igloi and Kossel, 1992)

	Maize	Pea	Spinach
Peptide components and assigned subunits	180 kDa (β'' subunit)		
		150 kDa	150 kDa (β'' subunit) 145 kDa
		130 kDa	
	120 kDa (β subunit)		120 kDa (110 kDa) (β subunit)
		115 kDa	
		110 kDa	102 kDa
		95 kDa	
		(90 kDa)	
	85 kDa (β' subunit)	85 kDa	78 kDa (80 kDa) (β' subunit)
		75 kDa	75 kDa
			64 kDa
			58 kDa
		48 kDa	48 kDa
		39 kDa	
	38 kDa (α subunit)		38 kDa (α subunit)
		(34 kDa)	34 kDa
		(32 kDa)	
		(27 kDa)	27 kDa
			17 kDa

number of polypeptides could not be reduced by any chromatographic procedures. The promoter affinity chromatography, using Sepharose to which the cloned pea chloroplast rDNA promoter in the form of long head to tail concatamers was covalently coupled, failed to further reduce the number of polypeptides. The purified soluble enzyme was found to faithfully transcribe rRNA, tRNA and mRNA genes. The question remains whether there is one enzyme which transcribes all three types of RNAs or there are three different types of RNA polymerases each specific for a given type of RNA as found in eukaryotic cells. At present, there is no answer to this important question.

A further problem in the understanding of the structure of RNA polymerase is the presence of *E. coli* RNA polymerase like genes in the chloroplast genome designated as *rpoA, rpoB, rpoC and rpoC2.* Hu and Bogorad (1990) have identified the products of these genes as a 180 kDa (β'' subunit), 120 kDa (β subunit), 85 kDa (β' subunit) and 38 kDa (α subunit) polypeptide. However, the RNA polymerase activity has not been demonstrated using these polypeptides alone.

In the absence of any direct approach to identify the functional polypeptide of chloroplast RNA polymerase, Khanna et al. (1991, 1992) and Lakhani et al. (1993) have utilized photoaffinity labeling technique. UV irradiation of lysed chloroplasts or the isolated transcriptionally active chromosome photocross linked nascent radiolabeled transcripts to a 48 kDa protein. The nascent transcript binding protein appears to be involved in transcription of all three classes of chloroplast genes. Since 48 kDa protein has been found to copurify with RNA polymerase activity isolated from chloroplasts of pea as well as spinach and the antibody raised against this protein has been shown to block *in vitro* transcription of the chloroplast 16S rRNA gene (Khanna et al., 1991), it can be concluded that 48 kDa protein is a part of the chloroplast RNA polymerase. In a similar approach, Khanna et al. (1992) have shown that a polypeptide of 150 kDa binds to both, chloroplast ribosomal (16S) and messenger RNA (*psbA*) promoters. The 16S rRNA and *psbA* promoters were amplified from chloroplast DNA by the polymerase chain reaction and labeled with a photoactive analogoue to TTP, 5 bromodeoxy UTP, as well as with α^{32}P-dCTP. The RNA polymerase was incubated with the labeled promoters, photocrosslinked and the mixture resolved in denaturing gel electrophoresis. The polypeptide of 150 kDa was the only labeled polypeptide. The results demonstrated that the 150 kDa polypeptide is a functional template binding polypeptide of the pea chloroplast transcription complex.

An interesting paper by Mache (1993) raises yet another question on the nature of RNA polymerase in chloroplasts. He has reported that highly purified RNA polymerase which contains a number of polypeptides can be separated under defined conditions into three different fractions by Heparin-Sepharose chromatography. Immunological studies revealed that the first fraction contained RNA polymerase activity associated with all seven major polypeptides. However, similar analysis of the remaining fractions showed that activity associated only with the 110 kDa polypeptide. The 110 kDa polypeptide purified by SDS-PAGE actively synthesized RNA in a reaction dependent on a supercoiled DNA template and the four ribonucleoside triphosphates. The fact that 110 kDa enzyme is only separated from other polypeptides prepared from very young plants indicates that the abundance of the two types of RNA polymerases changes during plant development. Mache (1993) speculates that 110 kDa enzyme preferentially transcribes the plastid genome during early stages of plant development until the other multi-subunit RNA polymerase starts functioning.

The steady-state level of a given mRNA is determined by its rates of transcription and degradation. In chloroplast of higher plants, mRNA stability is known to vary at different developmental stages (Deng and Gruissem, 1987). The control of chloroplast mRNA degradation is therefore likely to play a significant role in regulation of gene expression. The half-lives of chloroplast mRNAs have been estimated to range from 3 h to more than 40 h in land plants (Rapp et al., 1992). In spinach chloroplasts, the post-transcriptional addition of poly(A)-rich sequences to the plastid-encoded *psbA* mRNA as well as to several other transcripts has been described (Kudla et al., 1996). Unlike eukaryotic nuclear-encoded and bacterial RNAs, the poly(A) moiety in the chloroplast was not found to be ribohomopolymer of adenosine residues but rather clusters of adenosines bounded mostly by guanosines and rarely by cytidines and uridines (Lisitsky et al., 1997). The tails might be several hundred nucleotides long (Lisitsky et al., 1996). Poly(A)-rich tails play a major role in the rapid degradation of intermediary products of mRNA decay in the chloroplast, targeting the cleavage products to rapid degradation due to their high affinity to chloroplast exonuclease(s) (Kudla et al., 1996; Lisitsky et al., 1996). Lisitsky et al. (1997) observed that blocking the polyadenylation of RNA inhibits RNA degradation.

The two plastid RNA polymerases of higher plants, the eubacterial type, plastid-encoded plastid RNA polymerase (PEP) and the phage-type, nuclear-encoded plastid RNA polymerase (NEP), act sequentially, forming a cascade during chloroplast development (Maliga, 1998). NEP is active in proplastids of meristematic cells transcribing housekeeping genes and PEP takes over the transcription of housekeeping genes and initiates the transcription of photosynthesis-related genes in developing chloroplasts (Hajdukiewicz et al., 1997; Kapoor et al., 1997; Maliga, 1998).

The RNA editing in chloroplast of higher plants which is a post-transcriptional process and is responsible for specific C-to-U conversions, leading to RNA sequence different from the corresponding DNA sequences, is not covered in this Chapter. For a more detailed study please see Maier et al. (1996).

5.1 Promoters

For proper transcription, chloroplast genes at upstream regions also contain consensus which are almost similar to bacterial –35 (TTGACA) and –10 (TATAAT) promoter sequences. Usually by using *in vitro* transcriptions system and deleted or mutated genes, the promoters of chloroplast genes have been determined (Link, 1984; Gruissem and Zurawski, 1985). Although –35 region is not always essential for promoter activity and is not always identifiable in ct-genes (Gruissem and Zurawski, 1985). In *atpB* promoter the deletion of 5′ sequences up to -12 from transcription start, still shows transcription (Klein et al., 1992). On the other hand a class of chloroplast tRNA gene from spinach was shown to be not requiring their 5′ upstream promoter elements for transcription (Gruissem et al., 1986). In mustard (*Sinapis alba*) the *psbA* gene also contains TATA box, similar to nuclear, in between –35 and –10 region which is required for correct transcription (Eisermann et al., 1990). The rate of transcription varies among promoters of ct-genes and range up to 50-fold or greater in single chloroplast type (Deng and Gruissem, 1987, 1988; Blowers et al., 1990; Rapp et al., 1992; Salvador et al., 1993). The relative transcription rates of a single gene can differ between plant species, for example *psbA* transcription is about 30% that of the 16S rRNA in spinach (Deng and Gruissem, 1987) and 150% in barley (Rapp et al., 1992). The transcription rates from chimeric genes, in which foreign mRNAs are transcribed from chloroplast promoters, show that these chimeric RNAs are transcribed at similar rates to the corresponding endogenous RNA and that this transcription appears to be independent of the position of the gene within the chromosome (Blowers et al., 1990; Goldschmidt-Clermont et al., 1990; Salvadon et al., 1993; Staub and Maliga, 1993). The sequences downstream of the –10 and –35 elements may also have some role in determining the rate of transcription from individual promoter as shown in case of *atpB* promoter (Klein et al., 1992). Perhaps these elements act as transcription enhancers, as have been identified for the *psbA* gene in cyanobacteria (Li and Golden, 1993).

The chloroplast promoter regions are AT rich similar to prokaryotic consensus (Whitfeld and Bottomley, 1983). Usually promoter of the genes of the Rubisco large subunit (*rbc*L), the β/ε-subunits of ATPase (*atp*BE), and the 32-kDa PS II core protein (*psbA*) are highly conserved in monocotyledonous and dicotyledonous plants. It is shown that the 49 bp promoter region derived from the maize *rbcL* gene is recognized less efficiently by the maize RNA polymerase in a chloroplast extract than a template which includes an additional 26 bp 5′ proximal to the promoter region and 34 pb of 3′ DNA (Hanley-Bowdoin and Chua, 1987). The addition of these sequences does not affect the transcription initiation frequency of heterologous pea RNA polymerase at this promoter region. Base changes in the conserved promoter regions of chloroplast genes from different plants may be of functional importance for the chloroplast RNA polymerase. For example, pea *trn*M2 is a less efficient template for the spinach chloroplast RNA polymerase (Gruissem and Zurawski, 1985). Thus, chloroplast promoter regions having DNA sequences not defined by the canonical prokaryotic promoter may be important for the species-specific modulation of transcription in chloroplasts.

The promoter activity of chloroplast genes may also be regulated by local changes in superhelical densities of the DNA template (Thompson and Mosig, 1987; Lam and Chua, 1987). The light-induced barley *psbD-psbC* RNAs arise from an unusual promoter which mediates transcription initiation from a 23 nucleotide region (Sexton et al., 1990). The *psbD* operon of higher plant chloroplasts is regulated transcriptionally through the activity of an upstream light-responsive promoter (Allison and Maliga, 1995).

Recently, PEP (plastid-encoded RNA polymerase) promoter and NEP (nuclear-encoded RNA polymerase) promoter were also reported in the transcriptional apparatus of chloroplasts (Hajdukiewicz et al., 1997; Maliga, 1998). The PEP is prokaryotic type promoter and NEP is housekeeping gene promoter (Stern et al., 1997). In spinach, the NEP promoter is located between two –35/–10 promoters, which can be blocked *in vitro* by the putative transcription factor CDF2; in tobacco, the NEP promoter is located further downstream (Stern et al., 1997).

5.2 Transcription of rRNA Genes

The chloroplast rRNA genes are probably transcribed as large precursor molecules that subsequently undergo many reactions to generate the mature rRNAs (Delp et al., 1991). The 5′- and 3′- terminal precursor sequences of the 16S and 23S RNAs can form double-stranded stem structures similar to those of the *E. coli* rRNA genes, leading to the suggestion that these molecules may be processed by an RNase III-like endonucleolytic cleavage (Delp et al., 1991). However, tobacco chloroplast ribosomes contain a minor fraction of 16S rRNA molecules in which a 30-nt leader sequence containing the putative RNaseIII site is still present, suggesting that the final maturation of the 16S rRNA may take place within the ribosome (Verma et al., 1993). Whether the chloroplast rRNA genes are transcribed by a different RNA polymerase, is still a confliction (Tiller et al., 1991; Igloi and Kossel, 1992; Gruissem and Tonkyn, 1993; Lerbs-Mache, 1993). Chloroplast genome contains genes homologous to the *rpo*A, *rpo*B and *rpo*C genes, which encode RNA polymerase subunits of bacteria, but appear to lack a gene for *rpo*D, which encodes the principal σ factors of the bacterial RNA polymerase (Bergsland and Haselkorn, 1991; Fong and Surzycki, 1992; Troxler et al., 1994). However, there is also a report of σ-like factor in chloroplast RNA polymerase (Tiller et al., 1991; Troxler et al., 1994). It is shown in *Euglena* that the expression of chloroplast encoded RNA polymerase genes resulted in soluble fraction which transcribes tRNA and mRNA genes whereas the membrane-bound fraction is responsible for transcription of rRNA genes (Greenberg et al., 1984; Little and Hallick, 1988). In the case of spinach chloroplast both the rRNA and protein-coded genes are transcribed by soluble RNA polymerase (Briat et al., 1987).

The 16S rRNA and mRNAs for many chloroplastic ribosomal proteins usually accumulate during early development as shown in spinach (Bisanz-Seyer et al., 1989). The 16S rRNA, which is also present in seeds, starts increasing during seed germination. However, most of the mRNA of ribosomal protein starts appearing at the beginning of germination. For example *rps*19 and *rpl*23 mRNA appeared 2 days earlier. On the other hand the mRNAs of *rps*3 and *rpl*16 appear only after germination.

Although the presence of more than one type of chloroplast transcription machinery has been suggested but there is still uncertainty about the existence and characteristics of a second RNA polymerase which is distinct from the *E. coli* like enzyme. To address this question, recently the group of Pal Maliga has deleted the gene (*rpo*B) for one of the essential subunits of the *E. coli*-like enzyme from tobacco chloroplast genome and checked whether transcription was maintained in the mutant chloroplasts (Allison et al., 1996). Their data showed that in the absence of the chloroplast-encoded *E. coli*-like RNA polymerase, expression of photogenes was reduced significantly. In contrast, the transcription levels for the ct-genes encoding the gene expression apparatus were similar to levels in wild-type plants. It showed the existence of second chloroplast transcription system which selectively transcribed a subset of ct-genes by non-*E. coli*-like RNA polymerase. In contrast to the *E. coli*-like RNA polymerase, this novel transcription machinery preferentially transcribes genetic system rather than photosynthetic genes. This second transcription apparatus is designated as the Genetic System RNA polymerase, which also recognizes a novel promoter sequence. Although the mutant plants were photosynthetically defective but the transcription by this polymerase was sufficient for chloroplast maintenance and plant development (Allison et al., 1996).

5.3 Transcription of ct-genes for Ribosomal Proteins

The gene products of chloroplast and nucleus which are mainly controlling chloroplast biogenesis, are found within the multimeric thylakoid complexes or chloroplast ribosomes. The ct-proteins encoded genes, which are present in single copies or small gene families, are regulated mainly at the transcription level (Gilmartin et al., 1990; Kuhlemeier, 1992). However, most ct-genes transcribed at all times during chloroplast development and mainly regulated post transcriptionally. All the ribosomal protein genes in higher plants are known to transcribe but only some have been shown to be translated into the corresponding proteins. However, the ribosomal protein S12, S16, S19, L2, L20, L32, L33 and L36 of spinach chloroplasts are known to be products of their corresponding ct-genes (Schmidt et al., 1992). Many transcripts are also reported for individual ribosomal protein genes which probably arise from processing of larger polycistronic transcripts (Ohto et al., 1988;

Subramaniam, 1991). In *Euglena gracilis* the primary transcript of large ribosomal protein operon contain 11 genes encoding ribosomal proteins, a tRNA gene, and an ORF encoding a highly basic protein (Christopher and Hallick, 1990; Hallick et al., 1993). An 8.3 kb mRNA from which all introns have been spliced, is then processed stepwise into transcripts containing one or more genes. A mixed ct-gene cluster in maize was also reported which contains genes encoding 4 subunits of the ATP synthase (*atp*I, *atp*H, *atp*F, and *atp*A) and the gene encoding ribosomal protein S2 (*rps*2) produces a total of 12 transcripts including a major species of 6,200 nt containing mRNAs of all five genes (Stahl et al., 1993). They suggested that this ct-gene cluster is functionally organized as an operon with additional regulatory features to allow for increased accumulation of mRNA for the thylakoid components.

In rice the *psa*A, *psa*B and *rps*14 genes are organised into a single transcriptional unit (Chen et al., 1983). All these three genes recognized a 5.2 kb transcript.

5.4 Terminators

The RNA molecule also contains the signal sequences for termination of transcription, which supports the theory that IR allows the formation of intramolecular stem-loop structures in the mRNA, creating a pause in transcription elongation by RNA polymerase (Platt, 1986). Usually series of Uridine residues are present at 3′-end of transcript which help in disassociation of the transcript from template, because RNA-DNA hybrids with rU-dA pairing are not stable (Platt, 1986). The IR sequences in chloroplast genome have a general function as transcription terminators (Steinmetz et al., 1983; Deno et al., 1984; Sugita and Sugiura, 1984; Kirsch et al., 1986). The IR of different genes of chloroplast have different termination frequency which was inferred from the ratio of IR-containing RNA to tRNA produced in the chloroplast transcription extract (Stern and Gruissem, 1987). The termination frequencies of chloroplast 3′ IR sequences of some of the genes like *petD, rpoA, rbcL, psbA, trnH* and *trnS* are reported to be 38%, 0%, 0%, 0%, 85% and 86% respectively (Stern and Gruissem, 1987; Gruissem et al., 1986). A rapid and precise RNA processing by chloroplast RNA polymerase also corrects the ineffective termination at the IR sequence.

In plants *in vitro* transcription systems have also been reported which provide the best way to study the individual steps in transcription, e.g. initiation, elongation and termination. Furthermore, *in vitro* functional assays allow the biochemical isolation of various factors involved in each step (Sugita and Sugiura, 1996). Recently, Zhu et al. (1995) reported *in vitro* transcription system using monocotyledonous and dicotyledonous whole-cell extracts and circular DNA templates. They have also proved for the first time that the TATTTAA sequence (positions –35 to –28) is an authentic TATA box essential for RNA polymerase II-dependent transcription. Some more *in vitro* transcription systems have been reported, which include a RNA polymerase

I-dependent cell extract from *Vicia faba* (broad bean) (Yamashita et al., 1993), a nuclear extract from tobacco cells (Yamaguchi et al., 1994), a light-responsive system from parsley (*Petroselium crispum*) cells (Frohnmeyer et al., 1994) and a basal, activatable *in vitro* system from tobacco cultured-cell nuclei, which supports transcription by RNA Polymerase I, II, and III (Fan and Sugiura, 1995).

References

Akkaya, M.S. and Breitenberger, C.A. 1992. Light regulation of protein synthesis factor EF-G in pea chloroplast. *Plant Mol. Biol.* **20:** 791–800.

Alfano, C. and McMacken, R. 1989. Ordered assemblies of nucleoprotein structures at the bacteriophage and replication origins during the initiation of DNA replication. *J. Biol. Chem.* **264:** 10699–10708.

Allet, B. and Rochaix, J.D. 1979. Structure analysis at the end of the intervening DNA sequences in the chloroplast 23S orbosomal genes of *C. reinhardtii. Cell* **18:** 55–60.

Allison, L.A. and Maliga, P. 1995. Light-responsive and transcription enhancing elements regulate the plastid *psbD* core promoter. *EMBO J.* **14:** 3721–3730.

Allison, L.A., Simon, L.D. and Maliga, P. 1996. Deletion of *rpoB* reveals a second distinct transcription system in plastids of higher plants. *EMBO J.* **15:** 2802–2809.

Apel, K. 1979. Phytochrome-induced appearance of mRNA activity for the apoprotein of light-harvesting chlorophyll a/b-protein. *Eur. J. Biochem.* **97:** 183–188.

Baur, E., 1909. Das Wesen und die Erblichkeitsverhältnisse der "varietates albomarginatachort" von *Pelargonium zonale. Z. Vererbungsl,* **1:** 330–351.

Bedbrook, J.R. and Bogorad, L. 1976. Endonuclease recognition sites mapped on Zea mays chloroplast DNA. *Proc Natl. Acad. Sci. USA,* **73:** 4309–4313.

Bendich, A.J. 1987. Why do chloroplasts and mitochondria contain so many copies of their genome? *BioEssays,* **6:** 279–282.

Bergsland, K.J. and Haselkorn, R. 1991. Evolutionary relationship among eubacteria, cyanobacteria and chloroplasts. *J. Bacteriol.* **173:** 3446–3455.

Bisanz-Seyer, C., Li, Y.-F., Seyer, P. and Mache, R. 1989. The components of the plastid ribosome are not accumulated synchronously during the early development of spinach plant. *Plant Mol. Biol.* **12:** 201–211.

Blanc, M. Briat, I.F. and Laulhere, J.P. 1981. Influence of the ionic environment on the *in vitro* transcription of the spinach plastid DNA. *Biochim. Biophys. Acta.* **655:** 347–382.

Blowers, A.D., Ellmore, G.S., Klein, U. and Bogorad, L. 1990. Transcriptional analysis of endogenous and foreign genes in chloroplast transformants of *Chlamydomonas. Plant Cell* **2:** 1059–1070.

Bottomley, W., Smith, M.J. and Bogorad, L. 1971. RNA polymerase of maize: partial purification and properties of the chloroplast enzyme. *Proc. Natl. Acad. Sci. USA* **68:** 2412–2416.

Bramhill, D. and Kornberg, A. 1988. A model for initiation at origins of DNA replication. *Cell,* **54:** 915–918.

Breitenberger, C.A., Graves, M.C. and Spremulli, L.L. 1979. Evidence for the nuclear location of the gene for chloroplast elongation factor *G. Arch. Biochem. Biophys.* **194:** 265–270.

Briat, J.-F., and Mache, R., Properties and characterization of a spinach chloroplast RNA polymerase isolated from a transcriptionally active DNA-protein complex. *Eur. J. Biochem.* **111:** 503–509.

Briat, J.-F., Bisanz-Seyer, C. and Lescure, A.-M. 1987. *In vitro* transcription initiation of the rDNA operon of spinach chloroplast by the highly purified soluble homologous RNA polymerase. *Curr. Genet.* **11:** 259–263.

Bryant, D.A. and Stirewalt, V.L. 1990. The cyanella genome of *Cyanophora paradoxa* encodes ribosomal proteins not encoded by chloroplast genome of land plants. *FEBS Lett.* **259:** 273–280.

Bubunenko, M.G., Schmidt, J. and Subramanian, A.R. 1994. Protein substitution of chloroplast ribosome evolution. *J. Mol. Biol.* **240:** 28–41.

Burke, J.M. 1988. Molecular genetics of group I introns: RNA structures and protein factors required for splicing. *Gene* **73:** 273–294.

Carrasco, J.L., Chueca, A., Prado, F.E., Hermoso, R., Lazaro, J.J., Ramos, J.L., Sahrawy, M. and Gorgi, L. 1994. Cloning structure and expression of a pea cDNA clone coding for a photosynthetic furctose-1, 6-bisphosphatase with some features different from those of the leaf chloroplast enzyme. *Planta* **193:** 494–501.

Cech, T.R. 1988. Conserved sequences and structures of group I introns. *Gene,* **73:** 259.

Chen, S.G., Cheng, M., Chung, K., Yu, N. and Chen, M. 1992. Expression of the rice chloroplast psaA-psaB-rps14 gene cluster. *Plant Sci.* **81:** 93–102.

Christopher, D.A. and Hallick, R.B. 1990. Complex RNA maturation pathway for a chloroplast ribosomal protein operon with an internal tRNA cistron. *Plant Cell* **2:** 659–671.

Chu, N.M., Oishi, K.K. and Tewari, K.K. 1981. Physical mapping of the pea chloroplast DNA and localization of the ribosomal RNA genes. *Plasmid* **6:** 279–292.

Clarke, A.K., Gustafsson, P. and Lidholm, J A. 1994. Identification and expression of the chloroplast clp P gene in the conifer *Pinus contorta. Plant Mol. Biol.* **26:** 851–862.

Correns, C. 1909. Vererbungsversuche mit Plaβ(gelb) grunen und buntblattrigen sippen bei *Mirabilis jalapa, Urtica pilulifera* und *Lunaria annua. Z. Vererbungsl* **1:** 291.

Daniell, H., Vivekananda, J., Nielsen, B.L., Ye, G.N., Tewari, K.K. and Sanford, J.C. 1990. Transient foreign gene expression in chloroplasts of cultured tobacco cells after biolistic delivery of chloroplast vectors. *Proc. Natl. Acad. Sci. USA*, **87:** 88–92.

Danon, A. and Mayfield, S.P.Y. 1991. Light regulated translational activators: identification of chloroplast gene specific mRNA binding proteins. *EMBO J.* **10:** 3993–4001.

Dean, C., Favreau, M., Bond-Nutter, D., Bedbrook, J. and Dunsmuir, P. 1989. Sequences downstream of translation start regulate quantitative expression of two *petunia* rbcS genes. *Plant Cell* **1:** 201–208.

Deno, H., Shinozaki, K. and Sugiura, M. 1984. Structure and transcription of a tobacco chloroplast gene for the III subunit of proton-translocation ATPase. *Gene* **32:** 195–201.

Delp, G. and Kosseel, H. 1991. rRNAs and rRNA genes of plastids. *Cell Cult. Somatic Cell Genet. Plants* **7A:** 139–167.

Deng, X.W. and Gruissem, W. 1987. Control of plastid gene expression during development: the limited rate of transcriptional regulation. *Cell* **49:** 379–387.

Deng, X.W. and Gruissem, W. 1988. Constitutive transcription and regulation of gene expression in non-photosynthetic plastids of higher plants. *EMBO J.* **7:** 3301–3308.

Dormann-Przybyl, D., Strittmatter, G. and Kossel, H. 1986. The region distal to the rRNA operon from chloroplasts of maize contains genes coding for tRNA Arg (ACG) and tRNA Asn (GUU). *Plant Mol. Biol.* **7:** 419–431.

Drager, R.G. and Hallick, R.B. 1993. A complex twintron is excised as four individual introns. *Nucleic Acids Res.* **21:** 2389–2394.

Dujon, B. 1989. Group I introns as mobile genetic elements. *Gene* **82:** 91–114.

Durrenberger, F. and Rochaix, J.D. 1991. Chloroplast ribosomal intron of *Chlamydomonas reinhardtii. EMBO J.* **10:** 3495–3501.

Edelman, M., Cowan, C.V., Epstein, G.T. and Schiff, J.A. 1964. Studies of chloroplast development in Euglena VIII. Chloroplast associated DNA. *Proc. Natl. Acad. Sci. USA* **52:** 1214–1219.

Edwards, K., Bedbrook, J., Dyer, T. and Kossel, H. 1981. 4.5S rRNA from *Zea Mays* chloroplasts shows structural homology with the 3'end of prokaryotic 23S rRNA. *Biochem. Int.* **2:** 533–538.

Eisermann, A., Tiller, K. and Link, G. 1990. *In vitro* transcription and DNA binding characteristics of chloroplast and etioplast extracts from mustard. *EMBO J.* **9:** 3981–3987.

Fan, H. and Sugiura, M. 1995. A plant basal *in vitro* system supporting accurate transcription of both RNA polymerase II and III dependent genes: supplement of green leaf component(s) drives accurate transcription of a light-responsive *rbcS* gene. *EMBO J.* **14:** 1024–1031.

Fong, S.E. and Surzycki, S.J. 1992. Chloroplast RNA polymerase genes of C. *reinhardtii* exhibit an unusual structure and arrangement. *Curr. Genet.* **21:** 485–497.

Fox, L., Erion, J., Tarnowski, J., Spremulli, L.L., Brot, N. and Weissbach, H. 1980. *Euglena gracilis* chloroplast EF-Ts: evidences that it is nuclear encoded gene product. *J. Biol. Chem.* **255:** 6018–6019.

Frohnmeyer, H., Hahlbrock, K. and Schafer, E. 1994. A light-respective *in vitro* transcription system from evacuolated parsley protoplasts. *Plant J.* **5:** 437–449.

Fujiwara, S., Kawachi, M., Inouye, I. and Someya, J. 1994. The gene for ribosomal protein L27 is located on the plastid rather than the nuclear genome of chlorophyll c-containing alga. *Plant Mol. Biol.* **24:** 253–257.

Fukata, H., Mochida, A., Maruyama, N. and Fukasawa, H. 1991. Chloroplast DNA topoisomerase I from cauliflower. *J. Biochem.* (Tokyo) **109:** 127–131.

Gamble, P.F. and Mullet, J.E. 1989. Translation and stability of proteins encoded by the plastid *psbA* and *psbB* genes are regulated by a nuclear gene. *J. Biol. Chem.* **264:** 7236–7243.

Garriga, G. and Lambowitz, A.M. 1984. RNA splicing in Neurospora mitochondria: Self splicing of a mitochondrial intron *in vitro. Cell* **38:** 631–641.

Gellert, M. 1981. DNA topoisomerases. *Ann. Rev. Biochem.* **50:** 879–910.

Gilmartin, P.M., Sarokin, L., Memelink, J. and Chua, N-H. 1990. Molecular light switches for plant genes. *Plant Cell* **2:** 369–378.

Gold, J.C. and Spremulli, L.L. 1985. *Euglena gracilis* chloroplast initiation factor 2: identification and initial characterization. *J. Biol. Chem.* **260:** 14897–14900.

Goldschmidt-Clermont, M., Girard-Bascou, J. Choquet, Y. and Rochaix, J.D. 1990. Transsplicing mutants of *C. reinhardtii. Mol. Gen. Genet.* **223:** 417–425.

Gott, J.M., Shub, D.A. and Belfort, M. 1986. Multiple self-slicing introns in bacteriophage T4: Evidence from autocatalytic GTP labeling of RNA *in vitro. Cell* **47:** 81–87.

Gray, M.W. 1988. Organelle origins and ribosomal RNA. *Biochem. Cell Biol.* **66:** 325–348.

Gray, J.C., Hird, S.M., Dyer, T.A. 1990. Nucleotide sequence of the wheat chloroplast gene encoding the proteolytic subunit of an ATP-dependent protease. *Plant Mol. Biol.* **15:** 947–950.

Greenberg, B.M., Narita, J.O., Deluca-Flaherty, C.R. and Hallick, R.B. 1985. In: Molecular Biology of the Photosynthetic Apparatus Steinback, K.E., Bonitz, S., Arntzen, C.J. and Bogorad, L., (eds), Cold Spring Harbor, N.Y., pp 303–309.

Greenberg, B.M., Narita, J.O., DeLuca-Flaherty, C., Gruissem, W., Rushlow, K.A. and Hallick, R.B. 1984. Evidence for two RNA polymerase activities in *Euglena gracilis* chloroplasts. *J. Biol. Chem.* **259:** 14880–14887.

Gruissem, W., Elsner-Menzel, C., Latshaw, S., Narita, J.O., Schaffer, M.A. and Zurawski,

G. 1986. A sub-population of spinach chloroplast tRNA genes does not require upstream promoter elements for transcription. *Nucleic Acids Res.* **14:** 7541–7556.

Gruissem, W., Greenberg, B.M., Zurawski, G. and Hallick, P.B. 1986. Chloroplast gene expression and promoter identification in chloroplast extracts. *Meth. Enzymol.*, **118:** 253–270.

Gruissem, W. and Tonkyn, J.C. 1993. Control mechanisms of plastid gene expression. *Crit. Rev. Plant Sci.* **12:** 19–55.

Gruissem, W. and Zurawski, G. 1985. Identification and mutational analysis of the promoter for a spinach chloroplast tRNA gene. *EMBO J.* **4:** 1637–1644.

Gutell, R.R., Larson, N. and Woese, C.R. 1994. Lessons from an evolving rRNA: 16S and 23S rRNA structures from a comparative perspective. *Microbiol. Rev.* **58:** 10–26.

Hajdukiewicz, P., Allison, L.A. and Maliga, P. 1997. The two plastid RNA polymerases encoded by the nuclear and plastid compartments transcribe distinct groups of genes in tobacco plastids. *EMBO J.* **16:** 4041–4048.

Hallick, R.B. 1989. Proposals for the naming of chloroplast genes. *Plant Mol. Biol. Rep.* **7:** 266–275.

Hallick, R.B., Hong, L., Drager, R.G., Favreau, M.R., Monfort, A., Orsat, B., Spielman, A. and Stutz, E. 1993. Complete sequence of *Euglena gracilis* chloroplast DNA. *Nucleic Acids Res.* **21:** 3537–3544.

Hallick, R.B., Lipper, C., Richards, O.C. and Rutter, W.J. 1976. Isolation of a transcriptionally active chromosome from chloroplasts of *Euglena gracilis. Biochemistry* **15:** 3039–3045.

Hanley-Bowdoin, L. and Chua, N.H. 1987. Chloroplast promoters. *Trends Biochem. Sci.* **12:** 67–70.

Harris, E.H., Boyntron, J.E. and Gillham, N.W. 1994. Chloroplast ribosomes and protein synthesis. *Microbiol. Rev.* **58:** 700–754.

Harris, S.A. and Ingram, R. 1991. Chloroplast DNA and biosystematics. *Taxon* **40:** 393–412.

Hartz, D., Binkley, J., Hollingsworth, T. and Gold, L. 1990. Domains of initiator tRNA and initiation codon crucial for initiator tRNA selection by *E. coli* IF3. *Genes Dev.* **4:** 1790–1800.

Hatfield, P.M., Shoemaker, R.C. and Palmer, R.G. 1985. Maternal inheritance of chloroplast DNA within the geneus *Glycine,* subgenus *Soja. J. Hered.* **76:** 373–374.

Heinhorst, S., Cannon, G.C. and Weissbach, A. 1985. Plastid and nuclear DNA synthesis are not coupled in suspension cells of *N. tabacum. Plant Mol. Biol.* **4:** 3–12.

Heinhorst, S., Cannon, G.C. and Weissbach, A. 1990. Chloroplast and mitochondrial DNA polymerases from cultured soybean cells. *Plant Physiol.* **92:** 939–945.

Henze, K., Schnarrenberger, C., Kellermann, J. and Martin, W. 1994. Chloroplast and cytosolic triosephosphate isomerases from spinach. *Plant Mol. Biol.*, **26:** 1961–1973.

Herrmann, R.G. and Kowallik, K.V. 1970. Multiple amounts of DNA related to the size of chloroplasts. *Protoplasma* **60:** 365–372.

Herrmann, R.G. 1972. Do chloroplasts contain DNA? II. The isolation and characterization of DNA from chromoplasts, chloroplasts, mitochondria and nuclei of *Narcissus. Protoplasma* **74:** 7–17.

Hiratsuka, J., Shimada, H., Wittier, W.F., Isibashi, T. and Sakamato, M., 1989. The complete sequence of the rice (*Oryza sativa*) chloroplast genome. *Mol. Gen. Genet.* **217:** 185–194.

Hu, J. and Bogorad, L. 1990. Maize chloroplast RNA polymerase: the 180-, 120- and 38-kDa polypeptides are encoded in chloroplast gene. *Proc. Natl. Acad. Sci.* **87:** 1531–1535.

Hubscher, U. and Spadari, S. 1994. DNA replication and chemotherapy. *Physiol. Rev.* **74:** 259–304.

Igloi, G.L. and Kossel, H. 1992. The transcriptional apparatus of chloroplasts. *Crit Rev Plant Sci.* **10:** 525–558.

Kanakari, S. Timmler, G., Knoblauch, K.V. and Subramanian, A.R. 1992. Nucleotide sequence, map position and transcript pattern at the intron-containing gene for maize chloroplast ribosomal protein S16. *Plant Mol. Biol.* **18:** 419–422.

Kapoor, S., Suzuki, J.Y. and Sugiura, M. 1997. Identification and functional significance of a new class of non-consensus-type plastid promoters. *Plant J.* **11:** 327–337.

Kato, A., Takaiwa, F., Shinozaki, K. and Sugiura, M. 1985. Location and nucleotide sequence of the genes for tobacco chloroplast tRNA Arg(ACG) and $tRNA^{Leu}$ (UAG). *Curr. Genet.* **9:** 405–409.

Kawashima, N. and Wildman, S.G. 1972. Studies on fraction I protein IV. Mode of inheritance of primary structure in relation to whether chloroplast or nuclear DNA contains the code for a chloroplast protein. *Biochim. Biophys. Acta* **262:** 42–49.

Khanna, N.C., Lakhani, S. and Tewari, K.K. 1991. Photoaffinity labelling of the pea chloroplast transcriptional complex by nascent RNA *in vitro. Nucleic Acids Res.* **19:** 4849–4855.

Khanna, N.C., Lakhani, S. and Tewari, K.K. 1992. Identification of the template binding polypeptide in the pea chloroplast transcriptional complex. *Nucleic Acids. Res.* **20:** 69–74.

Kieber, J.J., Tissier, A.F. and Signer, E.R. 1992. Cloning and characterization of an *Arabidopsis thaliana* topoisomerase I gene. *Plant Physiol.* **99:** 1493–1501.

Kirsch, W., Seyer, P. and Herrmann, R.G. 1986. Nucleotide sequence of the clustered genes for two P_{700} chlorophyll. *Curr. Genet.* **10:** 843–855.

Klein, R.R., Mason, H.S. and Mullet, J.E. 1988. Light-regulated translation of chloroplast proteins. *J. Cell Biol.* **106:** 289–301.

Klein, R.R. and Mullet, J.E. 1990. Light-induced transcription of chloroplast genes. *J. Biol. Chem.* **265:** 1895–1902.

Klein, U., De Camp, J.D. and Bogorad, L. 1992. Two types of chloroplast gene promoters in *Chlamydomonas reinhardtii. Proc. Natl. Acad. Sci. USA* **89:** 3453–3457.

Ko, K., Bornemisza, O., Kourtz, L., Ko, Z.W., Platon, W.C. and Cashmore, A.R. 1992. Isolation and characterization of a cDNA clone encoding a cognate 70 kDa heat shock protein of the chloroplast envelope. *J. Biol. Chem.* **267:** 2986–2993.

Kolodner, R.D. and Tewari, K.K. 1972. Molecular size and conformation of chloroplast DNA from pea leaves. *J. Biol. Chem.* **247:** 6355–6364.

Kolodner, R.D. and Tewari, K.K. 1975a. The molecular size and conformation of the chloroplast DNA from higher plants. *Biochim. Biophys. Acta* **402:** 372–390.

Kolodner, R.D. and Tewari, K.K. 1975b. Denaturation mapping studies on the circular chloroplast DNA from pea leaves. *J. Biol. Chem.* **250:** 4888–4895.

Kolodner, R.D. and Tewari, K.K. 1975c. Chloroplast DNA from higher plants replicates by both the Cairns and rolling circle mechanism. *Nature* **256:** 708–711.

Kolodner, R.D., Warner, R.C. and Tewari, K.K. 1975. The presence of covalently linked ribonucleotides in the closed circular DNA from higher plants. *J. Biol. Chem.* **250:** 7020

Kolodner, R.D. and Tewari, K.K. 1979. Inverted repeats in chloroplast DNA from higher plants. *Proc. Natl. Acad. Sci.* **76:** 41.

Kornberg, A. 1988. DNA replication. *Biochim. Biophys. Acta* **951:** 235–239.

Kornberg, A. and Baker, T.A. 1991. DNA Replication, 2nd edn. W.H. Freeman and Co., New York.

Kraus, B.L. and Spremulli, L.L. 1986. Chloroplast initiation factor 3 from *Euglena gracilis*: purification and initial characterization. *J. Biol. Chem.* **261:** 4781–4784.

Kruger, K., Grabowski, P.J., Zaug, A.J., Sands, J., Gottschling, D.E. and Cech, T.R. 1982.

Self-splicing RNA: auto-excision, auto-cyclization or rRNA intervening sequences of *Tetrahymena. Cell* **31:** 147–157.

Kudla, J., Hayes, R., and Gruissem, W. 1996. Polyadenylation accelerates degradation of chloroplast mRNA. *EMBO J.* **15:** 7137–7146.

Kuhlemeier, C. 1992. Transcriptional and post-transcriptional regulation of gene expression in plants. *Plant Mol. Biol.* **19:** 1–14.

Lakhani, S., Khanna, N. and Tewari, K.K. 1993. Nascent transcript-binding protein of the pea chloroplast transcriptionally active chromosome. *Plant Mol. Biol.* **23:** 963–979.

Lam, E. and Chua, N.-H. 1987. Chloroplast DNA gyrase and *in vitro* regulation of transcription by template topology and novobiocin. *Plant Mol. Biol.* **8:** 415–424.

Lammpa, G.K. and Bendich, A.J. 1979. Changes in chloroplast DNA level during development of pea (*Pisum sativum*). *Plant Physiol.* **64:** 126–130.

Lawrence, M.E. and Possingham, J.V. 1986. Microspectrofluorometric measurement of chloroplast DNA in dividing and expanding leaf cells of *S. oleracea. Plant Physiol.* **81:** 708–710.

Lebrum, M., Briat, J.-F., Laulhere, J.P. 1986. Characterization and properties of the spinach chloroplast transcriptionally active chromosome isolated at high ionic strength. *Planta*, **169**: 505–512.

Lemieux, B., Turmel, M. and Lemieux, C. 1990. Recombination of *Chlamydomonas* chloroplast DNA occurs more frequently in the large inverted repeat sequence than in the single copy regions. *Theor. Appl. Genet.* **79:** 17–27.

Lerbs, S., Brautigam, E. and Mache, R. 1988. DNA-dependent RNA polymerase of spinach chloroplast. *Mol. Gen. Genet.* **211:** 459–464.

Lerbs, S., Briat, J.F. and Mache, R. 1983. Chloroplast RNA polymerase from spinach. *Plant Mol. Biol.* **2:** 67–74.

Lerbs-Mache, S. 1993. The 110-kDa polypeptide of spinach plastid DNA-dependent RNA polymerase. *Proc. Natl. Acad. Sci. USA* **90:** 5509–5513.

Liere, K. and Link, G. 1994. Structure and expression characteristics of the chloroplast DNA region containing the split gene for tRNA from mustard. *Curr. Genet.* **26**: 557–563.

Li, R. and Golden, S.S. 1993. Enhancer activity of light-responsive regulatory elements in the untranslated leader regions of cyanobacterial psbA genes. *Proc. Natl. Acad. Sci. USA* **90:** 11678–11682.

Lin, Q., Ma, L., Burkhart, W. and Spremulli, L.L. 1994. Isolation and characterization of cDNA clones for chloroplast translational initiation factor-3 from *Euglena gracilis. J. Biol. Chem.* **269:** 9436–9444.

Link, G. 1984. DNA sequence requirements for the accurate transcription of protein coding plastid gene in a plasmid *in vitro* system from mustard. *EMBO J.* **3:** 1697–1704.

Lisitsky, I., Klaff, P., and Schuster, G. 1996. Addition of poly(A)-rich sequences to endonucleolitic cleavage sites in the degradation of spinach chloroplast mRNA. *Proc. Natl. Acad. Sci. USA* **93:** 13398–13403.

Lisitsky, I., Klaff, P., and Schuster, G. 1997. Blocking polyadenylation of mRNA in the chloroplast inhibits its degradation. *Plant J.* **12:** 1173–1178.

Little, M.C. and Hallick, 1988. Chloroplast *rpoA, rpoB* and *rpoC* genes specify at least three components of a chloroplast DNA dependent RNA polymerase. *J. Biol. Chem.* **263**: 14302–14307.

Luan, S., Albers, M.W. and Schreiber, S.L. 1994. Light-regulated, tissue-specific immunophilins in a higher plant. *Proc. Natl. Acad. Sci. USA* **91:** 984–988.

Lyttleton, J.W. 1962. Isolation of ribosomes from spinach lamellar proteins in *Vicia faba. Exp. Cell Res.* **26:** 312–317.

MacAllister, T.W., Kelley, W.K., Miron, W.K., Stenzel, T.T. and Bastia, D. 1991. Replication of plasmid R6K origin γ *in vitro*. *J. Biol. Chem.* **266:** 16056–16062.
Mache-Lerbs, S. 1993. The 110-kDa polypeptide of spinach plastid DNA-dependent RNA polymerase. *Proc. Natl. Acad. Sci.* USA **90:** 5509.
Maier, R.M., Zeltz, P., Kossel, H., Bonnard, G., Gualburto, J.M. and Grienenberger, J.M. 1996. RNA editing in plant mitochondria and chloroplasts. *Plant Mol. Biol.* **32:** 343–365.
Maliga, P. 1998. Two plastid RNA polymerases of higher plants: An evolving story. *Trends in Plant Science,* 3, **1:** 4–6.
Masoud, S.A., Johnson, L.B. and Sorensen, E.L. 1990. High transmission of paternal plastid DNA in alfalfa plants. *Theor. Appl. Genet.* **79:** 49–55.
McKnown, R.L. and Tewari, K.K. 1984. Purification and properties of pea chloroplast DNA polymerase. *Proc. Natl. Acad. Sci.* USA **81:** 2354–2358.
Mecsas, J. and Sugden, B. 1987. Replication of plasmids derived from bovine papilloma virus type I and Epstein-Barr virus in cells in culture. *Ann. Rev. Cell. Biol.* **3:** 87.
Meeker, R. and Tewari, K.K. 1980. Transfer RNA genes in the chloroplast DNA of pea leaves. *Biochemistry* **19:** 5973–5981.
Meeker, R., Nielsen, B.L. and Tewari, K.K. 1988. Localisation of replication origins in pea chloroplast DNA. *Mol. Cell. Biol.* **8:** 1216–1223.
Michel, F. and Dujon, B. 1983. Secondary structure in two intron families including mitochondrial-chloroplasts and nuclear encoded member. *EMBO J.* **2:** 33–38.
Mukherjee, S.K., Reddy, M.K., Kumar, D. and Tewari, K.K. 1994. Purification and characterization of eukaryotic type I topoisomerase from pea chloroplasts. *J. Biol. Chem.* **269:** 3793–3801.
Narita, J.O., Rushlow, K.E. and Hallick, R.B. 1985. Characterization of a *Euglena gracilis* chloroplast RNA polymerase specific for ribosomal RNA genes. *J. Biol. Chem.* **260:** 11194–11199.
Nielsen, B.L., Lu-Zhun and Tewari, K.K. 1993. Characterization of the pea chloroplast DNA *OriA* region. *Plasmid* **30:** 197.
Nielsen, B.L., Rajshekhar, V.K. and Tewari, K.K. 1991. Pea chloroplast DNA primase. *Plant Mol. Biol.* **16:** 1019–1034.
Nielsen, B.L. and Tewari, K.K. 1988. Pea chloroplast topoisomerase: purification characterization and role in replication. *Plant Mol. Biol.* **11:** 3–14.
Ohta, N., Sato, N., Kawano, S. and Kuroiwa, T. 1994. The *trpA* gene on the plastid genome of *Cyanidium caldarium* strain RK-I. *Curr. Genet.* **25:** 357–361.
Ohyama, K., Fukuzawa, H., Kohchi, T., Sano, T., Sano, S., Shirai, H., Umesno, K., Yasuhiko, S., Takeuchi, M., Chang, Z., Aota, S., Inokuchi, H. and Ozeki, H. 1988. Structure and organization of *Marchantia polymorpha* chloroplast genome I. cloning and gene identification. *J. Mol. Biol.* **203:** 281–298.
Ohto, C., Torazawa, K., Tanaka, M., Shinozaki, K. and Sugiura, M. 1988. Transcription of ten ribosomal protein genes from tobacco chloroplasts. *Plant Mol. Biol.* **11**: 589–600.
Ohyama, K., Fukuzawa, H., Kohchi, T., Shirai, H., Sano, T., et al., 1986. Chloroplast gene organization deduced from complete sequence of liverwort *M. polymorpha* chloroplast DNA. *Nature* **322:** 572–574.
Oishi, K.K., Sumnicht, T. and Tewari, K.K. 1981. Messenger RNA transcripts of pea chloroplast DNA. *Biochemistry* **20:** 5701–5717.
Orozco, E.M.Jr., Mullet, J.E. and Chua, N. 1985. An *in vitro* system for accurate transcription initiation of chloroplast protein genes. *Nucleic Acids Res.* **13:** 1283–1302.
Owttrim, G.W., Mandel, T., Trachsel, H., Thomas, A.A.M. and Kuhlemeier, C. 1994. Characterization of the tobacco *eIF-4A* gene family. *Plant Mol. Biol.* **26:** 1747–1757.

Pakrasi, H.B. 1995. Genetic analysis of the form and function of Photosystem I and II. *Ann. Rev. Genetics* **29**: 755–776.

Palmer, J.D. 1985. Comparative organization of chloroplast genomes. *Ann. Rev. Genetics* **19:** 325–354.

Palmer, J.D. 1991. Plastid chromosomes: structure and evolution. *Cell. Cult. Somatic Cell Genet. Plants* **7A:** 55–92.

Pancic, P.G., Strotmann, H. and Kowallik, K.V. 1992. Chloroplast ATPase genes in the diatom *Odontella sinensis* reflect cyanobacterial character in structure and arrangement. *J. Mol. Biol.* **224:** 529–536.

Partono, S. and Lewin, A.S. 1988. Autocatalytic activities of intron 5 of the *cob* gene of yeast mitochondria. *Mol. Cell. Biol.* **8:** 2562–2571.

Plant, A.L. and Gray, J.C. 1988. Introns in chloroplast protein coding genes of land plants. *Photosynth. Res.* **16:** 23–39.

Platt, T. 1986. Transcription termination and the regulation of gene expression. *Ann. Rev. Biochem.* **55:** 339–372.

Polans, N.O., Corriveau, J.L. and Coleman, A.W. 1990. Plastid inheritance in *Pisum sativum. Curr. Genet.* **18:** 477–480.

Rajora, O.P. and Dancik, B.P. 1995. Chloroplast DNA variation in *Populus* III. Novel chloroplast DNA variants in natural *populus* × *canadensis* hybride. *Theor. Appl. Genet.* **90:** 331–334.

Rajshekhar, V., Sun, E., Meeker, R., Wu, D.W. and Tewari, K.K. 1991. Highly purified pea chloroplast RNA polymerase transcribes both rRNA and mRNA genes. *Eur. J. Biochem.* **195:** 215–228.

Randolph-Anderson, B.L., Boynton, J. E. and Gillham, N.W. 1989. J. Electrophoretic and immunological comparisons of chloroplast and prokaryotic ribosomal proteins. *Mol. Evol.* **29**: 68–88.

Rapp, J.C., Baumgartner, B.J. and Mullet, J. 1992. Quantitative analysis of transcription and RNA levels of 15 barley chloroplast genes. *J. Biol. Chem.* **267:** 21404–21411.

Rapp, J.C. and Mullet, J.E. 1991. Chloroplast transcription is required to express the nuclear genes *rbcS* and *cob. Plant Mol. Biol.* **17:** 813–823.

Reddy, M.K., Choudhary, N.R., Kumar, D. Mukherjee, S.K. and Tewari, K.K. 1994. Characterization and mode of *in vitro* replication of pea chloroplast OriA sequences. *Eur. J. Biochem.* **220:** 933–941.

Reiss, T. and Link, G. 1985. Characterization of transcriptionally active DNA protein complexes from chloroplast and etioplasts of mustard (*Sinbapis alba* + L). *Eur. J. Biochem.* **148**: 207–212.

Reiter, R.S., Coomber, S.A., Bourett, T.M., Bartley, G.E. and Scolnik P.A. 1994. Control of leaf and chloroplast development by the *Arabidopsis* gene pale cress. *Plant Cell* **6:** 1253–1264.

Ris, H. and Plaut, W. 1962. The ultrastructure of DNA-containing areas in the chloroplasts of *Chlamydomonas. J. Cell. Biol.* **13:** 383–391.

Rochaix, J.D. 1978. Restriction endonuclease map of the chloroplast DNA of *Chlamydomonas reinhardii. J. Mol. Biol.* **126:** 597–617.

Rochaix, J.D. 1992. Post-transcriptional steps in the expression of chloroplast genes. *Ann. Rev. Cell. Biol.* **8:** 1–28.

Rochaix, J.D. and Malnor, P. 1978. Anatomy of chloroplast ribosomal DNA of *C. reinhardii. Cell* **15:** 661–670.

Rochaix, J.D., Rahire, M. and Michel, F. 1985. The chloroplast ribosomal intron of *C. reinhardii* code for a polypeptide related to mitochondrial maturases. *Nucleic Acids Res.* **13:** 975–984.

Rushlow, K.E., Orozco, E.M., Jr., Lipper, C. and Hallick, R.B. 1980. Selective *in vitro* transcription of *Euglena* chloroplast ribosomal RNA genes by a transcriptionally active chromosomes. *J. Biol. Chem.* **255:** 3786–3792.

Sala, F., Amileni, A.R., Parisi, B. and Spadari, S. 1980. A gamma like DNA polymerase in spinach chloroplast. *Eur. J. Biochem.* **112:** 211–217.

Salvador, M.L., Klein, U. and Bogorad, L. 1993. 5' sequences are important positive and negative determinants of the longevity of *Chlamydomonas* gene transcription. *Proc. Natl. Acad. Sci USA*, **90:** 1556–1560.

Schmidt, J., Herfurth, E. and Subramanian, A.R. 1992. Purification and characterization of seven chloroplast ribosomal proteins. *Plant Mol. Biol.* **20:** 459–465.

Sexton, T.B., Christopher, D.A. and Mullet, J.E. 1990. Light-induced switch in barley *psbD-psbC* promoter utilization: a novel mechanism regulation chloroplast gene expression. *EMBO J.* **9:** 4485–4494.

Shapiro, D.R. and Tewari, K.K. 1986. Nucleotide sequences of tRNA genes in the *P. sativum* chloroplast DNA. *Plant Mol. Biol.* **6:** 1–12.

Shinozaki, K., Ohme, H., Tanaka, M., Wakasugi, T. and Hayashida, N., et al., 1986. The complete nucleotide sequence of the tobacco chloroplast genome: its gene organization and expression. *EMBO J.* **5:** 2043–2049.

Siedlecki, J., Zimmersmann, W. and Weissbach, A. 1983. Characterization of a prokaryotic topoisomerase I activity in chloroplasts from spinach. *Nucleic Acids Res.* **11:** 1523–1536.

Siu, C.H., Swift, H. and Chiang, K.-S. 1976. Characterization of cytoplasmic and nuclear genomes in the colorless alga *Polytoma. J. Cell. Biol.* **69:** 371–382.

Smith, H.J. and Bogorad, L. 1974. The polypeptide subunit structure of the DNA-dependent RNA polymerase of *Zea mays* chloroplast. *Proc. Natl. Acad. Sci. USA.* **71:** 4839–4842.

Sreedharan, S.P., Beck, C.M. and Spremulli, L.L. 1985. *Euglena gracilis* chloroplast elongation factor Tu: purification and initial characterization. *J. Biol. Chem.* **260:** 3126–3131.

Stahl, D.J., Roderme, S.R., Bogorad, L. and Subramanian, A.R. 1993. Cotranscription pattern of an introgressed operon in the maize chloroplast genome. *Plant Mol. Biol.* **21:** 1069–1076.

Staub, J.M. and Maliga, P. 1993. Accumulation as D1 polypeptide in tobacco plastids is regulated via the untranslated region of the *psbA* mRNA. *EMBO J.* **12:** 601–606.

Staub, J.M. and Maliga, P. 1994. Translation of *psbA* mRNA is regulated by light via the 5′-untranslated region in tobacco plastids. *Plant. J.* **6:** 547–553.

Staub, J.M. and Maliga, P. 1994. Extrachromosomal elements in tobacco plastids. *Proc. Natl. Acad. Sci. USA* **91:** 7468–7472.

Steinmetz, A.A., Krebbers, E.T., Schwarz, Z., Gubbins, E.J. and Bogorad, L. 1983. Nucleotide sequences of five maize chloroplast tRNA genes and their flanking region. *J. Biol. Chem.* **258:** 5503–5511.

Stern, D.B. and Gruissem, W. 1987. Control of plastid gene expression. *Cell* **51:** 1145–1147.

Stern, D.B., Higgs, D.C., and Yang, J. 1997. Transcription and translation in chloroplasts. *Trends in Plant Science*, 2, **8:** 308–315.

Strauss, S.H., Palmer, J.D., Howe, G.T. and Doerksen, A.H. 1988. Chloroplast genome of two conifers like a large inverted repeat and are extensively rearranged. *Proc. Natl. Acad. Sci. USA* **85:** 3898–3902.

Subramanian, A.R. 1993. Molecular genetics of chloroplast ribosomal proteins. *TIBS* **18:** 177–181.

Subramanian, A.R. 1991. Nuclear-coded chloroplast r-proteins, precursors cDNA clones and transit sequences. *NATO ASI Ser.* **H 55**: 95–105.

Sugita, M. and Sugiura, M. 1983. A putative gene of tobacco chloroplast coding for ribosomal protein similar to *E. coli* ribosomal protein S19. *Nucleic Acids Res.* **11:** 1913–1918.

Sugita, M., and Sugiura, M. 1996. Regulation of gene expression in chloroplasts of higher plants. *Plant Mol. Biol.* **32:** 315–326.

Sugiura, M. 1992. The Chloroplast genome. *Plant Mol. Biol.* **19:** 149–168.

Sugiura, M. and Wakasugi, T. 1989. Compilation and comparison of tRNA genes from tobacco chloroplast. *Crit. Rev. Plant Sci.* **8:** 89–101.

Svab, Z., Hajdukiewiez, P. and Maliga, P. 1990. Stable transformation of plastids in higher plants. *Proc. Natl. Acad. Sci. USA* **87:** 8526.

Takahashi, Y., Rahire, M., Breyton, C., Popot, J.-L., Joliot, P. and Rochaix, J.D. 1996. The chloroplast ycf7 (*pet* L) open reading frame of *C. reinhardtii* encodes a small functionally important subunit of the cytochrome b_6f complex. *EMBO J.* **15:** 3498–3506.

Tewari, K.K. 1979. Structure and replication of chloroplast DNA. In: Nucleic Acids in Plants, (eds.) Halls, T.C. and Danice, J.W., Vol. 1, CRC Press, pp. 41–108.

Tewari, K.K. 1986. Purification and properties of chloroplast DNA polymerase. *Methods Enzymol.* **118:** 186–201.

Tewari, K.K. 1987. Replication of Chloroplast DNA. In: DNA Replication in Plants Bryant, J.A. and Dunham, V.L., (eds), CRC Press, Boca Raton, pp. 69–116.

Tewari, K.K. and Goel, A. 1983. Solubilization and partial purification of RNA polymerase from pea chloroplasts. *Biochemistry* **22:** 2142–2148.

Tewari, K.K. and Wildman, A.G. 1968. Function of chloroplast DNA I. Hybridization studies involving nuclear and chloroplast DNA with RNA from cytoplasmic (80S) *Proc. Natl. Acad. Sci. USA* **59:** 569–576.

Tewari, K.K. and Wildman, S.G. 1969. Function of chloroplast DNA II. Studies on DNA-dependent RNA polymerase activity of tobacco chloroplast. *Biochim. Biophys. Acta* **186:** 358.

Tiller, K., Eisermann, A. and Link, G. 1991. The chloroplast transcription apparatus from mustard. *Eur. J. Biochem.* **198:** 93–99.

Thomas, J.R. and Tewari, K.K. 1974a. Ribosomal RNA genes in chloroplast DNAs of higher plants. *Biochim. Biophys. Acta* **361:** 73.

Thomas, J.R. and Tewari, K.K. 1974b. Conservation of 70S rRNA genes in chloroplast DNAs of higher plants. *Proc. Natl. Acad. Sci. USA* **71:** 3147.

Thommes, P. and Hubscher, U. 1990. Eukaryotic DNA replication. Enzymes and proteins acting at the fork. *Eur. J. Biochem.* **194:** 699–712.

Thompson, R.J and Mosig, G. 1987. Stimulation of a *Chlamydomonas* chloroplast promoter by novobiocin. *Cell* **48:** 281–287.

Tobin, E.M. and Silverthorne, J. 1985. Light regulation of gene expression in higher plants. *Ann. Rev. Plant Physiol.* **36:** 569–593.

Tomizawa, J. 1986. Control of ColE1 plasmid replication. *Cell* **47:** 89–97.

Troxler, R.F., Zhang, F., Hu, J. and Bogorad, L. 1994. Evidence that sigma factors are components of chloroplast RNA polymerase. *Plant Physiol.* **104:** 753–759.

Tsudzuki, J., Nakashima., K., Tsudzuki, T., Hiratsuka, J., Shibata, M., Wakasugi, T. and Sugiura, M. 1992. Chloroplast DNA of black pine retains a residual inverted repeat lacking rRNA genes. *Mol. Gen. Genet.* **232:** 206–214.

Tuteja, N. 1996. Unraveling DNA helicases from plant cells. *Plant Mol. Biol.* **33:** 947–952.

Tuteja, N. and Phan, T.-N. 1998. Inhibition of pea chloroplast DNA helicase unwinding and ATPase activities by DNA-interacting ligands. *Biochem. Biophys. Res. Commun.* **244:** 861–867.

Tuteja, N., Phan, T.-N., Tuteja, R., Ochem, A. and Falaschi, A. 1977. Inhibition of DNA unwinding and ATPase activities of human DNA helicase II by chemotherapeutic agents. *Biochem. Biophys. Res. Commun.* **236:** 636–640.

Tuteja, N. and Tuteja, R. 1996. DNA helicases: the long unwinding road. *Nature Genet.* **13:** 11–12.

Tuteja, N. Phan, T.N. and Tewari, K.K. 1996. Purification and characterization of DNA helicase from pea chloroplast that translocates in the 3' to 5' direction. *Eur. J. Biochem.* **238:** 54–63.

Ursin, V.M., Becker, C.K. and Showmaker, C.K. 1993. Cloning and nucleotide sequence of tobacco chloroplast translational elongation factor, EF-Tu. *Plant Physiol.* **101:** 333–334.

Vera, A., Yokoi, F. and Sugiura, M. 1993. The existence of the pre-mature 16S rRNA species in plastid ribosomes. *FEBS Lett.* **327:** 29–31.

Walbot, V. and Coe, E.H. 1979. Nuclear gene iojap conditions a programmed change to ribosomeless plastids in *Zea mays. Proc. Natl. Acad. Sci. USA* **76:** 2760–2764.

Wang, J.C. 1985. DNA topoisomerases. *Annu. Rev. Biochem.* **54:** 665–697.

Wang, Z.F., Yang, J., Nie, Z.Q. and Wu, M. 1991. Purification and characterization of a gamma-like DNA polymerase from *C. reinhardtii. Biochemistry* **30:** 1127–1131.

Watt, P.M. and Hickson, I.D. 1994. Structure and function of type II DNA topoisomerases. *Biochem. J.* **303:** 681–695.

Webber, A.N. and Malkin, R. 1990. Photosystem I reaction-centre proteins contain leucine zipper motifs: a proposed role in dimer formation. *FEBS Lett.* **264:** 1–4.

White, E.E. 1990. Chloroplast DNA in *Pinus monticola:* 1. Physical map. *Theor. Appl. Genet.* **79:** 119–124.

Whitfeld, P.R. and Bottomley, W. 1983. Organization and structure of chloroplast gene. *Ann. Rev. Plant Physiol.* **34:** 279–310.

Xu, M-Q., Kathe, S.D. Goodrich-Blair, H., Nierzwicki-Bauer, S.A. and Shub, D.A. 1990. Bacterial origin of a chloroplast intron: conserved self-splicing group I introns in cyanobacteria. *Science* **250:** 1566–1569.

Yamashita, J., Nakajima, T., Tanifuji, S. and Koto, A. 1993. Accurate transcription initiation of *Vicia faba* rDNA in a whole cell extract from embryonic axes. *Plant J.* **3:** 187–190.

Yamaguchi, Y., et al., 1994. In: Plant Molecular Biology Manual E2. Gelvin, S.B. and Schilperoort, R.A., (eds), Kluwer Acad. Publ., pp. 1–15.

Yang, L., Li, R., Mohr, I.J., Clark, R. and Botchan, M.R. 1991. Activation of BPV-I replication *in vitro* by transcription factor E2. *Nature* **353:** 628–632.

Yuan, W., Quinn, D.M., Sigler, P.B. and Gelb, M.H. 1990. Kinetic and inhibition studies of phospholipase A_2 with short chain substrates and inhibitors. *Biochemistry* **29:** 6082–6094.

Zhu, Q. 1996. RNA polymerase II dependent plant *in vitro* transcription systems. *Plant J.* **10:** 185–188.

Concepts in Photobiology: Photosynthesis and Photomorphogenesis
G.S. Singhal, G. Renger, S.K. Sopory, K-D Irrgang and Govindjee (Eds)

24. Regulation of Plastid Gene Expression

Akhilesh K. Tyagi
Centre for Plant Molecular Biology and Department of Plant Molecular Biology, University of Delhi South Campus, New Delhi-110 021, India

Summary
Plastids are known to contain prokaryote-like genome within a eukaryotic cell. The plastid genome is a circular molecule of about 150 kb and codes for functions related to genetic system of plastids as well as those related to photosynthesis. Due to limited coding potential of plastid genome, its products must interact with nuclear-encoded products to establish a functional form of plastids. It has been found that in addition to well-defined prokaryote-like promoter sequences, plastid genes also contain some novel regulatory elements. These interact with multiple forms of RNA polymerase(s) and other regulatory proteins which control expression of plastid genes in light-dependent and development-specific manner. These stimuli seem to exert their effect by multi-step signal transduction chain at transcriptional, post-transcriptional and translational levels. While post-transcriptional steps include processing of polycistronic RNA, editing and stability, translation is controlled at the steps of chain elongation and protein turn-over which in turn depends on chlorophyll. The components involved in these processes include a plethora of proteins whose identity is being revealed by recent investigations. It seems that important regulatory steps are dependent on redox state and phosphorylation status. Knowledge is expected to accumulate rapidly keeping in view concerted efforts in the area of signal transduction and gene expression and development of techniques for transforming plastids.

1. Introduction

Plastids perform various functions including the carbon fixation by way of photosynthesis, fatty acid production and amino acid biosynthesis. About 40 years back, discovery of DNA in plastids resulted in intensive study on plastid genome characterization at the level of electron microscopy, restriction endonuclease mapping, gene localization; and sequencing of complete genome of tobacco and a liverwort was completed in 1986. The plastid genome generally varies from 120–200 kb and contains a large and a small single copy regions separated by two inverted repeat regions. The genome has a capacity to code for 4 subunits of RNA polymerase, 4 ribosomal RNAs, 23 ribosomal proteins, 30 tRNAs, 30 thylakoidal proteins for light reactions and the large subunit of Rubisco (Tyagi et al., 1993; Sugita and Sugiura, 1996). In addition, there are over a dozen genes capable of coding NADH dehydrogenase subunits or other products and few other open reading frames have a capacity to code for proteins of unknown function. Both strands of plastid genome (plastome) can code for products and most of the genes are arranged in polycistronic operons.

While the products of several genes may be present in all the forms that plastids represent, e.g. proplastids, chloroplasts, chromoplasts, amyloplasts or leucoplasts, their expression is highly regulated (Link, 1996). The level of different chloroplast proteins in barley has been shown to vary 1000-fold and out of this up to 300-fold can be accounted for by variation in transcription (Rapp et al., 1992). Stability of RNA, translation control and protein turn-over are other significant levels of control (Mayfield et al., 1995). In this article is presented an overview of the facts related to regulation of plastid gene expression at various levels (Fig. 1). One may also consult the

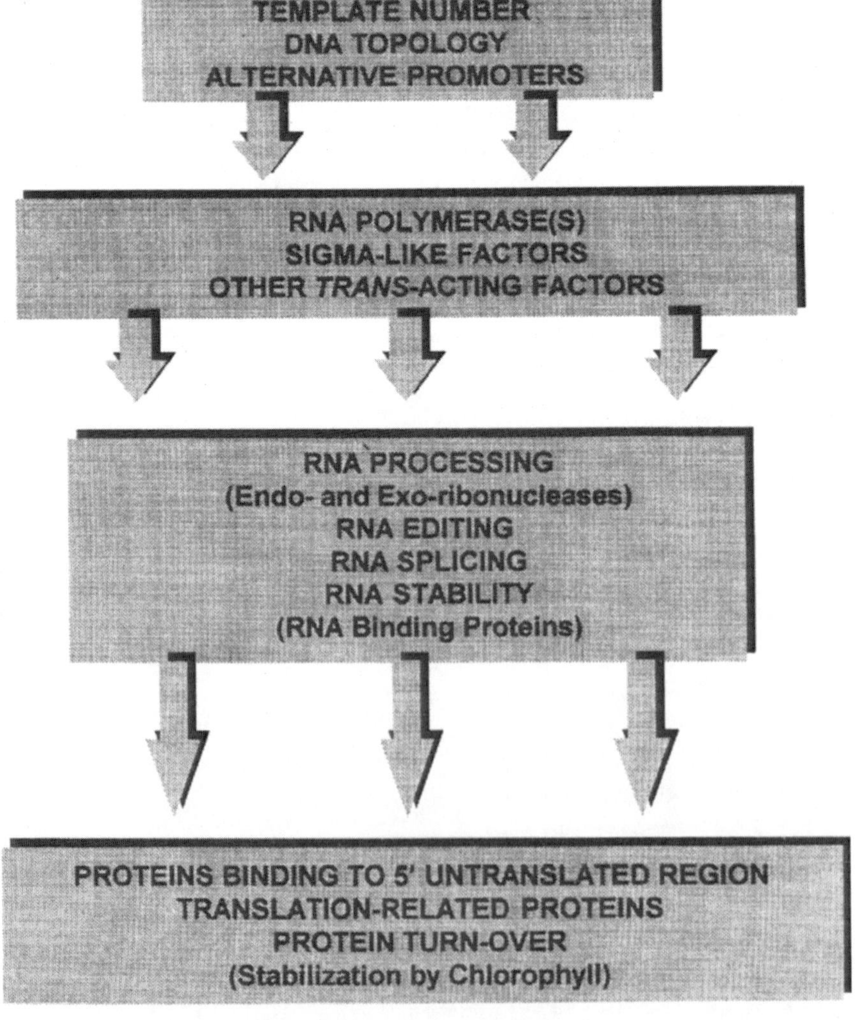

Fig. 1 Major steps on which light and developmental cues may act to control chloroplast gene expression.

book by Andersson et al. (1996) for further details on various aspects discussed here.

2. How to Study Regulation of Plastid Gene Expression?

While it is straight-forward to study the steady-state transcript levels or proteins at any specified state of plastid differentiation by northern and western blotting, one needs to do more to understand the mechanism controlling these levels. The soluble protein extracts of chloroplasts have been developed that support synthesis of mRNA or tRNA from exogenous templates and allow delineation of promoter and regulatory sequences for plastid genes. Other system allows quantification of transcription rates by measuring the amount of radio-label incorporated into nascent RNA by RNA polymerase bound to DNA in a very short period. It is called as "run-on" assay. When these rates are compared to steady-state levels, the influence of post-transcriptional control of mRNA stability is revealed. Alternatively, one can feed the system with plastid transcription inhibitor, tagetitoxin, and analyze the relative stability of various transcripts by detecting the decay rates. Some groups have also attempted to transcribe genes with purified preparations of RNA polymerase(s) (Link, 1996). More recently, it has become possible to transform chloroplasts of *Chlamydomonas* and tobacco (Maliga, 1993) and several questions related to transcription rate, mRNA/protein stability, RNA editing and co-translation of multicistronic mRNA are being addressed to with the aim of understanding plastid regulation biology in a better way.

3. Number and Topology of Template Control Transcript Levels

The copy number of plastome per plastid has been shown to change spatially and during light-dependent greening. This, in turn, can influence the level of transcripts but the contribution of control mechanisms like replication level or degradation remains to be defined. The topology of plastome has also been shown to influence transcription. In one line of investigation, calf thymus topoisomerase I was employed to generate topoisomers of a plasmid containing two divergently transcribed genes (*rbcL* and *atpB*) and, in another investigation, an inhibitor of topoisomerase II (novobiocin) was used under *in vitro* transcription conditions. The rate of *in vitro* transcription was higher with super-coiled form than with relaxed form (Stirdivant et al., 1985; Lam and Chua, 1987). Further, existence of gyrase and topoisomerases in plastids has also been demonstrated giving credence to the idea that plastome topology and accessibility to transcription machinery are relevant for regulation of plastid gene expression. That torsional stress of cpDNA does change on exposure to light or depending on developmental state has also been shown by UV cross-linking assay based on intercalating agent which gets altered reflecting on change in superhelical tensions (Davies et al., 1991).

4. Transcription Machinery of Plastids

Most of the genes in plastome show the presence of consensus elements similar to prokaryotic –10 and –35 sequences (TATAAT and TTGACA), although –35 sequence may show functional redundancy. On the other hand, some tRNA genes may have internal promoters and for others, like *atpB/E*, *rrn* and *rpoB*, a consensus sequence is not discernible. Additionally, novel promoter elements functional in light (Allison and Maliga, 1995; Kim and Mullet, 1995) have been identified.

The prokaryotic-type of promoters are used by a RNA polymerase containing α, β, β'-like polypeptides encoded by the plastome. But, sigma polypeptide(s) are not encoded in the plastome although several sigma-like factors (SLFs) have been identified which show differential activity during plastid differentiation and are coded by nuclear genome. Genes for sigma factors of chloroplast RNA polymerase have been isolated and found to be regulated by light and development (*see* Maliga, 1998). There is evidence that such an enzyme is involved in regulation of the differential activity of photosynthesis-related genes (Link, 1996). Besides the sigma-like factors interacting with RNA polymerase, other DNA binding proteins have been identified by employing DNA sequences from *psbD–C* (Kim and Mullet, 1995) and 16S rDNA (Iratni et al., 1994). The factor binding to 16S rDNA promoter, CDF2, has been shown to be a repressor while functional role of others remains to be worked out. Furthermore, it has also been shown that phosphorylation status of RNA polymerase and other regulatory proteins may influence the DNA protein interactions since an etioplast-like state is achieved by phosphorylation and chloroplast-like interactions can be achieved by dephosphorylation in an *in vitro* system (Tiller and Link, 1993). Other RNA polymerase activities defined from the plastids include multi-protein enzyme (Pfannschmidt and Link, 1994), transcriptionally active chromosome, TAC (Hallick et al., 1976) and a single polypeptide activity (Lerbs-Mache, 1993) which may be involved in maintaining plastid expression machinery as evident by transcription of plastid genes in deletion mutants of plastome that lack the genes for prokaryote-type RNA polymerase or the mutants lacking plastid ribosomes. Recently, a nuclear gene for single polypeptide RNA polymerase for plastids has been cloned (Hedtke et al., 1997). In mustard, an RNA polymerase activity A (ca. 13 polypeptides) is predominant in chloroplasts and another activity B (4 polypeptides) is predominant in etioplasts (Pfannschmidt and Link, 1994) and their relative activities are known to change during development of etioplasts into chloroplasts.

In short, although multiple promoters and RNA polymerase activities appear to exist for the transcription in plastids, more insight is required both on composition and mechanistic aspects of plastid transcription machinery.

5. Transcript Levels Change During Plastid Development

While transcript levels for various plastid genes show global changes during development of plastids, transcript for some genes—specifically those related to photosynthesis—change significantly on exposure to light, i.e. during conversion of etioplasts into chloroplasts. Transcript levels are also known to change during development of proplastids into chloroplasts although these can be demarcated more easily in monocots where meristematic region containing proplastids is present near the leaf base and a gradient of differentiated states of chloroplast follows towards the leaf-tip (Link, 1996). Spatial control of plastid gene expression also becomes apparent by the fact that root cells do not accumulate transcripts to any significant level despite exposure to light (Fig. 2., Kapoor et al., 1993).

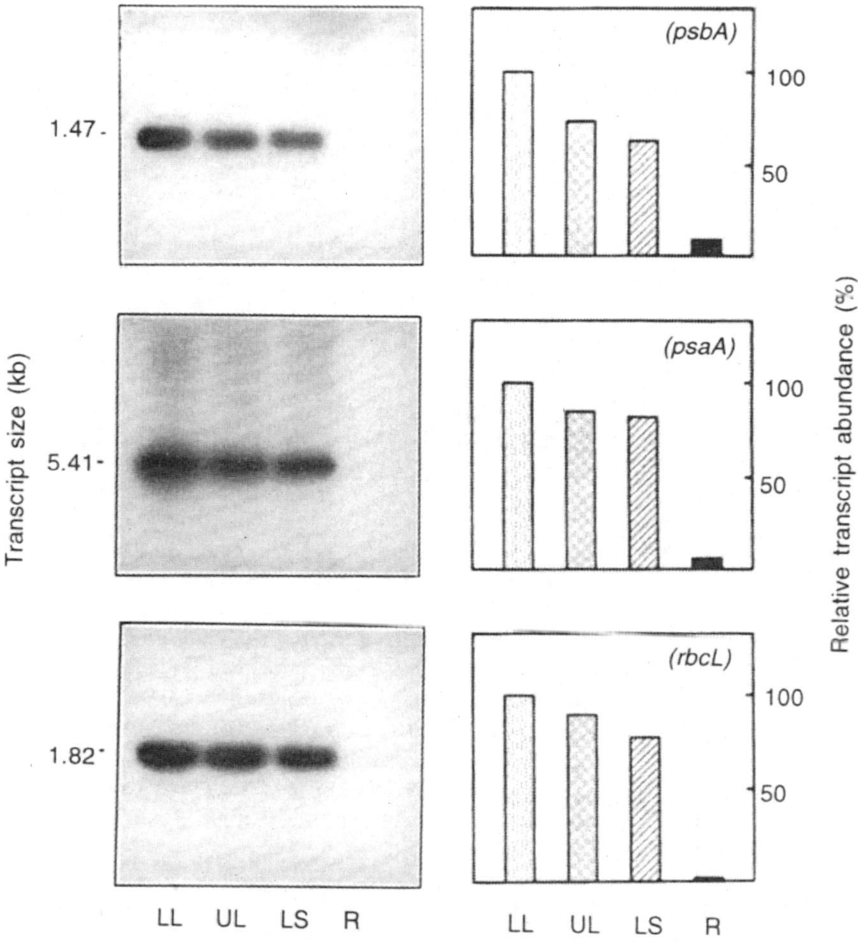

Fig. 2 **Spatial control of mRNA accumulation for plastid genes in rice. Total RNA from lower (LL) and upper (UL) halves of leaf blade, leaf sheath (LS) and roots (R) was analyzed by northern hybridization. Gene designations and their products: *psbA*, 32 kDa, D1 polypeptide of Photosystem II; *psaA*, P700 chlorophyll *a* apoprotein A of Photosystem I; *rbcL*, large subunit of ribulose-1, 5-bisphosphate carboxylase/oxygenase. The size of transcripts is given in kb on left (from Kapoor et al., 1993).**

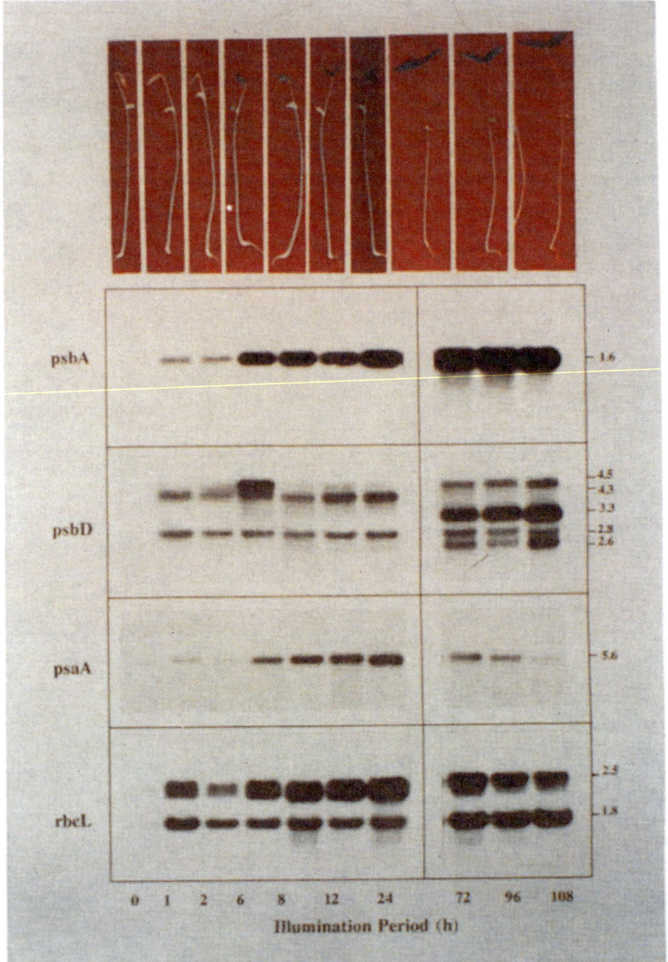

Fig. 3 Light-dependent accumulation of transcripts for plastid genes in moth bean. Total RNA was isolated from 5-day-old dark-grown seedlings and from dark-grown seedlings exposed to varying duration of light (upper panel) and used for northern hybridization (lower panel). Size of transcripts is given in kb on right. Multiple transcripts appearing with gene probes, *psaD* and *rbcL*, are known to result from use of alternative promoters and/or RNA processing. Gene designations: as in Fig. 1; *psbD*, 34 kDa, D2 polypeptide of Photosystem II.

Light-induced changes have been reported for over a dozen photosynthesis-related genes (Fig. 3) in several plants like maize, rice, wheat, barley, sorghum, mustard, moth bean and pea. Some contradictory reports may simply reflect the variation in the experimental plan since sensitivity to light is known to vary depending on the age of the seedlings (Fig. 4., Kelkar et al., 1993) and both the developmental state-dependent and light-dependent cues seem to contribute for establishing the steady-state levels of plastid gene transcripts (Kapoor et al., 1994). Light seems to be influencing the steady-state level both by its intensity and quality. Among the various wavelengths, red/far-red regions working via phytochrome and blue light acting via cryptochrome seem to be most important.

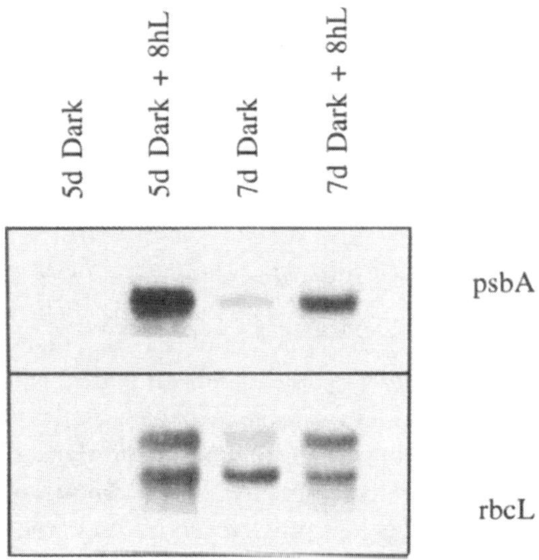

Fig. 4 Quantum of light-inducibility of plastid genes depends on the age of the seedlings. Total RNA was isolated from 5 and 7 days old dark-grown moth bean seedlings or those illuminated for 8 h and used for northern hybridization (from Kelkar et al., 1993).

It has been proposed that chloroplast transcription may be controlled via phosphorylation and redox status (Link, 1996) both of which can be controlled by light either via signal transduction, involving components like kinases/ phosphatases or via alterations generated in redox due to light-dependent electron transfer. Recently, role of phosphatases and kinases has been shown in signaling pathways that regulate light-dependent plastid gene transcription in rice and barley (Tyagi and coworkers, unpublished, Christopher et al., 1997). While inhibitors of phosphatases suppress the gene expression, an inhibitor of protein kinases has been shown to enhance transcription. Other components of signal transduction chain operative in control of plastid transcript levels are being defined by mutagenesis in *Arabidopsis* leading to identification of gene loci that influence plastid development (Chory, 1993).

6. Transcription Versus RNA Stability

The changes in steady-state levels of transcripts may depend either on rate of transcription or on RNA stability. Recent studies employing *in vitro* transcription assays have shown that the rates of transcription change during development of chloroplasts from etioplasts as well as proplastids. Also, different organs as well as different plastid genes (Rapp et al., 1992) show variable rates of transcription. Changes in transcription rates have also been found to occur when activity of *psbA* promoter was evaluated in transplastomic tobacco (Staub and Maliga, 1993). Certain plastid genes show differential transcription qualitatively because exposure to light results in activation of novel promoters, e.g. *psbD-psbC* (Berends-Sexton et al., 1990) and *ORF31-petG-ORF42* (Haley and Bogorad, 1990). Similar to nuclear genes, a role of methylation in regulation of plastid gene transcription was proposed. But, later studies did not support this view. It, therefore, remains yet to be understood as to how transcription rates of plastid genes change?

The changes in rate of transcription do not match the alterations in steady-state level of RNA for several plastid genes. When a comparison for six photosynthesis-related genes (*psbA, psbD, psaA, petB, atpB* and *rbcL*) was made in barley, mRNA stability was found to vary up to 14-fold (Rapp et al., 1992). Similarly, use of transcription inhibitors like tagetitoxin revealed that half-life of mRNA for photosynthesis-related plastid genes may range from 6–40 h (Kim et al., 1993).

Similar to the mechanism of regulation of transcription, the control mechanism of RNA stability is not very clear. Some components involved in the process have been identified recently, however, and are discussed below.

7. Processing of Transcripts

Before one considers the stability of plastid transcripts, it would be worth learning about their post-transcriptional processing. The primary transcript produced in plastids contains 5′-triphosphates and can be labelled specifically by guanylyltransferase and [α-^{32}p] GTP—similar to cap formation in nuclear mRNA—in variance to transcripts containing processed 5′ ends as a result of endonucleolytic activity. About 60 primary transcription units have been identified in tobacco (Sugita and Sugiura, 1996). But, most of the times, the primary transcript represents a dicistronic or polycistronic transcription unit and it is further processed resulting into transcripts representing a single or double genes. Thus, in some cases as many as 20 transcripts may be produced from a primary transcript, e.g. *psbB* operon. Sometimes, alternatively processed transcripts may turn out to be non-functional, e.g. treatment of barley with jasmonate resulted in a 35 nucleotide longer mRNA for *rbcL* which could not be translated because 5′ extension possibly forms a complex with 3′ end of 16S rRNA (Reinbothe et al., 1993).

The primary transcript is also processed by removal of introns which may vary from 503–2526 bp and are found to be present in 18 genes of tobacco (Sugita and Sugiura, 1996). These may fall into classes called as Group I and II and are prevalent in fungal mitochondria and ribosomal RNA genes in *Tetrahymena.* The identification is based on conserved sequence elements and potential folding patterns. However, no self-splicing introns, like rRNA genes of *Tetrahymena*, have yet been reported in plastids and, most likely, splicing involves protein-RNA components. Some nuclear mutations influencing splicing of specific-genes in chloroplast have recently been identified in maize to understand the mechanism of splicing (Jenkins et al., 1997). Interestingly, plastid transcripts synthesized separately can also be spliced to produce functional transcript by *trans*-splicing e.g. for *rps12* (Koller et al., 1987). Other important modification of transcripts is their editing which mostly results in substitution of C to U. The editing may occur before splicing and cleavage and could have regulatory role since both developmental state-specific and light-dependent splicing has been reported. Editing of RNA may involve site-specific factors since presence of an extra-site in transplastomic tobacco reduces editing of native RNA for *psbL* (Chaudhuri et al., 1995).

Perhaps the most crucial *cis*-acting motifs involved in RNA processing and stability are present in 3′ untranslated regions. These are mostly unique short inverted repeats capable of forming a stem-and-loop structure. Earlier, it was considered to be termination motif but recent studies show its involvement in RNA processing and stability (Sugita and Sugiura, 1996). It seems that a cleavage near stem-and-loop structure is followed by polyadenylation of transcripts even in plastids. Interestingly, polyadenylation has been shown to accelerate mRNA degradation specifically in dark (Kudla et al., 1996). Various internal processing sites have been shown to interact with endonucleolytic activities whose protein components have already been defined and may contain RNA component as well like RNAse P of *E. coli*. Several other RNA binding proteins have also been identified in tobacco and spinach. One of the spinach 28 RNP has been shown to bind with variable avidity to RNA species depending on its phosphorylation status (Lisitsky et al., 1995) thereby invoking the regulatory importance of RNA-binding proteins in maintaining steady-state levels.

Genetic determinants of components involved in RNA processing and stability are revealed by nuclear mutants in maize and barley (see Sugita and Sugiura, 1996). One such *h*igh *c*hlorophyll *f*luorescence mutant (*hcf109*) of *Arabidopsis* has recently been characterized which specifically controls the stability of selected transcripts (Meurer et al., 1996). Since it is only < 0.1 centimorgans from a defined marker gene, it should be possible to do positional cloning. Thus, we may soon see the developments that correlate RNA stabilizing/processing activities with their genetic determinants that will help understand molecular nature/mechanism of these post-transcriptional regulatory events.

8. Translational Control

Most of the polypeptides encoded by plastid genes accumulate even in dark. But, some photosynthesis-related polypeptides binding to chlorophyll do not accumulate in etioplasts even in plants like barley and spinach where their RNA are present in significant amount. These polypeptides are components of Photosystem II (47 and 43 kDa chlorophyll *a* apoproteins, 32 kDa, D1 and D2 polypeptides binding to P680 chlorophyll *a*) and Photosystem I (68 and 65 kDa P700 chlorophyll *a* apoproteins) and are encoded by plastid genes, namely *psbA, B, C, D* and *psaA, B*. Since chlorophyll deficient *Xantha1* mutant also does not accumulate these polypeptides and addition of chlorophyllide *a* in the presence of phytylpyrophosphate to lysed etioplasts of barley—which leads to sythesis of chlorophyll—resulted in accumulation of these polypeptides, it has been proposed that protochlorophyllide/chlorophyllide system acts as the photoreceptor for the process responsible for chlorophyll-mediated translational control (Laing et al., 1988; Eichacker et al., 1990). The lack of polypeptide accumulation has been considered due to control at translational step, chain elongation or apoprotein stabilization. It seems that cotranslational binding of molecules like chlorophyll *a* may enhance protein stability (Kim et al., 1994). Although large subunit of Rubisco does not bind chlorophyll, lack of a strict correlation between mRNA and polypeptide levels during dark/light transition has been considered to point out the role of translational control.

As far as the components involved in translation are concerned, several proteins binding to 5′ untranslated region of mRNA have been shown to be important in *Chlamydomonas* (Mayfield et al., 1995). In higher plants also, the leader region for *psbA* has been shown to control light-dependent accumulation of D1 polypeptide (Staub and Maliga, 1993, 1994). It seems that 5′-untranslated region interacts with proteins variably depending on phosphorylation status (Danon and Mayfield, 1994a) or redox state (Danon and Mayfield, 1994b) which can be controlled by light. Thus translation of mRNAs related to photosynthesis can be linked to the photosynthetic process and chlorophyll biosynthesis *per se*. In addition, light-dependent activity of chloroplast's elongation factor or differential organ-specific expression of its genes, e.g. in case of tobacco (Sugita et al., 1994), may play a significant role in translation level control of plastid gene expression.

9. Conclusions

The unique features of plastid genome are also reflected in the control of its expression. The earlier belief about truly prokaryotic-type control system has been taken over by the complexity of components and their interaction responsible for gene expression. Thus, multiple RNA polymerases seem to be involved in controlling state-specific transcription. Post-transcription events like transcript processing and stability have turned out to be major players and, in some cases, translational control may still act supreme. Interestingly, however, some common elements seem to be operational at all levels. These include phosphorylation and redox control both of which can in turn depend on light, a crucial factor for the process of photosynthesis, although details of signal transduction still remain to be worked out (Link, 1996). The semi-autonomous nature of plastids should also be kept in view and, therefore, nuclear components controlling plastid development have to be identified and their role needs to be defined. Recent work on mutants has led to identification of nuclear gene loci controlling plastid gene expression and their molecular characterization would definitely add to our knowledge of gene regulation. Same is true for application of plastid transformation system (Maliga, 1993) for gaining insights into regulatory aspects.

Acknowledgements

The work in our lab is supported by DBT, DST and the Rockefeller Foundation.

References

Allison, L.A. and Maliga, P. 1995. Light-responsive and transcription-enhancing elements regulate the plastid *psbD* core promoter. *EMBO J.* **14:** 3721–3730.

Andersson, B., Slater, A.H. and Barber, J. eds 1996. Molecular Genetics of Photosynthesis. Oxford University Press, Oxford.

Berends-Sexton, T., Christopher, D.A. and Mullet, J.E. 1990. Light-induced switch in barley *psbD-psbC* promoter utilization: a novel mechanism regulating chloroplast gene expression. *EMBO J.* **9:** 4485–4494.

Chaudhuri, S., Carrer, H. and Maliga, P. 1995. Site-specific factor involved in the editing of the *psbL* mRNA in tobacco plastids. *EMBO J.* **14:** 2951–2957.

Chory, J. 1993. Out of darkness: mutants reveal pathways controlling light-regulated development in plants. *Trends Genet* **9:** 167–172.

Christopher, D.A., Xinli, L., Kim, M. and Mullet, J.E. 1997. Involvement of protein kinase and extraplastidic serine/threonine protein phosphatases in signaling pathways regulating plastid transcription and the *psbD* blue-light responsive promoter in barley. *Plant Physiol.* **113:** 1273–1282.

Danon, A. and Mayfield, S.P. 1994a. ADP-dependent phosphorylation regulates RNA-binding *in vitro:* implications in light-modulated translation. *EMBO J.* **13:** 2227–2235.

Danon, A. and Mayfield, S.P. 1994b. Light-regulated translation of chloroplast messenger RNAs through redox potential. *Science* **266:** 1717–1719.

Davies, J.P., Thompson, R.J. and Mosig, G. 1991. Intercalation of psoralen into DNA of plastid chromosomes decreases late during barley chloroplast development. *Nucl. Acids Res.* **19:** 5219–5225.

Eichacker, L.A., Soll, J., Lauterbach, P., Rudiger, W., Klein, R.R. and Mullet, J.E. 1990. *In vitro* synthesis of chlorophyll *a* in the dark triggers accumulation of chlorophyll *a* apoproteins in barley etioplasts. *J. Biol. Chem.* **265:** 13566–13571.

Haley, J. and Bogorad, L. 1990. Alternative promoters are used for genes within maize chloroplast polycistronic transcription units. *Plant Cell* **2:** 323–333.

Hallick, R.B., Lipper, C., Richards, O.C. and Rutter, W.J. 1976. Isolation of a transcriptionally active chromosome from chloroplasts of *Euglena gracilis*. *Biochemistry* **15:** 3039–3045.

Hedtke, B., Börner, T. and Weihe, A. 1997. Mitochondrial and chloroplast phage-type RNA polymerases in *Arabidopsis*. *Science* **277:** 809–811.

Iratni, R., Baeza, L., Andreeva, A., Mache, R. and Lerbs-Mache, S. 1994. Regulation of rDNA transcription in chloroplasts: promoter exclusion by constitutive repression. *Genes Develop.* **8:** 2928–2938.

Jenkins, B.D., Kulhanek, D.J. and Barkan, A. 1997. Nuclear mutations that block group

II RNA splicing in maize chloroplasts reveal several intron classes with distinct requirements for splicing factors. *Plant Cell* **9:** 289–296.

Kapoor, S., Maheshwari, S.C. and Tyagi, A.K. 1993. Organ-specific expression of plastid-encoded genes in rice involves both quantitative and qualitative changes in mRNA. *Plant Cell Physiol.* **34:** 943–947.

Kapoor, S., Maheshwari, S.C. and Tyagi, A.K. 1994. Developmental and light-dependent cues interact to establish steady-state levels of transcripts for photosynthesis-related genes (*psbA, psbD, psaA* and *rbcL*) in rice (*Oryza sativa* L.). *Curr. Genet.* **25:** 362–366.

Kelkar, N.Y., Maheshwari, S.C. and Tyagi, A.K. 1993. Light dependent accumulation of mRNAs for chloroplast-encoded genes in *Vigna aconitifolia. Plant Sci.* **88:** 55–60.

Kim, M., Christopher, D.A. and Mullet, J.E. 1993. Direct evidence for selective modulation of *psbA, rpoA, rbcL* and *16S* mRNA stability during barley chloroplast development. *Plant Mol. Biol.* **22:** 447–463.

Kim, J., Eichacker, L.A., Rudiger, W. and Mullet, J.E. 1994. Chlorophyll regulates accumulation of the plastid-encoded chlorophyll proteins P700 and D1 by increasing apoprotein stability. *Plant Physiol.* **104:** 907–916.

Kim, J. and Mullet, J.E. 1995. Identification of a sequence-specific DNA binding factor required for transcription of the barley chloroplast blue light-responsive *psbD-psbC* promoter. *Plant Cell* **7:** 1445–1457.

Koller, B., Fromm, H., Galun, E. and Edelman, M. 1987. Evidence for *in vivo trans* splicing of pre-mRNAs in tobacco chloroplasts. *Cell* **48:** 111–119.

Kudla, J., Hayes, R. and Gruissem, W. 1996. Polyadenylation accelerates degradation of chloroplast mRNA. *EMBO J.* **15:** 7137–7146.

Lam, E. and Chua, N.H. 1987. Chloroplast DNA gyrase and *in vitro* regulation of transcription by template topology and novobiocin. *Plant Mol. Biol.* **8:** 415–424.

Laing, W., Kreuz, K. and Apel, K. 1988. Light-dependent, but phytochrome independent, translational control of the accumulation of the P700 chlorophyll-*a* protein of Photosystem I in barley (*Hordeum vulgare* L.). *Planta* **176:** 269–276.

Lerbs-Mache, S. 1993. The 110-kDa polypeptide of spinach plastid DNA-dependent RNA polymerase: single subunit enzyme or catalytic core of multimeric enzyme complexes? *Proc. Natl Acad. Sci. USA* **90:** 5509–5513.

Link, G. 1996. Green life: control of chloroplast gene transcription. *Bio Essays* **18:** 465–471.

Lisitsky, I., Liveanu, V. and Schuster, G. 1995. RNA-binding characteristics of a ribonucleoprotein from spinach chloroplast. *Plant Physiol.* **107:** 933–941.

Maliga, P. 1993. Towards plastid transformation in flowering plants. *Trends Biotechnol.* **11:** 101–107.

Maliga, P. 1998. Two plastid RNA polymerases of higher plants: an evolving story. *Trends Plant Sci.* **3:** 4–6.

Mayfield, S.P., Yohn, C.B., Cohen, A. and Danon, A. 1995. Regulation of chloroplast gene expression. *Annu. Rev. Plant Physiol. Plant Mol. Biol.* **46:** 147–166.

Meurer, J., Berger, A. and Westhoff, P. 1996. A nuclear mutant of *Arabidopsis* with impaired stability on distinct transcripts of the plastid *psbB, psbD/C, ndhH,* and *ndhC* operons. *Plant Cell* **8:** 1193–1207.

Pfannschmidt, T. and Link, G. 1994. Separation of two classes of plastid DNA-dependent RNA polymerases that are diferentially expressed in mustard (*Sinapis alba* L.) seedlings. *Plant Mol. Biol.* **25:** 69–81.

Rapp, J.C., Baumgartner, B.J. and Mullet, J.E. 1992. Quantitative analysis of transcription and RNA levels of 15 barley chloroplast genes: transcription rates and mRNA levels vary over 300-fold; predicted mRNA stabilities vary 30-fold. *J. Biol. Chem.* **267:** 21404–21411.

Reinbothe, S., Reinbothe, C., Heintzen, C., Seidenbecher, C. and Parthier, B. 1993. A methyl jasmonate induced shift in the length of the 5′ untranslated region impairs translation of the plastid *rbcL* transcript in barley. *EMBO J.* **12:** 1505–1512.

Sugita, M., Murayama, Y. and Sugiura, M. 1994. Structure and differential expression of two distinct genes encoding the chloroplast elongation factor Tu in tobacco. *Curr. Genet.* **25:** 164–168.

Sugita, M. and Sugiura, M. 1996. Regulation of gene expression in chloroplasts of higher plants. *Plant Mol. Biol.* **32:** 315–326.

Staub, J.M. and Maliga, P. 1993. Accumulation of D1 polypeptide in tobacco plastids is regulated via the untranslated region of the *psbA* mRNA. *EMBO J.* **12:** 601–606.

Staub, J.M and Maliga, P. 1994. Translation of *psbA* mRNA is regulated by light via the 5′-untranslated region in tobacco plants. *Plant J.* **6:** 547–553.

Stirdivant, S.M., Crossland, L.D. and Bogorad, L. 1985. DNA supercoiling affects *in vitro* transcription of two maize chloroplast genes differently. *Proc. Natl Acad. Sci. USA* **82:** 4886–4890.

Tiller, K. and Link, G. 1993. Phosphorylation and dephosphorylation affect functional characteristics of chloroplast and etioplast transcription systems for mustard (*Sinapis alba* L.). *EMBO J.* **12:** 1745–1753.

Tyagi, A.K., Kelkar, N.Y., Kapoor, S and Maheshwari, S.C. 1993. The chloroplast genome: genetic potential and its expression. In Abrol, Y.P., Mohanty, P. and Govindjee (eds) Photosynthesis—Photoreactions to Plant Productivity, Oxford and IBH Publishing Co. Pvt. Ltd., New Delhi and Kluwer Academic Publishers, Dordrecht, pp. 3–47.

PART C

Gene Regulation and Photomorphogenesis

Section VII: Photoreceptors and Functions

Concepts in Photobiology: Photosynthesis and Photomorphogenesis
G.S. Singhal, G. Renger, S.K. Sopory, K-D. Irrgang and Govindjee (Eds)

25. Phytochromes: molecular structure, photoreceptor process and physiological function

Vitally A. Sineshchekov
Biology Faculty, M.V. Lomonosov
Moscow State University, 119899 Moscow, Russia

Summary

Phytochrome (P)—a unique photochromic pigment—mediates a great variety of photoregulation reactions in plants thus determining the strategy of their optimum light development, i.e. photomorphogenesis. The pigment molecule is a dimer whose identical monomers consist of a linear tetrapyrrole covalently bound to a ca 125 kDa water soluble protein comprising ca 1200 amino acid residues. The monomer is structurally divided into an N-terminal chromophore bearing domain responsible for light perception and a C-terminal domain responsible for dimerization and light stimulus transduction. The spectroscopical and photochemical properties of the pigment are determined by the specific chromophore-apoprotein interaction. Its photoperception function is based on the photoreversible phototransformation of the initial red-light—absorbing form (Pr) into a physiologically active far-red-light—absorbing form (Pfr). The Pr $\rightarrow$ Pfr conversion includes an energy storing photoreaction, which is cis-trans photoisomerization of the chromophore proceeding within several tens ps with participation of the singlet excited state, and a following branched dark stepwise protein relaxation leading to its conformationally restructured form. The functional activity of the Pfr form is connected with the rearrangement of a number of amino acid residues in the C-terminal. The transduction of the signal from the Pfr form includes activation of a putative primary receptor (G-proteins) and changes of the concentration of Ca^{2+} and cyclic guanosine 5'-phosphate (cGMP) leading eventually to the modulation of the rate of gene expression. Action of P as a light-regulated enzyme is also considered as a possible mechanism of signal transduction. The complex behaviour of plants under different light conditions and the multiple functions of P are believed to be connected, at least in part, with heterogeneity of the pigment in the cell and first of all with the existence of a small family of Ps (five or more depending on plant species) which are encoded by different genes but possess the same chromophore. The main ones, phytochromes A and B (phyA and phyB), perform different functions or complement each other in some responses as shown by the investigations of P mutants. PhyA itself is found to be heterogeneous: it consists of two spectroscopically, photochemically and functionally distinct species: the major soluble and labile phyA' and a minor presumably membrane-associated and relatively stable phyA". Dimeric structure of the photoreceptor is also suggested to be a potential source of the multiple action of the pigment. And finally, it is believed that photoreactions of the intermediates of the P cycle can modify the P functions. Thus, investigation of the structure-function relationship in the P molecule acquires now a new dimension for understanding of the physiological polyfunctionality of the pigment.

1. Introduction

The autotrophic organisms entirely depend on light in their growth and development. To make optimal use of it, they have attained a highly sophisticated apparatus to monitor surrounding light conditions and to adjust to them the strategy of their development. The apparatus comprises at least three photoinformation pigment systems operating with maximum efficiency in UV-B, UV-A/blue and red/far-red regions. The former two are mediated by flavins and, possibly, pterins, the latter, by phytochrome, P. From the time of the P discovery until quite recently, it was widely accepted that all the varieties of the photoresponses in plants were mediated by one molecular species of the pigment. Now it is becoming increasingly evident that the multiplicity of the P functions may be connected with different molecular types of the pigment and that different photoresponses may develop already at the level of the initial photoreactions of the pigment and intramolecular transmission of the light signal. Discovery of Ps and investigation of their structure, photochemical and photophysiological properties are thus becoming the dominant trend of development in the area of photomorphogenesis and these aspects of research are discussed in the present chapter.

2. General

Discovery of P is connected with its unique property-functionally important photoreversible photochromic transformation of the pigment from the initial inactive dark-adapted red-absorbing form (Pr) into the physiologically active far-red-absorbing form (Pfr). In etiolated seedlings, all P is in the Pr form but, upon illumination, a certain concentration ratio establishes between Pr and Pfr which depends on the dose and spectral composition of the actinic light.

$$\text{Pr} \underset{\lambda_{max} = 730\text{ nm}}{\overset{\lambda_{max} = 666\text{ nm}}{\rightleftarrows}} \text{Pfr} \rightarrow \text{photoresponses}$$

With the development of spectroscopic, immunochemical and molecular biological techniques, P was found in angiosperms, gymnosperms, ferns, mosses, and algae (see Rüdiger and López-Figueroa, 1992; Thümmler et al., 1992; Wada et al., 1997; Wu and Lagazias, 1997 and the literature cited therein). Recently a photoreceptor similar to P was also detected in a slime mould (Staroztzik and Marwan, 1995) and in the cyanobacterium *Synechocystis* (Hughes et al., 1997; Yeh et al., 1997).

The influence of the P system on growth and development of higher plants is observed on all stages of their life including induction of seed germination, deetiolation and seedling establishment, floral induction and scenescence thus endowing them with competitive advantage. In particular, the fundamental function of P is to perceive relative amounts of red (R) and far-red (FR) in natural light (Smith, 1995; Smith and Whitelam, 1997) to provide plants with capacity to anticipate potential neighbour competition (shade avoidance syndrome). On the biochemical level, perception of light stimulus by P leads to activation of the expression of a number of light-regulated nuclear and chloroplast-encoded genes which in their integrity constitute a certain photophysiological response.

Based on the quantity of red light, which is necessary to produce photoresponses (intensities in the range from 10^{-10} mol m^{-2} to > 10^{-4} mol m^{-1}), they are divided into three categories: (1) very low fluence response (VLFR) (which are irreversible due to the fact that even under FR illumination Pfr concentration sufficient for physiological effect is achieved); (2) low fluence responses (LFR) (typical photoreversible P reactions) and (3) high irradience responses (HIR) dependent on the rate of cycling between Pr and Pfr. The latter are subdivided into the responses most effectively induced by the red and far-red light, R-HIR and FR-HIR. The photoresponses differ also by the time required to achieve the effect and, for LFR, by the time after illumination, when the reaction can be reversed by the subsequent illumination with FR. It is believed that the fast reactions (sec. and min.) operate at the level of biophysical and biochemical processes while the long term photoregulation phenomena (hours, days) involve regulation of gene expression.

To explain the two modes of the P action, two hypotheses have been advanced with respect to the mechanism of the light signal transduction from P to the effector systems. First is *the membrane hypothesis.* Its major point is an assumption that P interacts with the membrane and upon transformation to Pfr modifies, directly or via certain messengers, its properties (see Jordan, 1992; Volotovsky, 1992). This in turn leads to the changes in the membrane potentials and eventually to the mechanical and other related phenomena. The second is *the gene hypothesis.* The data indicate that derepression of the potentially active genes, the key point of the mechanism, proceed as a result of the activation by P of the biochemical cascades which include putative initial receptor (presumably, heterotrimeric G-proteins), Ca^{2+}, Ca^{2+}/calmodulin, and cyclic guanosine 5′-phosphate (cGMP). Increase of the concentration of these intermediates in the cytoplasm leads to the development of certain patterns of P photoresponses (Millar et al., 1994).

The possibility that photoresponses of plants to light are triggered by only two signalling intermediates, Ca^{2+} and cGMP, stresses the importance of the discovery of the heterogeneity of P in vivo which may be the source, at least partially, of the complexity of their photoregulatory functions. There is evidence from several experimental lines suggesting the existence of a number of P species in the cell: (1) Certain photophysiological effects could not be explained without postulating heterogeneity of the pigment—the existence of the "bulk" and "active" Ps (Hillman, 1967). (2) Two P pools were distinguished with respect to the kinetics of their light-induced destruction—a light-labile (Type 1) and a light-stable form (Type 2) (Heim et al., 1981; Brockmann and Schäfer, 1982). (3) Several species of P, which differ by the structure of their apoprotein, were detected with the use of immunochemical technique (Pratt et al., 1991; Pratt, 1994). (4) A small family of genes coding for P was discovered (see Quail, 1991, 1994, 1997; Boylan et al., 1992; Furuya, 1993; Smith, 1997 for review) and lastly (5) Existence of the heterogeneous population of P was shown by the *in vivo* low-temperature fluorescence spectroscopy and photochemistry of P (Sineshchekov, 1995a) The photophysiology of the P system is currently being successfully dissected by the use of the *photomorphogenetic mutants* (see reviews Kendrick and Nagatani, 1991; Reed et al., 1992; Peters, 1992; Chory, 1993; Cherri and Vierstra, 1994; Koornneef and Kendrick, 1994; Smith, 1997 and also recent original publications below) which fall into two classes: (1) *pigment mutants* (affected either in the synthesis or attachment of the chromophore, *chromophore mutants,* or in the synthesis of the respective apoprotein, *apoprotein mutants*) and (2) *transduction chain mutants.* The mutants provide a powerful instrument to follow structure-function relationship in the phytochrome molecule and attribute distinct photoresponses to a concrete P species and to identify intermediates in the transduction chain as is discussed in detail in Chapter 31.

3. Phytochrome Structure and Primary Photoprocesses

3.1 The Molecule

The notion that P is a chromoprotein with an open tetrapyrrole as a chromophore covalently bound to apoprotein has been recognized already in the early works of its discoverers. However, the size of the molecule (so called "native P") and its structure have been determined much later along with the development of protein extraction and gene sequencing technique. With the introduction of other methods of the structural protein analysis (such as analytical micromethods, ^{1}H-NMR spectroscopy, CD-spectroscopy, X-ray scattering under low angles, electron microscopy) one can speak now with a certain degree of confidence of P structure at all levels of its organization.

3.1.1 The chromophore

The chromophore (phytochromobilin, PΦB) is a linear tetrapyrrole with the system of the conjugated double bonds in a 3E-PΦB configuration (Fig. 1) (Rüdiger and Thümmler, 1991, 1994; Rüdiger, 1992). It is synthesized by reduction of biliverdin IX-α (BV) which is a product of the enzymatic haem formation from the first specific precursor 5-aminolevulinate (ALA) ('C-5 pathway') and its subsequent degradation via two enzymatic steps. The reduction occurs in plastids with participation of the enzyme PΦB synthase to form 3Z-PΦB which eventually isomerizes to 3E-PΦB (Terry et al., 1995). Rüdiger and Thümmler (1994) however put forward a question on the possible existence of a second tetrapyrrole synthesis pathway outside the plastids because the P chromophore is found after the photodestruction of plastids and in mutants which lack the C-5 pathway. The P formation in the cell is governed primarily by the level of the apoprotein synthesis, i.e. the limiting step in the formation of the chromophore-protein complexes is the protein part. This is suggested, in particular, from the fact (Schendel et al., 1996) that the ALA feeding to the etiolated seedlings does not change the level of P, [Ptot], in the cell as well as the level of the photoactive protochlorophyllide (Pchlid^{655}, where the superscript is the position of the emission maximum), i.e. of the pigments whose chromophore makes a complex with a protein. At the same time, the level of the inactive Pchlid^{633}, which does not associate with the enzymatic complex, increases by 1-2 orders of magnitude.

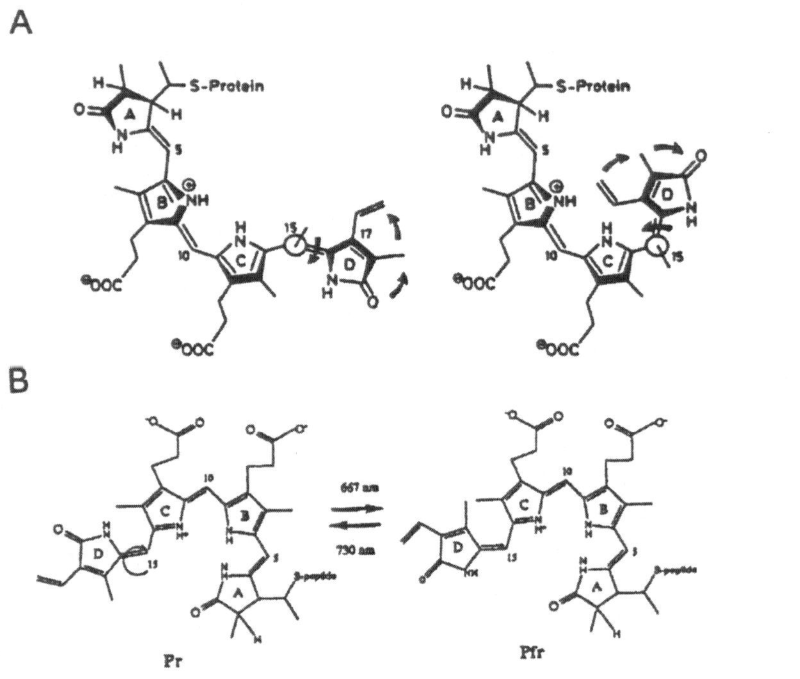

A

S-Protein

S-Protein

B

667 nm

730 nm

Pr

Pfr

Fig. 1. Proposed structure of the phytochrome chromophore with extended conformation in native protein and its changes in the Pr ↔ Pfr photoconversion: Pr form (left), Pfr form (right). (A) The movement of ring D and of the methine bridge for conrotation of bonds C14-C15 and C15-C16 in the photoreaction. (After Rüdiger and Thümmler, 1991; Rüdiger, 1992.) (B) The mechanism of the photoreaction suggesting overall conservation of the chromophore conformation. (After Song, 1994; Parker et al., 1994.) The arrows indicate the direction of movement.

3.1.2 The apoprotein

The primary structure is obtained from the analysis of the nucleotide sequences of genes encoding for several molecular types of the pigment and of the sequences of the amino acid residues in respective proteins (Furuya, 1989, 1993; Quail, 1991, 1994; Boylan et al., 1992). It suggests that the native P is an approx. 125 kDa protein with NH_2- and COOH-terminals comprising about 1200 amino acid residues. The size and composition of the molecule vary depending on P type and plant species. The higher proportion of the charged hydrophilic residues than the hydrophobic ones in the chain and the character of their distribution make the pigment molecule water-soluble, cytosolic. However, the existence of the amphiphilic regions detected both in the C- and N-terminal parts suggests that the pigment may interact with membrane (Partis and Grimm, 1990; Jordan, 1992; Furuya and Song, 1994; Song, 1994; Song et al., 1997).

The secondary structure includes primarily α-helices (approx. 50%), random coils (25–30%) and β-turns (20–25%). The content of β-sheets is low in holochrome but apoprotein has a higher proportion of it suggesting a certain role of the chromophore in the organization of the molecular structure (Song, 1988; Sommer and Song, 1990; Furuy and Song, 1994; Song et al., 1997).

The tertiary structure, which was determined with the use of proteolytic cleavage, comprises two major domains: the globular N-terminal domain of 70 kDa and the C-terminal domain of 55 kDa. The former contains a 10 kDa subdomain which in its turn is subdivided into 6 kDa and 4 kDa fragments. Several subdomains were found within the C-terminal domain (see below). Early extractions dealt with these partially degraded pigments: 'small' with molecular mass of ca 65 kDa and 'large' with molecular mass of 114 and 118 kDa.

The quaternary structure has been deduced from indirect physico-chemical experiments since the data on X-ray crystallographic structure of P are lacking. Electron microscopy, X-ray scattering and linear dichroism experiments provide information for the visualization of its structure and orientation of the chromophore (Fig. 2). P molecule is a dimer which consists of two identical subunits. It has an 'Y' shape with joint C-terminal segments and separate N-terminal segments and a two-fold symmetry axis (Nakasako et al., 1990; Tokutomi et al., 1992).

3.1.3 The chromophore-apoprotein interaction

The chromophore and the apoprotein assemble into a mature holoprotein: a single bilin chromophore binds with cystein-323 (in oats) in the hydrophobic pocket within the N-terminal domain. The process proceeds autocatalytically as shown with the use of *E. coli* and yeast systems to express and assemble photoactive holophytochrome (Furuya and Song, 1994; Li and Lagarias, 1994). Recently holophytochrome have been reconstituted in living cells of the yeast *Saccharomyces cerevisiae.* The cells expressing recombinant oat apophytochrome A were able to take up exogenous linear tetrapyrroles and these pigments combine with the apoprotein to form photoactive holophytochrome *in situ.* Unlike phytochrome A (phyA) in higher plant tissue the reconstituted photoreceptor appears to be light stable in yeast (Li and Lagarias, 1994; Kunkel et al., 1995). Bilin attachment to apophytochrome involves two steps, an initial formation of a reversible non-covalent complex followed by thioether bond formation (Li et al., 1995). The yield of photoreversible pigment assembled in the dark could be increased by far-red/red irradiation cycles, indicating that the presence of the chromophore promotes the correct folding of the binding site (Hill et al., 1994).

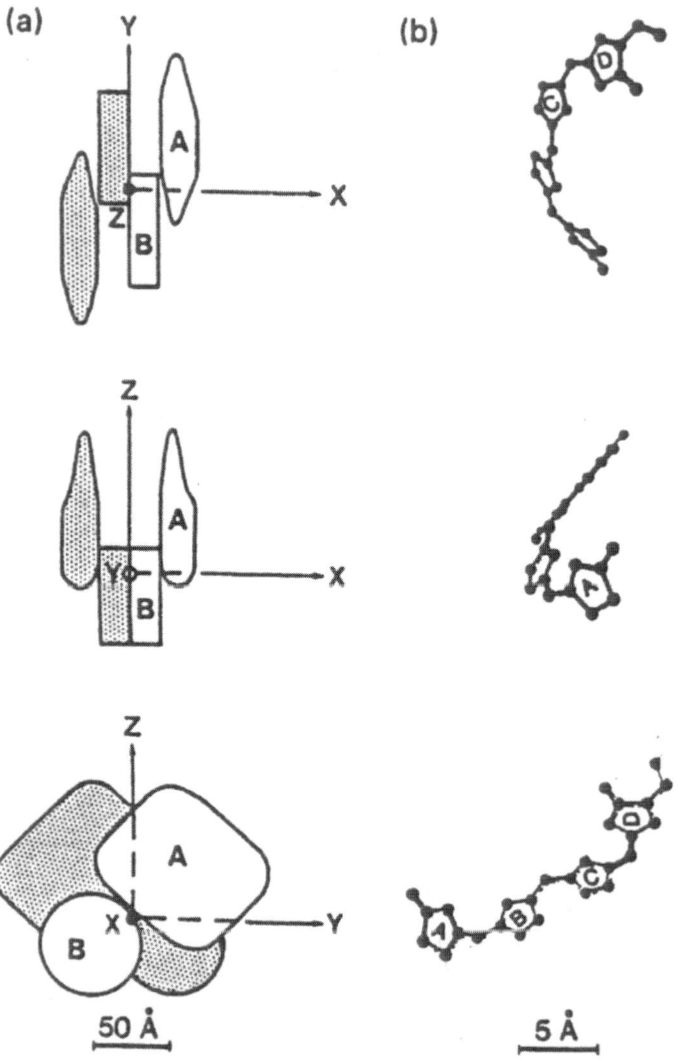

Fig. 2. Dimeric molecular model for pea phytochrome 1 in the Pr form (Nakasako et al., 1990) (a) and images of the Pr chromophore (Tokutomi et al., 1992) (b). A and B in (a) indicate the N-terminal 59 kDa chromophoric and the C-terminal 55 kDa domains; A through D in (b) indicate the four pyrrole rings in the chromophore of phytochrome. (From Tokutomi et al., 1992.)

Upon attachment both the chromophore and the apoprotein acquire new properties. On the other hand, denaturing leads to the loss of photochemical activity and changes of spectroscopic characteristics. It is believed that the photophysical and photochemical properties of the chromophore in its protein pocket are due to (1) stretching the chromophore (see Fig. 1); (2) protonation of the chromophore in the Pr form; and (3) interaction of the chromophore with charged groups near the chromophore (Rüdiger, 1992). A model is suggested (Parker et al., 1994) for *Avena* P in which Arg-318, Ala-319, Ser-322, Leu-325, and Tyr-327 form a pocket where one side of the chromophore is anchored, while the other side remains relatively free for movement. This allows rotation and subsequent conformational changes of the chromophore in the photoreaction.

4. Primary Photoprocesses and Phytochrome Cycle

Absorption of light quanta generates singlet excited states in the pigment whose energy is then either used in the photochemical cycle or dissipated as heat production and fluorescence. The quantum yields of the Pr ↔ Pfr conversion are close to one another and are estimated to be within the range 0.06–0.17 depending on the size of the pigment. Since there are no back thermal reactions from the intermediates of the P cycle these values can be considered as the quantum yields of the initial photoreactions.

4.1 Absorption and Emission of Light

Light quanta are absorbed by the P chromophore in the UV, whole visible and near infrared region ranging from ca 300 nm to 800 nm, the most effective being the red-far-red bands. *Absorption coefficient* in the main maxima for Pr and Pfr are, according to different research groups, 70–130 and 40–50 mM^{-1} cm^{-1}, respectively. This is comparable with the absorption coefficients of chlorophylls and phycobiliproteins. The *absorption spectra* of P *in vivo* cannot be obtained, however, because of its very low concentration and high light scattering of plant tissues. Its characteristics are deduced from the red-far-red light induced difference absorption spectra *in vivo* and from the data on the extracted native full lenght 124 kDa P. In the Pr form it has a maximum at 665–666 nm, a shoulder at 608 nm and a small maximum in the Soret band at 379 nm. They are attributed to Q_y, Q_x and B_y electronic transition, respectively (Song, 1988). The maxima in the Pfr form are at 730 and 400 nm (see Fig. 3). Deep freezing of the samples (in particular, in experiments on the initial photoreaction and photochemical cycle, see below) causes a 5–10 nm red shift and narrowing of the main bands. The *oscillator strength ratio* for the red and Soret band is approximately unity which is typical for extended conformation of the chromophore in the pigment. *Linear dichroism spectra* measurements in the region 500–700 nm suggest that the band at 608 nm may belong, in contrast to the accepted view, to a vibrational satellite of the Q_y transition (Tokutomi and Mimuro, 1989; Tokutomi et al., 1992). And finally, the *spectra of circular dichroism* with strong negative band for Pr and much weaker positive signal for Pfr suggest chromophore chirality due to the interaction of the chromophore with the protein (Eilfeld and Eilfeld, 1988; Björling et al., 1992; Chen et al., 1993). It should be noted

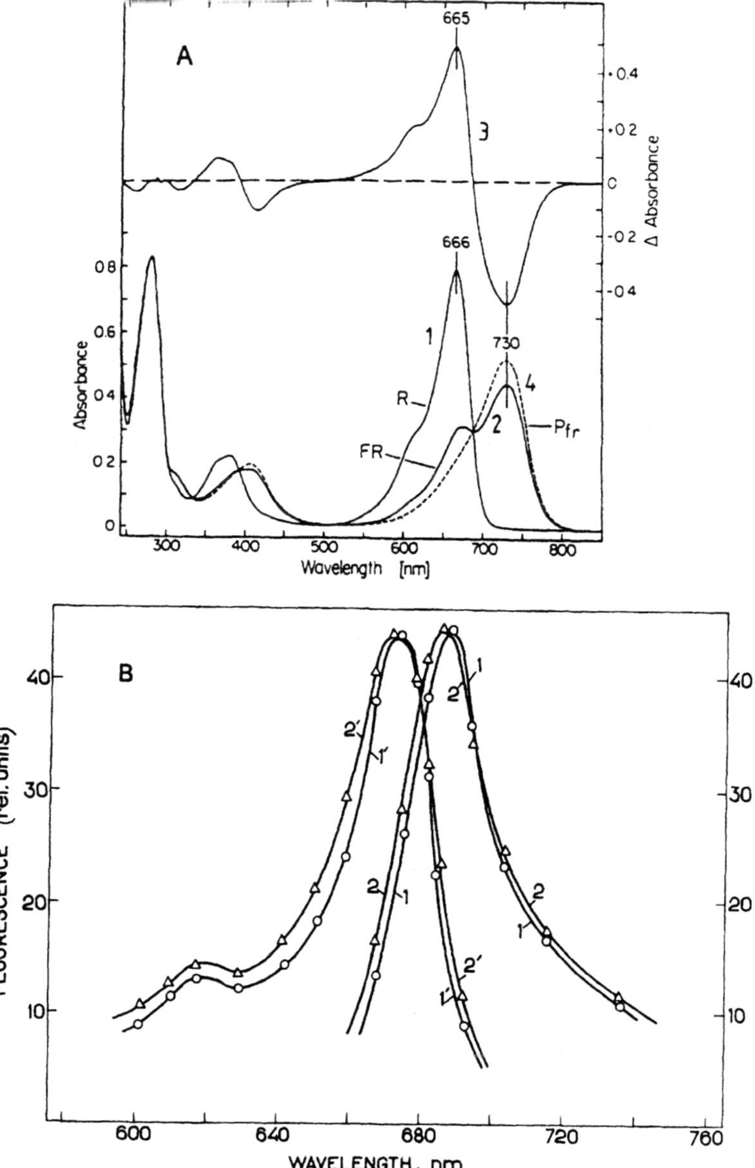

Fig. 3. **(A) Initial (1, 2), difference (3) and absolute (1, 4) absorption spectra of 124 kDa *Avena* phytochrome in the Pr (1) and Pfr (4) forms. Initial spectra were measured after far-red (1) and red (2) illumination. Difference spectrum 3 is spectrum 1 minus spectrum 2. Initial spectrum 1 represents absolute spectrum of Pr because practically all phytochrome is in the Pr form after far-red illumination; the absolute Pfr spectrum (4) was calculated from spectrum 2 after correction for incomplete photoconversion of Pr to Pfr assuming relative Pfr content of 0.86 at photoequilibrium. (From Vierstra and Quail, 1983a, b.) (B) Emission (1, 2, $\lambda_e = 630$ nm) and excitation (1', 2', $\lambda_m = 720$ nm) fluorescence spectra of the red phytochrome form in the cells of coleoptiles of etiolated *Avena* seedlings (1, 1') and 124-kDa phytochrome isolated from them (2, 2') at 85 K. (From Sineshchekov et al., 1990.)**

that the above absorption properties are presented for the light-labile P from etiolated seedlings (Type 1). The data on green-tissue light-stable P (Type 2) (which is believed to comprise phyB and other minor Ps) are rather controversial: according to different sources the spectra are either close to those of Type 1 (in pea, *Arabidopsis*) or blue-shifted by approx. 10–15 nm (in oat) (see Sineshchekov, 1995a and the literature cited therein). It has been found (Wagner et al., 1991) that in dicots phyA and phyB have similar spectral properties while in monocots they differ. This conclusion agrees with the observation (Sineshchekov et al., 1998b) that overexpression of rice phyB apoprotein (PHYB) in *Arabidopsis* (RBO in Wagner et al., 1991) brings about ca 3 nm short-wavelength shift of the low-temperature fluorescence emission spectrum of the etiolated *Arabidopsis* stems as compared with the wild-type plants while overexpression of *Arabidopsis* PHYB (ABO in Wagner et al., 1991) does not practically affect the position of the spectrum (λ_{max} = 683 nm). Recent experiments with recombinant phyB from potato with derivative phycocyanobilin (PCB), which differ from phytochromobilin by the substituent at C-18 (vinyl for phytochromobilin vs. ethyl for phycocyanobilin), reveal extrema in the difference absorption spectra at 659 nm (for Pr) and at 713 nm (for Pfr) (Gärtner et al., 1995). In comparison, PCB-substituted phyA from oat give, respectively, 651 nm and 715 nm to the extrema of both forms. (Incorporation of phytochromobillin yielded maxima nearly identical to native oat phytochrome, 663 and 728 nm (Gärtner, 1995).) It should be noted that the comparison of the spectroscopic properties of phyA and phyB is complicated by the fact that phyA is heterogeneous and comprises two spectroscopically distinct species (see Sineshchekov, 1995a and below). *Fluorescence* of P (Type 1, phyA) in its Pr form was detected both *in vitro* and *in vivo* with *emission and excitation maxima* at 680–687 nm and 666–673 nm depending on plant species and tissues used and experimental conditions (Fig. 3) (see Sineshchekov, 1995a for a review). PhyB *in vivo* has an emission maximum at 683 nm as shown on transgenic *Arabidopsis* (Sineshchekov et al., 1998b) and potato (V. Sineshchekov, O. Ogorodnikova, M. Herold and C. Gatz, unpublished results) and phyA-deficient mutant of pea (V. Sineshchekov, O. Ogorodnikova and J. Weller, unpublished results). *The quantum yield* (ϕ_f) is pretty low at ambient temperature (ϕ_f is in the range ca (1–5) × 10^{-3} according to different sources) but it increases up to ca 0.3 and more upòn freezing at 77 K and below. This reflects very fast radiationless processes of energy degradation at room temperature and their slowing down upon the temperature decrease. The Pr *fluorescence decay kinetics* reveal a number of components with lifetimes (τ_f) ranging from several ps to 1 ns. The components do not depend on the degree of intactness of the P molecule suggesting that the properties of the Pr excited states are governed primarily by the immediate protein surrounding of the chromophore. The *time-resolved Pr fluorescence spectra* of the kinetic components had different position of the maxima (λ_{max}) which varied within 10 nm. Lumi-

R, the first intermediate, stable at low temperatures, fluoresces with a yield similar to that of Pr (ca 0.3) while Pfr does not fluoresce even at the liquid helium temperature (1.5–4 K), $\phi_f < 10^{-4}$.

4.2 Photoreaction and Internal Conversion

There are three main experimental approaches to the investigation of the various competing radiationless processes of the dissipation of the Pr excited states: (1) Flash photolysis of the Pr phototransformation; (2) Low-temperature absorption and fluorescence spectroscopy and kinetics; and (3) Time-resolved optoacoustic spectroscopy.

Picosecond kinetics measurements on large rye P reveal a decay of the Pr excited states (Pr*) within ca 15–40 ps to prelumi-R which in its turn decays with a delay (up to 100 ps) to the photoproduct lumi-R (Lippitsch et al., 1988). It was also reported (Kandori et al., 1992; Tokutomi et al., 1995) that depletion of the absorption of Pr of pea large P upon excitation is followed by recovery of the absorption and appearance of a product in 24 ps. The maximum of the absorption of the photoproduct formed directly from Pr (named batho-R) is at 677 nm and decays to lumi-R in the time range from 0.4 to 100 ns. Presumably, batho-R corresponds to prelumi-R in Lippitsch et al. (1988). The question arises, however, if the appearance of lumi-R proceeds via an intermediate prelumi-R or directly because Lippitsch et al. (1993) have followed later a complex kinetics of the light-induced changes of the Pr → lumi-R and presented a kinetic model which include five stages in the time range from several ps to over 180 ps where no prelumi-R formation takes place as shown in Fig. 4, see below.

Under intense illumination the reverse photoreaction lumi-R → Pr is observed which may explain peculiarities of the laser-induced physiological photoresponses. Investigations with laser double-flash/double-color technique have shown that the amount of Pfr, which is formed under the first flash, is reduced by the second flash absorbed by lumi-R during its lifetime (see Inoue et al., 1990; Scurlock et al., 1993 and the literature cited therein). *At low temperatures* (below 150–170K), the photoprocess of the Pr → Pfr conversion is stopped at the stage of lumi-R and the direct and reverse photoreactions (Pr ↔ lumi-R) can be investigated with the use of the *conventional absorption and fluorescence technique* (see Kendrick and Spruit, 1977; Rüdiger and Thümmler, 1991, 1994; Sineshchekov, 1995a for a review). Absorption (and fluorescence excitation) spectrum of lumi-R is characterised by the main maximum at 696–698 nm, a shoulder at 630 nm, by the better resolved structure, and higher oscillator strength than Pr. It has an emission maximum at 702–705 nm. The kinetics of the light induced absorption and fluorescence intensity changes in the Pr ↔ lumi-R conversion is complex. It reveals several components whose rate and relative amplitudes change depending on temperature and direction of transformation, Pr → lumi-R or reverse. This was interpreted as the existence of several Pr ground state populations which differ by the activation barrier of the photoreaction (from several hundreds to several thousands J mol^{-1}) and thus by its quantum yields at low temperatures (see Sineshchekov and Sineshchekov, 1990; Sineshchekov and Akhobadze, 1992; Sineshchekov, 1995a and below).

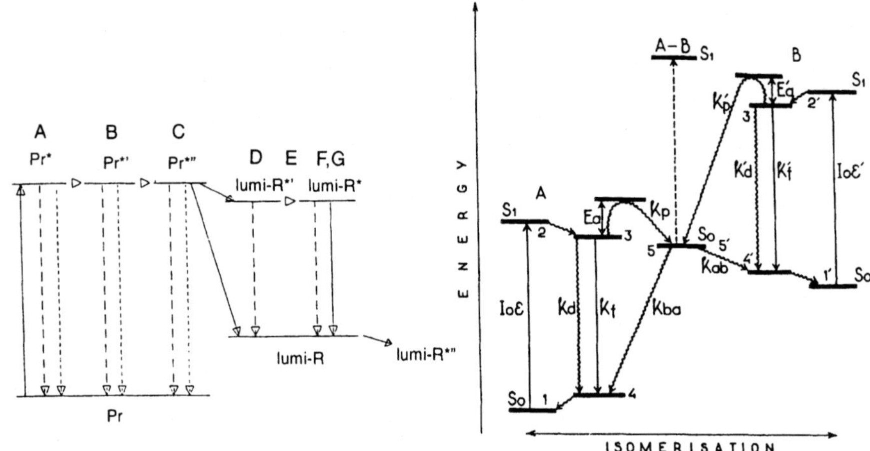

Fig. 4. Alternative schemes of the Pr → lumi-R photoreaction: energy levels and transitions between them. 1. Model based on the time-resolved absorption and fluorescence measurements. Stages at different times (ps) after excitation: (A) 10, (B) 20, (C) 35, (D) 50, (E) 130, (F) 170 and (G) 270. Long-dashed arrows: fluorescence; short-dashed arrows: internal conversion; full arrows: photoconversion. Pr*: excited state Pr; Pr*': orthogonal Pr*; Pr*": fully rotated Pr*; lumi-R*': orthogonal excited-state lumi-R; lumi-R*: excited state lumi-R. (From Lippitsch et al., 1993.) (left panel) 2. Model based on low-temperature measurements of the Pr and lumi-R fluorescence and phototransformation. A, initial state, Pr; B, the first stable photoproduct, lumi-R; A-B, the shortliving unstable intermediate, possibly, prelumi-R or batho-R; k, rate constants of fluorescence (k_f), temperature-independent degradation of excitation (k_d), primary photoreaction (k_p), phototransformation into the product (k_{ab}), and return to the initial state (k_{ba}); Ea and Ea', activation energies of the temperature-dependent forward and reverse primary photoreactions. (From Sineshchekov and Sineshchekov, 1990.) (right panel).

Laser-induced optoacoustic spectroscopy (LIOAS) has allowed an estimation of the upper value of the energy accumulated in lumi-R as a result of the photoreaction, 180 kJ mol^{-1}, what is close to the energy of the 0-0 transition in Pr determined from the emission spectra (see above), and the sum of the constants of the initial photoprocesses of Pr was evaluated to be of 0.3×10^{10} s^{-1} (see Braslavsky, 1990; Schaffner et al., 1990 for a review). Recent experiments with phyA from oat have obtained a value for the quantum yield of the lumi-R formation ca 0.1 in agreement with the literature data for the total Pr → Pfr transformation and have shown that the chromophore and its protein environment undergo an expansion of 13 ml per mol of photoconverted molecules (Braslavsky, 1995).

4.3 The Scheme of the Photoprocesses

From the analysis of the structure of the chromophore in the Pr and Pfr forms with the use of analytical methods and ^{1}H-NMR, Raman and difference infra-red (IR) spectroscopy it is generally accepted that photoinduced isomerization about C15 double bond between rings C and D is the most likely mechanism of the photoreaction (configuration of the chromophore is 15Z and 15E in Pr and Pfr, respectively) (see for a review Rüdiger and Thümmler, 1991, 1994; Sineshchekov, 1995a). Rüdiger (1992) has proposed a model according to which conrotation of the C14/C15 single bond with the C15/C16 double bond would produce only a displacement of Ring D in the plane of the ring but in effect its rotation. This would require the least space in the protein pocket. The model with conserved chromophore conformation has been presented in Parker et al. (1994) and Song (1994) (Fig. 1). The mechanism of the initial events at the level of excited states leading to the structural changes in the molecule remains, however, unclear. In particular, the results of the picosecond and low-temperature measurements of the light-induced absorption and fluorescence changes in the Pr → lumi-R conversion have produced energy level schemes of the process (Fig. 4) which are controversial in some of their aspects (see discussion in Sineshchekov, 1995a). The most recent is the scheme of Lippitsch et al. (1993) (Fig. 4, Scheme 1) which was put forward to explain two major points: (1) several intermediates are involved in the ps phototransformation and (2) absorption changes cannot be described by a simple kinetic model involving sequential or parallel pathways with first-order transformation kinetics of the intermediates. Acording to their view, excited state Pr (Pr*) generated by light absorption transforms into orthogonal Pr*, Pr*', and the latter, into fully rotated Pr*, Pr*''. Pr*'' relaxes into the orthogonal excited-state lumi-R, lumi-R*', and ground states of lumi-R. Lumi-R*' produces excited state lumi-R* which in its turn relaxes into lumi-R. Thus, the transitions in the time scale up to 120 ps (stages A-D) are between the excited state intermediates of Pr and lumi-R and the first ground state intermediate appears only after 157 ps (the second species of lumi-R) and there is no place for prelumi-R which was initially proposed by the same group (see above). The authors of Scheme 1 believe that the chromophore excitation and subsequent rotation to form lumi-R proceeds without the activation barrier and without energy storage in the photoproduct lumi-R. The changes are only to trigger protein restructuring without providing energy for it. The scheme is also to explain the heterogeneity of the emitting Pr species with different lifetimes and slightly different spectra which are produced from the homogeneous ground state Pr population. The scheme with similar notion—existence of a Pr excited-state potential energy surface with minima along the isomerization reaction coordinate—was put forward earlier by Holzwarth et al. (1992) to explain the heterogeneity of the Pr fluorescence decay.

An alternative scheme significantly different was proposed by Sineshchekov and Sineshchekov (1990) to incorporate the following major features of the process as observed in the low-temperature experiments (Fig. 4, Scheme 2). (1) The photoreaction is an activation photoprocess, i.e. it has an energy barrier in the excited state determining the quantum yield of the photoreaction at low temperatures. (2) The photoreaction (completed or incompleted) is the main (if not the only) temperature dependent process for the deactivation of the excited state, i.e. the route, which leads via the activation barrier from the excited state of Pr, branches in the direction to the ground state lumi-R and back to that of Pr. (3) It is assumed that the branching takes place at the orthogonal 'hot' ground state between Pr and lumi-R. This unstable even at low temperature state may correspond to prelumi-R (or batho-R). (4) The photoreaction is the energy conserving process which is then used to drive the Pr → Pfr semicycle. (5) The photoconversion lumi-R → Pr proceeds along a symmetrical scheme (but may not via the same intermediate) since lumi-R is pretty close by its photophysical and photochemical properties to Pr (Sineshchekov and Akhobadze, 1992).

Photoprocesses in Pfr are much less investigated. This is connected with two major difficulties—the absence of the Pfr fluorescence and admixture of the Pr form which is always present in the sample under photoequilibrium conditions. The former prevents measurements of the excited states dynamics in Pfr with the use of time-resolved fluorometry and the latter interferes with the ps absorption kinetics of the Pfr photoconversions. The absence of the Pfr fluorescence can be interpreted either as an indication to the reordering of the Pfr excited states so that the state with a prohibited transition becomes the lowest or that the photoreaction and internal conversion very efficiently compete with fluorescence even at low temperatures. In this case the sum of the constants of the radiationless processes in Pfr were estimated to be two-three orders of magnitude higher in Pfr than in Pr. Very fast photoprocesses in Pfr demand that the potential curve in the excited state between Pfr and the first photoproduct lumi-F has a barrier-less profile. The latter is, however, at odds with the fact that photoreaction of Pfr is an activation process with the potential barrier similar to that of Pr (ca 5 kJ mol^{-1}) (see Sineshchekov, 1995a for discussion).

4.4 Dark Stages of the Pr ↔ Pfr Cycle

The light energy is likely to be stored in the photoreactions as a strain imposed by the chromophore on the apoprotein upon its rearrangement and changes of the distances between the chromophore and N-terminus (Rospendovski et al., 1989). The following dark events involve the protein moiety and formation of Pfr (and Pr in the reverse reaction) is then realized via stepwise protein rearrangements. These steps in the Pr ↔ Pfr conversions are associated with certain intermediates which are characterized by a number of parameters: absorption and emission maxima, extinction coefficients, lifetimes and temperature limits of stability and others. It is accepted that no intermediates on the two pathways are common.

Laser photolysis investigations reveal 4–5 kinetic intermediates in the *Pr → Pfr phototransformation* with lifetimes ranging form several μs to hundreds ms. Recent data by Song's group (Zhang et al., 1992; Song et al., 1997) obtained with the use of global anaylsis of the transient spectra of native 124 kDa oat P disclose five components with the lifetimes of 7.4 and 89.5 μs and 7.6, 42.4 and > 266 ms (lumi-R1, lumi-R2, meta-Ra1, meta-Ra2, and meta-Rc). The data agree with the earlier observations (see Schaffner et al., 1990 for a review) with respect to the first two short-lived intermediates and are compatible with the kinetic models involving two parallel pathways or a sequential pathway with equilibria at certain stages. The formation of Pfr at the last step involves an unbranched reaction (Zhang et al., 1992). The authors in Scurlock et al., 1993; Braslavsky, 1995; Braslavsky et al., 1997, from the experiments on native and recombinant (PΦB-derived) Ps, favour the parallel pathway model of the Pfr formation. The phycocyanobilin (PCB)-derived phyA revealed a Pfr formation with shorter times suggesting participation of the C-18 vinyl group (present in PΦB and absent in PCB) in the conversion. In the dark steps in the reverse *Pfr → Pr photoconversion* three species with lifetime of 320 ns, 265 μs, and 5.5 ms were identified (Chen et al., 1996). The results agree with the notion that the Pfr photoreversion follows a sequential pathway that does not share any intermediates with the Pr phototransformation pathway. The rate of the Pr → Pfr transformation is influenced by the surrounding medium of the P molecule, in particular, by microviscosity, suggesting that the process depends on the state of the chromophore domain and in general on the flexibility of the P molecule (Lindemann et al., 1992; Braslavsky, 1995).

The low-temperature absorption investigations produced the picture of the Pr ↔ Pfr phototransformation which in its major features is similar to that of kinetics measurements (Kendrick and Spruit, 1977). The results presented by the Rüdiger's group (Rüdiger and Thümmler, 1994) can be summarized as follows: dark relaxation of lumi-R produces meta-Ra, blue shifted and with lower extinction coefficient; in native P meta-Ra yields meta-Rc (meta-Rb is formed in degraded P). And meta-Rc is converted into Pfr. The dark reaction Pfr → Pr is much less investigated. Two intermediates, lumi-F and meta-F, were initially found and there were indications that the picture may be more complex: two species of meta-F exist, meta-Fa and meta-Fb (Kendrick and Spruit, 1977) and two subspecies of meta-Fa were reported (Volotovsky, 1992). The long-wavelength shift in the absorption spectrum to 740 nm occurs only at the last step in the Pr → Pfr conversion, the slowest and the most temperature dependent, suggesting that the changes in the protein structure take place at the latter stages. They may involve molecular reconstruction proceeding via a relatively high potential barrier (see energy level diagram of the P cycle in Fig. 5).

The intermediates in the Pr → Pfr photoconversion are also being investigated by the *circular dichroism spectroscopy* (low-temperature and

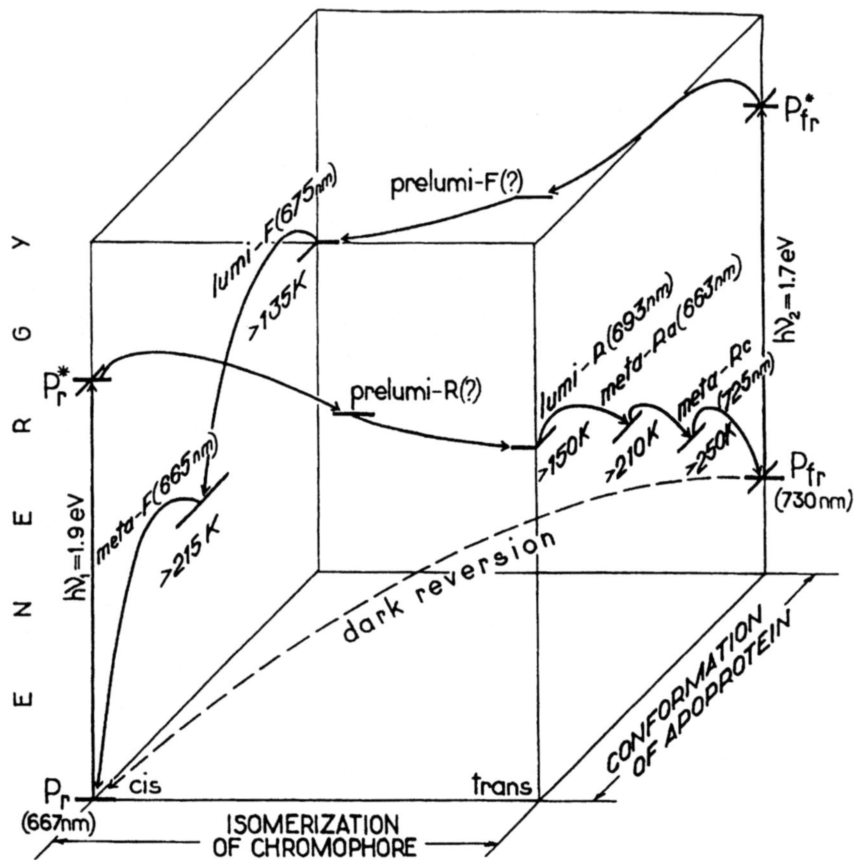

Fig. 5. Simplified energy level diagram of the phytochrome photocycle reflecting (i) changes of the chromophore configuration and energy storage in the photoreactions and (ii) dark relaxations of the photoproducts involving changes of the apoprotein conformation. Temperature presents approximate limits of stability of the intermediates, position of their absorption maxima are given in parentheses. The scheme is to be considered in the light of recent publications suggesting, in particular, formation of at least two lumi-R and, possibly, two meta-Ra states from homogeneous Pr pool and/or existence of heterogeneous Pr populations which initiate parallel cycles of Pr → Pfr phototransformations (see for discussion Section 4.4). (From Sineshchekov, 1995a.)

time-resolved) and by *vibrational spectroscopy* (Raman and infrared, IR) which provide information on the structural alterations in the pigment molecule. It was found (Eilfeld and Eilfeld, 1988; Björling et al., 1992) that the sign of the circular dichroic absorption bands changed upon formation of lumi-R and CD intensity dropped down upon generation of the subsequent intermediates (meta-Ra and meta-Rc) and Pfr pointing to a decrease in the interaction between the chromophore and the aromatic residue(s). It is believed (Eilfeld and Eilfeld, 1988) that the steps Pr → lumi-R → meta-Ra mainly involve configurational and conformational changes of the chromophore while further steps involve protein rearrangement. FT-IR difference spectroscopic studies (Siebert et al., 1990; Foerstendorf et al., 1995) suggest that the chromophore in the intermediates may differ from that of the main absorbing forms in the interaction of the carbonyl groups at rings A and D, respectively, with the protein.

It should be noted that the above data were obtained on P1 or Type 1 P. The data on photoreaction of PII or Type 2 P are scarce. According to Furuya (Furuya, 1989) the ratio of the absorbance at 724 nm at the red-light induced photostationary state to that at 667 nm of Pr was 0.135 (for PI it is ca 0.6). This low value was considered as artifactual and attributed, in particular, to a very rapid reversion of Pfr to Pr. In experiments on phototransformation kinetics of recombinant phytochromes from potato, Gärtner et al. (1995) have shown that phyB expressed in yeast and converted into chromoprotein upon incorporation of the chromophore derivative phycocyanobilin (PCB) the decay of I_{700} (lumi-R) was followed with a time constant of ca 75 μs and formation of Pfr revealed biexponential kinetics of ca 35 and 800 ms. In the same research group the PCB-derived oat phyA showed similarly a monoexponential decay of I_{700} with 90 μs and Pfr formation with two time constants, 13 and ca 500 ms (Braslavsky, 1995; Braslavsky et al., 1997). The authors (Gärtner et al., 1995), however, justifiably stress a necessity of a direct comparison of the data on potato phyA and phyB to yield information on the differences in the photochemical behaviour of the two phytochromes. In the experiments on transgenic *Arabidopsis* (Sineshchekov et al., 1998b) and potato (V. Sineshchekov, O. Ogorodnikova, M. Herold and C. Gatz, unpublished results) overexpressing phyB and on phyA-deficient mutant of pea (V. Sineshchekov, O. Ogorodnikova and J. Weller, unpublished results) it was shown that phyB *in vivo* differs from phyA by its photochemical properties: it has low extent of the Pr → lumi-R conversion at 85 K ($\gamma_1 \leq$ 0.05) and of the Pr → Pfr conversion at ambient temperatures ($\gamma_2 \approx$ 0.35–0.5).

5. Pfr → Pr Dark Reversion and Destruction of the Pfr Form

The physiological effect of Pfr is related to the absolute or relative concentration of this form so that the process which leads to [Pfr] changes would affect its action. Pr is shown to be relatively stable. However, the level of Pfr is subject to rapid changes due to its (1) degradation (see Clough and Vierstra, 1997 for a review) and (2) dark reversion into Pr. Furuya (1989) presents three patterns of Pfr transformation in the dark after red irradiation: 1. only dark reversion of Pfr (for instance, in light-grown cauliflower head, cauliflower type); 2. only Pfr decay (in etiolated tissues of oat, rice, monocot coleoptile type); 3. both Pfr reversion and decay (dicot tissues, dicot type). The reason for such a diverse picture remains largely unknown although one may seek explanation in different populations of light-labile phyA and light-stable phyB whose stability is believed (Wagner et al., 1991) to be due to determinants residing within the polypeptide. Existence of pigment pools with different activation barriers for the Pfr → Pr dark transformation could also be anticipated. The content of the different P population depends on plant species, pigment localization in tissues and degradation machinery in the cell. The latter, in particular, is supported by the data (Kunkel et al., 1995) that rice phyA adduct (phyA*) and the tobacco phyB adduct (phyB*) obtained by the *in vivo* reconstitution of phycocyanobilin with apophytochrome in living yeast cells, which does not possess the P degradation systems, show that the protein stability in yeast is independent of the form of the photoreceptor. After irradiation with red light, 25.6% of the far-red light-absorbing form of phyB* exhibited dark reversion with a half-life time of approximately 20 min. Control experiments with phyA* revealed no dark reversion. The data indicate that the molecular basis for this reaction is the formation of heterodimers between the red and the far-red light absorbing form of P.

6. Structure-function Relationship in the Phytochrome Molecule

The molecular nature of the changes leading to the appearance of the physiological activity of the pigment remains the major issue in the photochemistry of the pigment.

6.1 Differences between the Pr and Pfr Forms

The Pr → Pfr photoconversion induces changes both in the chromophore and the apoprotein as a system (see reviews Song, 1985, 1988; Furuya, 1989; Sineshchekov, 1995a) but for the sake of convenience we will consider them separately. The results summarized below are obtained primarily on phyA; however, data are being gathered indicating the phyB undergoes conformational changes in a red/far-red-light-dependent manner analogous to those of phyA (Wagner et al., 1991).

The chromophore in the apoprotein pocket—(1) The Pfr chromophore is different by its state of isomerization from that of Pr about the C15 double bond although configuration of the other two bonds is not yet agreed upon. (2) Possible involvement of proton transfer (at least intramolecular) from the chromophore in Pfr in the last step from Pr to Pfr is also debated (Furuya, 1989). (3) CD spectra measurements (Eilfeld and Eilfeld, 1988; Björling et al., 1992; Song et al., 1997) reveal lower CD signal in Pfr as compared to that of Pr, suggesting weaker interaction between the chromophore and the aromatic residue(s) in Pfr. (4) The chromophore moves within the protein (from its binding pocket) during photoconversion as shown by physiological experiments, linear dichroism studies and energy transfer methods and also suggested from experiments with oxidizing or reducing substances (Song, 1985; Sommer and Song, 1990) and molecular modelling of P (Parker et al., 1994). (5) These molecular alterations produce changes in the vibronic structure of the chromophore which show themselves in the observed photochromic absorption changes, fluorescence quenching of the Pfr form (Sineshchekov, 1995a) and in Raman and IR spectra (Farrens et al., 1989; Rospendovski et al., 1989).

The apoprotein—(1) An increase in the α-helical folding of the apoprotein takes place when Pr is converted to Pfr as shown with the time-resolved CD spectroscopy in the far-UV region (Chen et al., 1993; Deforce et al., 1994; Song et al., 1997). (2) Conformational changes in Pfr have been revealed also by differential sensitivity to proteolysis and binding studies with antibodies. They are within a 10 kDa subdomain at N-terminus, at the junction of the principal N- and C-terminal domains, and around glutamate-354. Different sites of cleavage are found in the two forms—in Pfr (arginine-746 to lysine-752, around glutamate-877 and arginine-1010 and only one in Pr (glutamate-38—arginine-62) (Nakazawa et al., 1993). (3) In Wells et al. (1994) differential accessibility of certain residues was shown by measuring dynamic quenching of the tryptophane fluorescence of *Pisum* phyA. It was concluded that the regions around four of the 10 tryptophanes may represent conformationally photoresponsive areas. The surface-exposed hydrophobic regions in Pfr are considered to be responsible for the biological activity of the pigment. Schematic representation of the changes in the Pr → Pfr conversion, see in Fig. 6.

6.2 Functional Roles of Different Structural Components in the P Molecule

Recent achievements in the field of site directed mutagenesis and deletion analysis allowed functional dissection of the P molecule domains and even individual amino acid residues in the organisation and functioning of the pigment.

Deletion of the N terminus to residue 46 and of the entire C-terminus is not critical to the chromophore attachment and photoreversibility; however, a deletion mutant lacking 222 amino acids failed to yield holophytochrome (Deforce et al., 1991). From the experiments on transgenic tobacco expressing truncated oats Ps, it is concluded that the N-terminal residues between 70 and 398 are necessary for chromophore attachment and residues 399-652, for spectral and photochemical integrity (Cherri et al., 1993). More specifically, among the residues in the vicinity of the chromophore (Arg-318, His-321, His-324 and His-326), only His-324 revealed an intimate involvement in the catalytic and photochemical process (Deforce et al., 1993; Song et al., 1996). Experiments (Gärtner, 1995) on mutant recombinant phyA from oat expressed in *E. coli* ('small' P-65 kDa segment of the native P plus 65 amino acids from the N-terminus with incorporated chromophore derived phycocyanobilin) have shown that substitution of leucine-glutamine for arginine-aspartate and tryptophane for tyrosine leads to a slowing down of the Pfr formation kinetics pointing to an important role of these residues. Exchange by a phenylalanin which gave unstable polypeptide suggested that the mutation affected a protein domain important for the stability of P. Tomizawa et al. (1995) have shown on rice phyA apoprotein expressed in yeast and reconstituted *in vitro* with the chromophore phycocyanobilin (PCB) that substitution of the first ten serine residues in the N-terminal domain by alanine residues (S/A mutation) gave very similar spectrophotometric peaks to the Pr and Pfr forms of wild-type product. However, the peak of the product, in which the amino acids of the C-terminal domain from 689 to 1128 were deleted (CD mutation), in the Pfr form was significantly shifted towards a shorter wavelength. This indicated

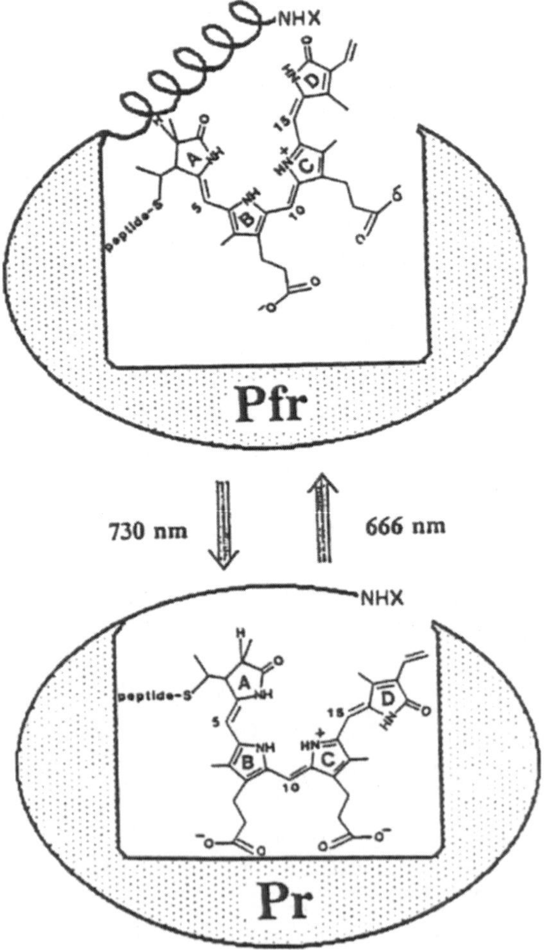

Fig. 6. Representation of the phototransformation of the Pr chromophore to the Pfr chromophore. The model incorporates: semi-extended conformation of the chromophore both in Pr and Pfr with conservation of the exocyclic dihedral angle at ring D by a chromophore-apoprotein interaction, protonation of ring C, reorientation and more exposed nature of the Pfr chromophore, increase in α-helical folding of the N-terminus. (From Rospendovski et al., 1989.)

that the C-terminal domain greatly influences the absorption maximum of Pfr. The S/A and CD mutations did not affect the dark reversion from Pfr to Pr. No photoreversibility was detected with the product where the first 80 N-terminal amino acids were deleted (mutation ND) suggesting that the missing residues are intimately involved in the photochemical process. Experiments with chromoproteins expressed in *E. coli* and assembled with phycocyanobilin (Hill et al., 1994) point of an essential role of the front 6-kDa region of the protein in the formation of the far-red absorbing chromophore-protein complex since the 118-kDa fragment, only lacking the 6-kDa N-terminus, exhibits a strong Pr band, but only a weak Pfr absorbance.

According to Quail's group (Boylan et al., 1994; Quail et al., 1995; Xu et al., 1995; Quail, 1997) sequences necessary for normal functional activity reside at both ends of the polypeptide within the terminal 110 residues or less. A subdomain within the N-terminal domain between residues 53 and 616 carries contact sites for interaction with downstream signalling molecules. The C-terminal domain between residues 617 and 1129 is believed to be necessary for completion of productive interactions under all irradiation conditions, while 52 residues at the N-terminus of phyA are necessary for its FR photosensory activity. From the experiments with chimeric molecules containing fused parts of phyA and phyB, it has been established that determinants of the photosensory specificities of phyA and phyB towards R and FR light reside in the N-terminal domains (see Fig. 7). Results of Emmler et al. (1995) suggest that the lack of the first 80 amino acids still allows a rice phyA to interact with the P transduction pathway, albeit non-productively in tobacco seedlings. At the same time, a rice phyA with a conversion of the first 10 serines into alanine residues (S/A) increased the response under both red and far-red light. Vierstra and co-authors (Jordan et al., 1996; Vierstra et al., 1996) have determined two distinct functional domains in the N-terminus (residues 1–70), one required for conformational stability and biological activity (residues 13–62) and the other, serine-rich domain (residues 6–12) that modulates phytochrome activity. Removal of or substitution within this second domain generate a hyperactive photoreceptor with 10–50 fold greater sensitivity to far-red light. Another region important both for biological activity and Pfr degradation was detected by the authors near the extreme C-terminus (residues 1095–1129). Implication of lysines within residues 740–790 in Pfr degradation was also shown. Nakazawa and Manabe (1996) have demonstrated that the part from residue 1 to 966 is physiologically active, the part from 967 to 1101 has suppressive or regulatory role while the part from residue 1102 to 1119 overcome the suppression effect. C-terminal domain carries determinants necessary for implementation of P action both in phyA and phyB the most essential being a 160-residue region between residues 680 and 840 (Quail et al., 1995; Wagner and Quail, 1995; Quail, 1997) (see Fig. 7). All phyB mutations and majority of phyA mutations are located in this region which is likely to be part of the "active site". Two regions in the

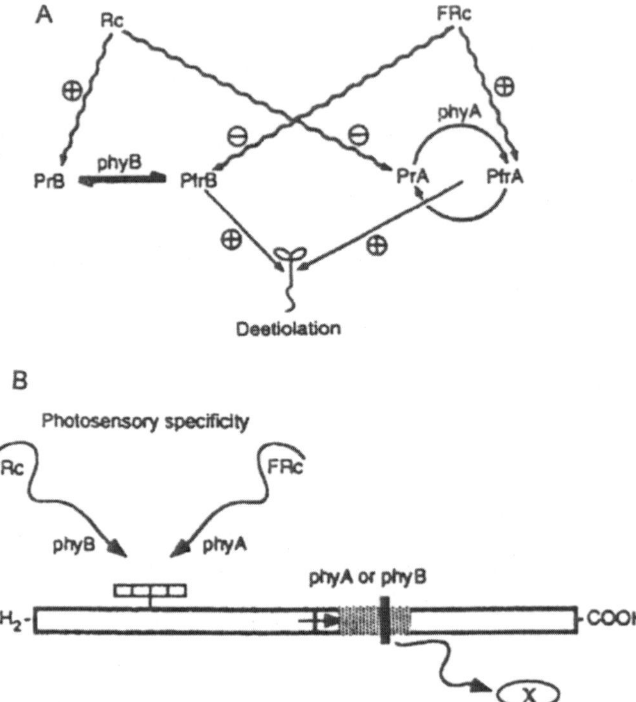

Fig. 7. Phytochromes A and B have discrete photosensory functions. In particular, they transduce mutually antagonistic signals to constant red (Rc) and far-red (FRc) light in deetiolation and early seedling development. (A) Schematic showing the action of Rc and Frc light (wavy lines) absorbed separately by phyA and phyB systems. Rc absorbed by phyB induces deetiolation (+) through maintenance of high amounts of its Pfr (PfrB) at low fluence rates of Frc. High fluence rates of FRc absorbed by phyB suppress (–) this induction at low fluence rates of Rc by reducing the amount of Pfr. Conversely, FRc absorbed by phyA induces deetiolation by means of the FR-HIR at low fluence rates of Rc. High fluence rates of Rc absorbed by phyA suppress this induction at low fluence rates of FRc by displacing the photoequilibrium toward its Pfr (PfrA). PrB and PrA: R-absorbing forms of phyB and phyA, respectively. (B) Schematic representation of the molecule of phytochromes A and B. Sequences determining the photosensory specificity and regulatory activity of phyA and phyB are spatially separated within the polypeptide. Sequence differences between phyA and phyB in the chromophore-bearing NH_2-terminal domain (chromophore indicated by series of four rectangles) result in contrasting interpretations of the same incoming Rc and FRc light signals by the two photoreceptors. Intramolecular information transfer (→) and successful transmission of the perceived signal to downstream transduction components (X) require a 160-residue COOH-terminal polypeptide segment (stippled, positions 681 to 840) (with indications that a subregion of 18 residues (solid) is particularly critical to this process. The activity specified by the COOH-terminal domain is common to both phyA and phyB, suggestive of a similar biochemical basis for the regulatory activity of the two photoreceptors. (Modified from Quail et al., 1995.)

molecule are believed to be involved in dimerization of the polypeptide near the center of the molecule and near the C-terminus (Cherri et al., 1993; Edgerton and Jones, 1993). And finally, Sakamoto and Nagatani (1997) have reported that the C-terminal part (in phyB) contains also putative localization signals. They have shown that most of the total phyB is located in the nucleus and that the nuclear localization of phyB is light-dependent.

7. Mechanisms of Signal Transduction

The mechanism of signal transduction is to mediate two phenomenological types of the photoresponses: (1) fast events at the membrane level and (2) slow processes which involve initiation and/or modulation of the protein synthesis.

The problem of the association of P with membrane and its direct modification is still open (Jordan, 1992). G-proteins, Ca^{2+}, calmodulin, protein kinases and inositol lipids are being considered as candidates for second messengers in this process (see Sopory and Munshi, 1998). Recent experiments of P activation of K^+ channels and chloroplast rotation in *Mougeotia* (Serlin et al., 1996) suggest, in particular, that it is unlikely that P acts directly on the channel. Within the classical "gene hypothesis", the signal is to be transduced from P in the cytoplasm to genes in the nucleus. The mechanism of light regulation of genes is believed to involve the binding of activators or repressors (protein factors, trans-acting elements) activated by P to regulatory elements of the gene (cis-acting DNA sequence elements). Traditional genetic approach which employs transduction chain mutants, in particular, those whose development in darkness resembles that of wild-type under illumination (see reviews Chory, 1993; Koornneef and Kendrick, 1994; Quail et al., 1995 and also Franzes et al., 1992 and the literature cited therein) revealed gene products of unknown nature which act negatively in the dark to repress photomorphogenesis and this repression is reversed by light (the negatively acting components).

The second effective approach is an original technique which uses a microinjection of P and various putative signalling molecules (the positively acting components) into the cells of the P-deficient tomato *aurea* mutants with impaired photomorphogenesis. These injections induced normal anthocyanin synthesis, chloroplast development, and other photomorphogenic processes (Bowler et al., 1994; Millar et al., 1994; Barnes et al., 1995). It was shown that P signalling involves heterotrimeric G-proteins, calcium, calmodulin, and cGMP.

Protein phosphorylation cascade is also being considered as a possible component of the transduction mechanism and this notion is compatible with several facts: (i) protein kinase activity is associated with purified P (probably connected with a polypeptide bound to P), in particular, a 124-kDa P in oat seedling extracts is tyrosine phosphorylated in its Pr form, but red light irradiation rapidly dephosphorylates it suggesting a possible role of tyrosine kinase in the P-mediated signal transduction cascade (Bhoo et al., 1995); phytochrome with a kinase domain that is catalytically active *in vitro* was obtained from the moss *Ceratodon purpureus* (Thümmler et al., 1992, 1995a, b); (ii) fast photoinduced phosphorylation/dephosphorylation events are observed in subcellular preparations; and (iii) the DNA binding characteristics change as a result of the light-induced phosphorylation of the G-box binding factor (see Harter et al., 1994; Thümmler et al., 1995a and the literature cited therein). It is suggested that phosphorylation of phytochrome may be involved in the modulation of its receptor activity and that phosphorylation of molecules downstream may be necessary to transduce and amplify the signal (Millar et al., 1994; Schäfer et al., 1997). Conclusion that P is a light-regulated enzyme is arrived at in Lagarias et al. (1996). Changes in pool size, distribution or sensitivity of hormones are also believed to be involved in the transduction pathways (Chory, 1993). At the same time, it is speculated that the Ps may use a previously unknown mechanism of primary signal transduction since there is no sequence similarities to other known proteins (Quail, 1991; Quail et al., 1994).

8. Phytochrome Heterogeneity in the Cell and its Functional Implication

As it was mentioned above, the problem of P heterogeneity as a means of monitoring complex environmental light signals acquires particular importance in the current phytochrome research.

8.1 Different Phytochromes

The major achievement of the recent years is the discovery of a small family of genes encoding for several Ps (phyA through phyE for *Arabidopsis* (see Smith and Whitlam, 1990; Quail, 1991, 1994; Furuya, 1993; Boylan et al., 1994 for a review). The chromophore is shown to be the same for all its members. However, the amino acid sequences may differ considerably: the homology between the three major Ps, phyA, B and C, for one and the same species (*Arabidopsis*) being approx. 50% while for the same Ps in different plant species it is higher. The phy gene family might be even more complex as shown in tomato and sorghum, they have more phytochrome genes than the five reported for *Arabidopsis.* Tomato contains two B-type P genes which are independently expressed in organ-specific fashion. In *Arabidopsis,* phyD is also considered to be a second B-type P (Pratt et al., 1995, 1997).

Discovery of the P family has prompted the development of the research of their physiological functions with the use of P mutants and transgenic plants. Now it is becoming clear that the Ps may play different roles in the photoregulation in plants, i.e. there is photosensory specialization among the family members (see reviews Smith and Whitelam, 1990, 1997; Kendrick and Nagatani, 1991; Reed et al., 1992; Peters, 1992; Chory, 1993; Cherri and Vierstra, 1994; Koornneef and Kendrick, 1994; Sineshchekov, 1995a; Casol et al., 1997; Kendrick et al., 1997; Whitelam and Devlin, 1997). The two major species, phyA and phyB, reveal contrasting functions for some processes while act complementary in the others. In particular, phyA and phyB monitor different facets of the light environment, they transduce mutually antagonistic signals to the seedling in response to continuous R and FR light (Ballare et al., 1994; Quail et al., 1995; Quail, 1997) (see Fig. 7). At the same time, both Ps, A and B, modulate the gravitropic orientation of hypocotyls (Robson and Smith, 1996) and the effect of phyB on hypocotyl extension growth of etiolated seedlings occurs via co-action with phyA (Casal, 1995).

PhyA participates in seed germination in FR and under very low fluences of light (see, in particular, Botto et al., 1996; Casal et al., 1997) and in deetiolation. Fluence requirements in photoinduction of seed germination for phyA is substantially lower than for phyB and phyA photoirreversibly triggers the germination upon irradiations with UV, visible and far-red light of very low fluence, while PhyB controls the photoreversible effects of low fluence (Shinomura et al., 1996). The phyA signalling system is believed to be important in early seedling establishment. In general, it makes higher contribution to VLFR than to LFR and has only minor influence on the R/FR ratio perception essential in the shade avoidance syndrome. PhyA is necessary for continuous far-red light (Frc) perception (FR-HIR). It is a primary receptor for red-light-induced enhancement of phototropism (Parks et al., 1996). PhyA induces CAB gene expression upon irradiation with VLFR and this induction is not red/far-red reversible (Hamazato et al., 1996). PhyA plays a role antagonistic to phyB in photoregulation of vegetative development in light-grown seedlings and in regulating flowering time (Devlin et al., 1996; Whitelam, 1995). It was shown that vascular tissue is a potential site of phyA action (Jordan et al., 1995).

PhyB promotes germination in the dark, inhibits germination in FR, regulates gravitropism and mediates photoperiodic control, participates in the shade avoidance syndrome and end-of-day FR response (Smith and Whitelam, 1997; Whitelam and Devlin, 1997). It is necessary for seedling growth and development under continuous red light (Rc). PhyB participates in the photoperiodic control of tuber formation in potato (Jackson et al., 1996). It induces CAB gene expression with red light of 2 to 3 order higher fluence than phyA (Hamazato et al., 1996). In Wester et al. (1994) it is shown that the expression level of this P gene is an important determinant of the intensity of light-induced plant responses. In general, the differential photosensory activity of phyA and phyB occurs largely through differences in their (i) intrinsic biochemical activities, (ii) relative abundances, and/or (iii) independent and separate reaction partners (Somers and Quail, 1995). The different photochemical properties of phyA and phyB (lower pho-chemical activity of phyB, see above) may also contribute to their functional specificity.

The role of minor Ps, phyC, -D, -E, is much less known. They are not necessary or sufficient for perception of constant R and FR (Quail et al., 1995) and much of traditional photomorphogenesis of higher plants can be explained on the basis of the actions of phyA and phyB (Smith, 1995, 1997). At the same time, there are data (Bagnal et al., 1995; Devlin et al., 1996) that at least one another species in addition to phyB is involved in mediating the shade avoidance responses and that neither phyA nor phyB exclusively regulated flowering in *Arabidopsis* and implication of the third P species is assumed. Hamazato et al. (1996) have shown that phytochrome species other than phyA and phyB photoreversibly regulates CAB gene expression. Involvement of gibberellins in the light-induced germination of *Arabidopsis* also appears to be regulated by Ps other than phyA and phyB (Yang et al., 1995).

Existence of multiple Ps is not unique for angiosperms. Recently, two different P types in the moss *Ceratodon purpureus* were characterized and expressed. One of them (PhyCer) has the C-terminus homologous to the catalytic domain of eucaryotic serine/threonine or tyrosine protein kinases (Thümmler et al., 1992) and is capable of phosphorylating serine and threonine residues (Thümmler et al., 1995b). The second P (PhyCer2) (Pasentsis et al., 1995) exhibits high sequence conservation with PhyCer within the chromophore domain while the C-terminal region is similar to those of conventional Ps. It is expressed at a much higher level than PhyCer.

8.2 Subpopulations of Phytochrome A

The P system seems to be even more complex. Along with the molecular biological and immunochemical data pointing to the heterogeneity of P population there were other experimental lines to this end. (1) Kinetics of light-induced P destruction revealed two pools of the pigment which differed by the rate constant of the process—bulk light-labile (half-time of destruction tens of minutes) and minor relatively light-stable (half-time of several hours) (Heim et al., 1981; Brockmann and Schäfer, 1982). In the recent observation (Poppe et al., 1994), red light pulses led to a P-dependent down-regulation of PHYA (phyA protein) mRNA abundance in etiolated parsley seedlings to a level of 10–20% compared with the dark control. The PHYA mRNA abundance in a parsley cell suspension culture was also down-regulated. The light-labile species is associated with phyA while the light-stable pool, with phyB and other P species (Furuya, 1989). However, within phyA there seem to be several subpopulations differing by spectroscopic and photochemical characteristics, lability and other properties (Sineshchekov, 1995a).

Biochemical experiments have detected along with the bulk soluble pigment a minor membrane-associated fraction (see Jordan, 1992; Lamparter et al., 1992; Terry et al., 1992 and the literature cited therein). Recently it was found that most of the total phyB upon illumination is located in the nucleus (Sakamoto and Nagatani, 1997). In this connection existence of the amphiphilic P sequences should be also mentioned which are exposed on the surface of the protein and which are potentially capable of interaction with other cellular macromolecules (Partis and Grimm, 1990; Jordan, 1992; Furuya and Song, 1994; Song, 1994).

Two P populations were found with the use of low-temperature fluorescence spectroscopy and photochemistry of the pigment *in vivo* (see Sineshchekov, 1995a for a review). They were distinguished by two major parameters—position of the emission (and absorption) maxima, λ_{max}, and extent of the Pr → lumi-R phototransformation at low temperatures, γ_1: Pr' with emission and absorption maxima at 686 and 673 nm, respectively, and $\gamma_1 = 0.5$ and Pr" with γ_{max} at 682 and 669 nm and $\gamma_1 \rightarrow 0$ (extent of the Pr → Pfr phototransformation, γ_2, was approx. the same for both species, 0.85 for the first and 0.70 for the second). Experiments with different plant species and tissues (Sineshchekov and Sineshchekov, 1990; Sineshchekov and Rüdiger, 1992; Sineshchekov, 1994, 1995b; Sineshchekov et al., 1995, 1996) have shown that Pr' is the bulk variable and light-labile component while Pr" is the minor saturable conserved and relatively light-stable one. Investigations carried out on P mutants of cucumber, *Arabidopsis* and pea deficient in phyB (*lh, hy3* and *lv5* respectively) have shown that the proportion of [Pr'] and [Pr"] remains the same as in respective wild-type plants what is compatible with the notion that both species reside within phyA (phyA' and phyA") (Sineshchekov, 1995b; Sineshchekov et al., 1998b; V. Sineshchekov, O. Ogorodnikova and J. Weller, unpublished results). Comparative analysis (Sineshchekov et al., 1994) of the fluorescence and photochemical properties of phyA' and phyA" in etiolated maize coleoptiles and of native (124 kDa) soluble and membrane-associated fractions extracted from them revealed close correlation of their parameters pointing to phyA" association with membrane (proteins) as a possible nature of their differences (what, presumably, is a result of post-translational modification of the pigment). In the light of recent observation on nuclear localization of phyB (Sakamoto and Nagatani, 1997) it is interesting to note that overexpression of rice and *Arabidopsis* phyB apoprotein (PHYB) in *Arabidopsis* (RBO and ABO in Wagner et al., 1991) and of oat phyA apoprotein (PHYA) (13k7 and 21k15 in Boylan and Quail, 1991) brings about decrease and increase in γ_1 values, respectively, as compared with the wild-type plants. Position of the emission maximum remains practically the same in the case of ABO or shifts by ca 3 nm to the blue in the case of RBO and by 2–3 nm to the red in the case of 13k7 and 21k15 (Sineshchekov et al., 1998b). These data suggest that phyB is close by its spectroscopic and photochemical properties to phyA", what could arise from similar localization of phyA" and phyB and/or their association with cellular particles, and also support earlier description of phyA' (see above). Interestingly, the recently discovered putative prokaryotic phytochrome from the cyanobacterium *Synechocystis* (Hughes et al., 1997; Yeh et al., 1997) does not show low-temperature Pr → lumi-R photoconversion, i.e. it belongs to the Pr" phenomenological type comprising phyA" and phyB

(Sineshchekov et al., 1998a). The two phyA species (phyA' and phyA") may perform different functions in the plant: the bulk light-labile phyA' operates in the processes of deetiolation while the minor relatively more light-stable phyA", under constant illumination (in agreement with the fact that phyA is functional in light-grown plants (Johnson et al., 1994; Clough et al., 1995)). This conclusion was arrived at in the experiments (Sineshchekov et al., 1996) on transgenic potato (Heyer et al., 1995) containing the P protein encoded by the PHYA gene cDNA (phyA) in sense or antisense orientation under the control of the 35S cauliflower mosaic virus promoter (see Fig. 8). These potato plants had increased and decreased phyA levels and were characterized by different phenotypes. And finally, within the phyA' molecular type there exist three subpopulations which differ by the energetics and kinetics of the photoreaction (Sineshchekov and Akhobadze, 1992). They are likely to be different conformers of the chromophore in the apoprotein pocket and may initiate parallel cycles of the formation of the physiologically active Pfr forms.

8.3 Possible Functional Implications of the Dimeric Structure of the P-molecule

Yet another source of the P polyfunctionality in the cell is the dimeric structure of its molecule. VanDer Woude (see VanDer Woude, 1987 and the literature cited therein) has introduced a model for explanation of the three types of photoresponses, VLFR, LFR and HIR. It recognizes interphototransformations of the three possible species of the dimer PrPr, PrPfr, and PfrPfr and interaction of the dimeric species with the next component of the transduction chain, a specific receptor, X, which is of very low abundance with respect to P. The model supports, in particular, the following: (1) PrPfr-X is established by fluences in the VLF range; (2) since concentration of PrPfr-X is very low, fluences in the LF range, 3 to 4 orders of magnitude greater than VLF, are required to produce PfrPfr-X; (3) responses to LF are related linearly to [PfrPfr-X] whereas VLF responses correlate well with log [PrPfr-X]; (4) formation and action of PrPfr-X in HIR is promoted by B or FR light, characteristic for this type of response, and cycling of PrPfr-X increases its action. PrPfr-X is assumed to be a membrane-associated complex whose activity and mobility are increased by physical, chemical, or hormonal sensitization. These characteristics of the model fit well into the real photophysiological behaviour of plants. Recent observations (Kunkel et al., 1997) suggest that phyA may act both as a Pfr-Pfr homodimer and as a Pfr-Pr heterodimer while phyB acts only as a Pfr-Pfr homodimer. The latter requires high Pfr/P concentration ratios thus explaining the lower signalling abilities of phyB. It should be mentioned that some authors (Quail, 1991; Jordan, 1992) suggest possible formation of heterodimers (from subunits of phyA and phyB, for instance) what widens the range of potential functions of the pigment. The notion that P functions as a dimer and that receptor

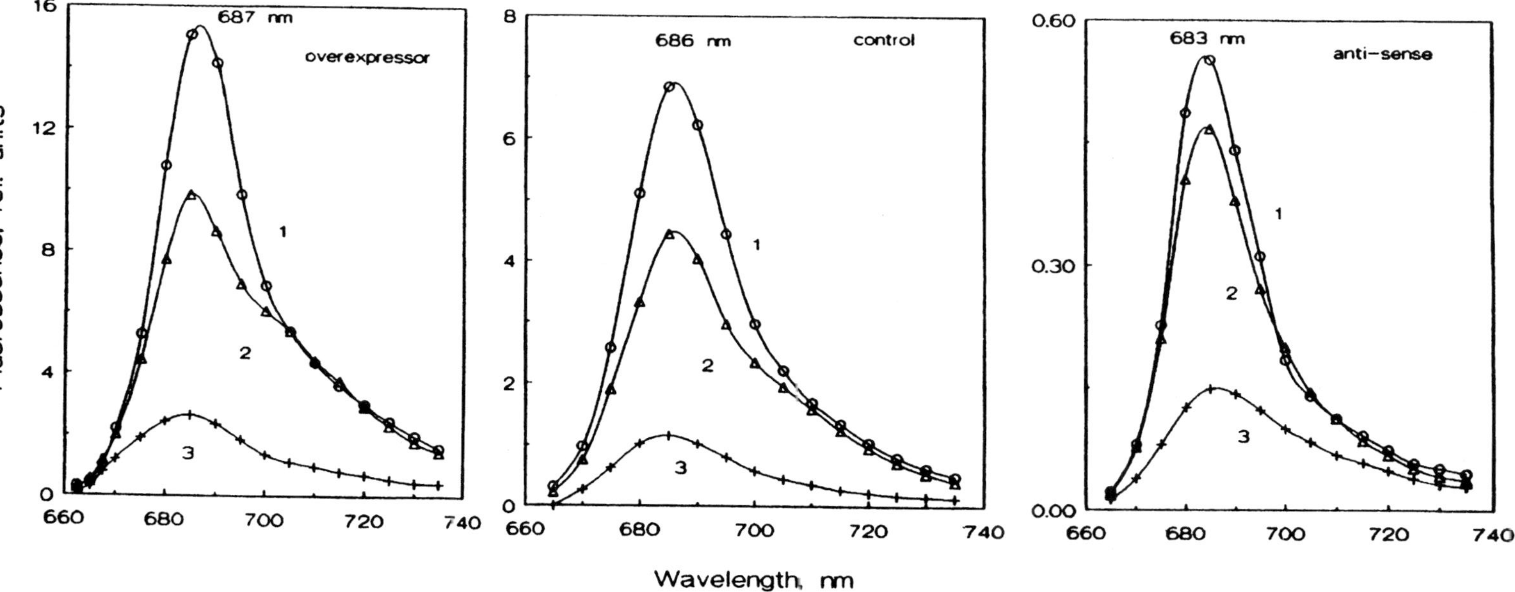

(a)

Fig. 8. Two phytochrome A populations, phyA' and phyA", in etiolated sprouts of transgenic potato with altered phyA levels and modified photoresponses due to expression of anti-sense RNA and constitutive expression of the potato PHYA (phyA protein) gene under the control of the cauliflower mosaic virus (CaMV) 35S promoter. Transgenic plants were obtained in Heyer et al., 1995; *in situ* fluorescent and photochemical analysis of the two species—major labile longer-wavelength phyA' and the minor constant shorter-wavelength phyA" with high and low extent of the Pr → lumi-R phototransformation, γ_1, respectively—was performed according to Sineshchekov, 1994. Low-temperature (85K) fluorescence emission spectra (λ_{ex} = 650 nm) of Pr in dark-grown sprouts (1) and after Pr phototransformation into lumi-R at 85K (2) and into Pfr at 273 K (3) show (i) the decrease of the total phytochrome concentration, proportional to the fluorescence intensity, in going from the overexpressor (Bin PS2) to the control (wild type) and to the anti-sense species (AP15/11), (ii) the short-wavelenght shift of the emission maximum and (iii) the reduction in the γ_1 values (a). Figure (b) presents the dependence of γ_1 on total phytochrome content in the anti-sense, control (WT) and overexpressor plants. This dependence and also the shift of the spectra reflects the changes of the concentration of the individual phyA' (Pr′) and phyA" (Pr″) forms with total phytochrome content (c). The phenotype of the gene-engineered potato plants (indicated in the figure) may be primarily connected with the variations in the phyA' content ([phyA"] remains relatively constant). (Modified from Sineshchekov et al., 1996.) (see next page).

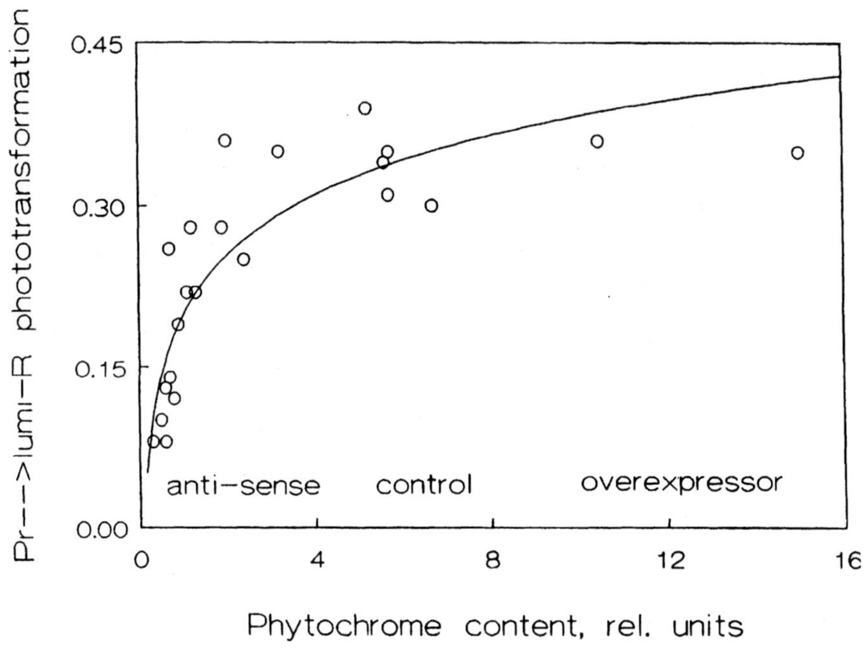

(b)

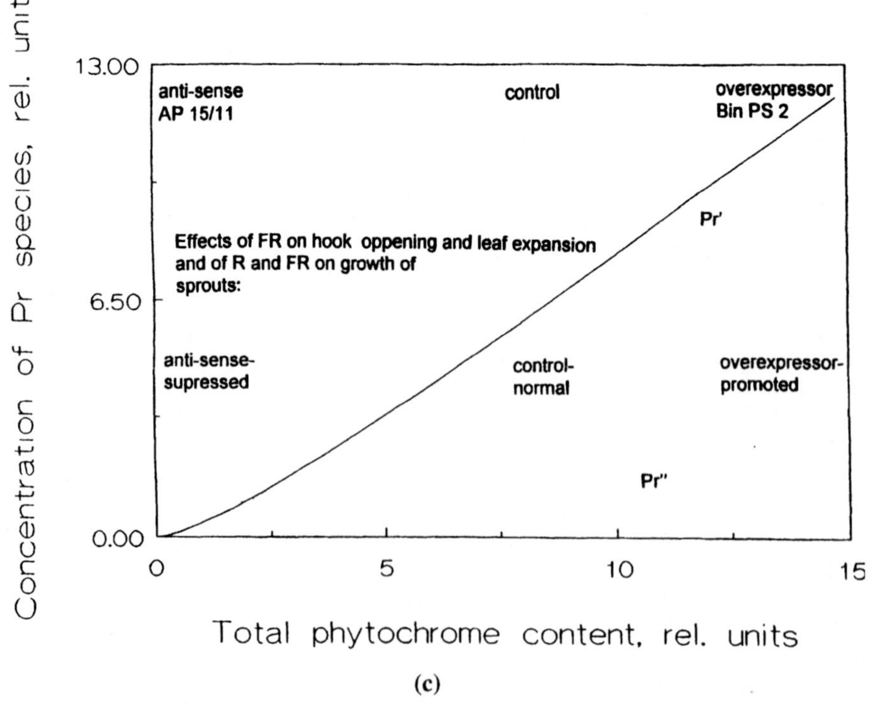

(c)

Fig. 8. **b, c: See legend on p. 783.**

recognition of P depends on its gross conformation was expressed in Jones and Edgerton (1994). It should be mentioned at the same time that there is an indication that monomer phytochrome also possesses biological function (Nakazawa et al., 1996).

8.4 Functional Role of the Photoreactions of the Intermediates

And finally, one cannot exclude participation in the photoresponses of the slow-transforming meta intermediates which are accumulated up to 50% of the total P under usual day light (Smith and Fork, 1992). They may modify the P action by undergoing photochemical transformation changing the photocycle of the Pr ↔ Pfr and ultimately the concentration of the physiologically active Pfr. It cannot be excluded either that they act by photoproducing physiologically active species different from Pfr or they themselves possess a physiological activity. The short-lived lumi-R state may also play a role in the processes induced by intense pulse illumination (for instance, in reduction of laser-induced seed and spore germination which is due the reverse lumi-R→ Pr photoreaction (Scheuerlein and Braslavsky, 1985; Scheuerlein et al., 1988). This is possible, in particular, because the quantum efficiency of the reverse lumi-R → Pr photoreaction is close to that of the direct one (Sineshchekov and Akhobadze, 1992; Sineshchekov et al., 1998b) or even higher by a factor of 2.5 (Gensch et al., 1996).

9. Conclusion

The native P in the cell is organized into a complex hierarchical and dynamic pigment system whose operation may determine to a large extent its regulatory polyfunctionality and provide a wide spectrum of opportunities in monitoring the environmental light conditions. The main block of this construction are different Ps, the products of P genes which vary by the amino acid composition of the apoprotein. The differences of the physiological functions of the two major Ps, phyA and phyB, are firmly established. Less defined is the functional role of the minor phytochromes, phyC and others, although the data are being gathered that they are not redundant. The second possible level of the diversity of the structurel organization is a post-translational modification of the pigment. Existence of the two spectroscopically and photochemically different P species is shown within phyA. Although the exact nature and functions of these populations—the major labile phyA' and the minor saturable and relatively constant phyA"—remain largely unknown, preliminary data suggest that they may differ by the state of association with proteins (soluble and membrance-bound) and operate in etiolated and green tissues, respectively. In spite of the fact that the major interest is drawn to the different phytochromes, the long standing problem of the P association with membranes and proteins deserves closer attention. Here new opportunities are being oppened with the development of the phytochrome extraction techniques in combination with the methods of *in situ* monitoring its state. And finally, dimeric structure of the pigment (homodimers and also presumed heterodimers consisting of monomers of different Ps) is considered as a possible basis for the multiple functions of the pigment. Here, however, direct experimental verifications of the existing hypotheses are needed.

The complex structure of the P system, on one hand, and the universality and limited number of the key intermediates in the transduction chain, on the other, actualise investigations of the molecular mechanisms of the photoreception because it is believed that the recognition of the character of the light information may take place already at the stage of the initial photoprocess in the pigment. PhyA and phyB are shown to differ primarily by the element in the molecule responsible for the perception of light and interpretation of its quality (photosensory function). One has also to mention in this connection the possibility of the existence of different pigment states within one molecular type (as shown for phyA'). They differ by the activation parameters of the initial photoreaction and are likely to be connected with different conformations of the choromophore in the apoprotein pocket. Although functional role of these states remains unclear, their investigations seem to be important for the understanding of the structure-function interrelations in the pigment molecule. It is also worth mentioning that, in spite of the fact that the nature of the P photoconversion is known (cis-trans photoisomerization), its molecular mechanism remains essentially unclear. This concerns both the initial photoreaction and the following relaxation steps. It should be also pointed out that the photochemical process is investigated on the bulk phyA and its mechanism cannot be readily applied to phyB and the other minor phytochromes. Further, the reverse Pfr → Pr photoconversion is much less elaborated than the direct one. Its investigation, which is important in itself, may help to understand the mechanism of the direct process.

As to the primary signal transduction pathways, in spite of the fact that the three key intermediates (G-proteins, calcium and cGMP) seem to be firmly established, the concrete mechanisms of their action needs to be elucidated (in particular, the number of G-proteins, their localization, interaction with P and activation of Ca^{2+} and cGMP and other steps). Still it is to be verified if the other possible processes such as light induced phosphoinositide pathway or phosphorylation/dephosphorylation or previously unknown mechanisms are operative in plants.

The major achievement in the area of P is, thus, that it starts to interpret both the photosensory functions and transduction of the light stimulus at the molecular and submolecular levels in terms of concrete intra- and intermolecular interactions, and first of all, of the chromophore with individual amino acid residues and of the latter with a putative primary receptors(s) in the early steps of the transduction of the perceived light stimulus. Combination of the methods of molecular biology and genetics of directional modifications of the P molecules and their functions with spectroscopic, photochemical and photophysiological approaches, which has already gained considerable progress in the area, promises its further fast development.

Acknowledgments

This work has been supported by the Russia Foundation for the Fundamental Investigations, Grant No. 96–04–50325.

Abbreviations

ALA—5-aminolevulinate; B—blue light; CAB—chlorophyll a/b binding protein; cGMP—cyclic guanosine 5'-phosphate; FR—far-red light; HIR—high irradience responses; LFR—low fluence responses; lumi-F, lumi-R (1700), meta-F, meta-R, prelumi-R—intermediates of the phytochrome cycle; PCB—phycocyanobilin; Pchlid—protochlorophyllide; P—phytochrome; P1, P2—Type 1 and Type 2 Ps; phyA, phyB, phyC—Ps A, B, C; PHYA, PHYB—apoproteins of phyA and phyB; phyA', phyA"—subpopulation of phyA; phyCer, phyCer2—moss Ps; Pfr, Pr—far-red- and red-light absorbing P forms; Pr', Pr"—subpopulations of Pr; P*Φ*B—phytochromobilin; R—red light; UV—ultraviolet; VLFR—very low fluence responses.

References

Bagnall, D.J., King, R.W., Whitelam, G.C., Boylan, M.T., Wagner D. and Quail P.H. 1995. Flowering responses to altered expression of phytochrome in mutants and transgenic lines of *Arabidopsis thaliana* (L.) Heynh. *Plant Physiol.* **108:** 1495–1503.

Ballare, C.L., Scopel, A.L., Jordan, E.T. and Vierstra, R.D. 1994. Signaling among neighboring plants and the development of size inequalities in plant populations. *Proc. Natl. Acad. Sci. USA* **91:** 10094–10098.

Barnes, S.A., Quaggio, R.B. and Chua, N.H. 1995. Phytochrome signal-transduction: characterization of pathways and isolation of mutants. *Philos. Trans. R. Soc. Lond., B, Biol. Sci.* **350:** 67–74.

Bhoo, S.H., Lapko, V.N., Wells, T.A., Sommer, D. and Song, P.-S. 1995. Phytochromes: structure and function. In: Tetrapyrrole Photoreceptors Freising, Germany, Book of Abstracts, O 22.

Björling, S.C., Zhang, C.-F., Farrens, D.L., Song, P. S. and Kliger, D.S. 1992. Time-resolved circular dichroism of native oat phytochrome intermediates. *J. Am. Chem. Soc.* **114:** 4581–4588.

Botto, J.F., Sanchez, R.A., Whitelam, G.C. and Casal, J.J. 1996. Phytochrome A mediates the promotion of seed germination by very low fluences of light and canopy shade light in *Arabidopsis. Plant Physiology* **110:** 439–444.

Bowler, C., Neuhaus, G., Yamagata, H. and Chua, N.H. 1994. Cyclic GMP and calcium mediate phytochrome phototransduction. *Cell* **77:** 73–81.

Bowler, C., Yamagata, H., Neuhaus, G. and Chua, N.H. 1994. Phytochrome signal transduction pathways are regulated by reciprocal control mechanisms. *Genes Dev.* **8:** 2188–2202.

Boylan, M.T. and Quail, P.H. 1991. Phytochrome A overexpression inhibits hypocotyl elongation in transgenic *Arabidopsis. Proc. Natl. Acad. Sci. USA* **88:** 10806–10810.

Boylan, M., Bruce, W., Dehesh, K., Deng, X.-W., Parks, B., Sharrock, R., Somers, D., Tepperman, J., Wagner, D. and Quail, P.H. 1992. Structure and expression of phytochrome genes. In: Plant Photoreceptors and Photoperception. British Photobiology Society, St John's College, Cambridge, September 20–23, 1992, 28 p.

Boylan, M., Douglas, N. and Quail, P.H. 1994. Dominant negative suppression of *Arabidopsis* photoresponses by mutant phytochrome A sequences identifies spatially discrete regulatory domains in the photoreceptor. *Plant Cell* **6:** 449–460.

Braslavsky, S.E. 1990. Phytochrome. In: Photochromism, Molecules and Systems (Eds. H. Dürr and H. Bouas-Laurent). pp. 738–755. Elsevier, New York.

Braslavsky, S.E. 1995. Chromophore protein interactions in phytochromes. In: Tetrapyrrole Photoreceptors, Freising, Germany, Book of Abstracts, 0 21.

Braslovsky, S.E., Gärtner, W. and Schaffner, K. 1997. Phytochrome photoconversion. *Plant Cell Environ.* **20:** 700–706.

Brockmann, J. and Schäfer, E. 1982. Analysis of Pfr destruction in *Amaranthus caudatus* L.-Evidence for two pools of phytochrome. *Photochem. Photobiol.* **35:** 555–558.

Casal, J.J. 1995. Coupling of phytochrome B to the control of hypocotyl growth in *Arabidopsis. Planta* **196:** 23–29.

Casal, J.J., Sanchez, R.A., and Yanovsky, M.J. 1997. The function of phytochrome A. *Plant Cell Environ.,* **20:** 813–819.

Chen, E.F., Parker, W., Lewis, J.W., Song, P.S. and Kliger, D.S. 1993. Time-resolved UV circular dichroism of phytochrome A: Folding of the N-terminal region. *J. Amer. Chem. Soc.* **115:** 9854–9855.

Chen, E.F., Lapko, V.N., Lewis, J.W., Song, P.S. and Kliger, D.S. 1996. Mechanism of native oat phytochrome photoreversion—a time-resolved absorption investigation. *Biochemistry* **35:** 843–850.

Cherri, J.R., Hondred, D., Walker, J.M., Keller, J.M., Hershey, H.P. and Vierstra, R.D. 1993 Carboxy-terminal deletion analysis of oat phytochrome A reveals the presence of separate domains required for structural and biological activity. *Plant Cell* **5:** 565–575.

Cherri, J.R. and Vierstra, R.D. 1994. The use of transgenic plants to examine phytochrome structure/function. In Kendrick and Kronenberg, 1994 pp. 271–300.

Chory, J. 1993. Out of darkness: mutants reveal pathways controlling light-regulated development in plants. *Trends Genet.* **9:** 167–172.

Clough, R.C. and Vierstra, R.D. 1997. Phytochrome degradation. *Plant Cell Environ.,* **20:** 713–721.

Clough, R.C., Casal, J.J., Jordan, E.T., Christou, P. and Vierstra, R.D. 1995. Expression of functional oat phytochrome A in transgenic rice. *Plant Physiol.* **109:** 1039–1045.

Deforce, L., Furuya, M. and Song, P.-S. 1993. Mutational analysis of the pea phytochrome A chromophore pocket: Chromophore assembly with apophytochrome A and photoreversibility. *Biochemistry* **32:** 14165–14172.

Deforce, L., Tokutomi, S. and Song, P.S. 1994. Phototransformation of pea phytochrome A induces an increase in alpha-helical folding of the apoprotein—comparison with a monocot phytochrome A and CD analysis by different methods. *Biochemistry* **33:** 4918–4922.

Deforce, L., Tomizawa, K.-I., Ito, N., Farrence, D., Song, P.-S. and Furuya, M. 1991. *In vitro* assembly of apophytochrome and apophytochrome deletion mutants expressed in yeast with phycocyanobilin. *Proc. Natl. Acad. Sci. USA* **88:** 10392–10396.

Devlin, P.F., Halliday, K.J., Harber, N.P. and Whitelam, G.C. 1996. The rosette habit of *Arabidopsis thaliana* is dependent upon phytochrome action: novel phytochromes control internode elongation and flowering time. *Plant J.* **10:** 1127–1134.

Edgerton, M.D. and Jones, A.M. 1993. Subunit interactions in the carboxy-terminal domain of phytochrome. *Biochemistry* **32:** 8239–8245.

Eilfeld, P.H. and Eilfeld, P.G. 1988. Circular dichroism of phytochrome intermediates. *Physiol. Plant.* **74:** 169–175.

Emmler, K., Stockhaus, J., Chua N.H. and Schäfer, E. 1995. An amino-terminal deletion of rice phytochrome A results in a dominant negative suppression of tobacco phytochrome A activity in transgenic tobacco seedlings. *Planta* **197:** 103–110.

Farrens, D.L., Holt, R.E., Rospendovski, B.N., Song, P.–S. and Cotton, T.M. 1989. Surface-Enhanced Resonance Raman Scattering (SERRS) spectroscopy applied to phytochrome and its model compounds. II. Phytochrome and phycocyanin chromophores. *J. Am. Chem. Soc.* **111:** 9162–9169.

Foerstendorf, H., Scheer, H., Rüdiger, W., Schäfer, E. and Siebert, F. 1995. FT-IR difference spectroscopic studies of intermediates of the photoreaction of phytochrome. In: Tetrapyrrole Photoreceptors,. Freising, Germany, Book of Abstracts, P 4.1.

Franzes, S., White, M.J., Edgerton, M.D., Elliott, R.C. and Thompson, W.F. 1992. Initial characterization of a pea mutant with light-independent photomorphogenesis. *Plant Cell* **4:** 1519–1530.

Furuya, M. 1989. Molecular properties and biogenesis of phytochrome I and II. *Adv. Biophys.* **25:** 133–167.

Furuya, M. 1993. Phytochromes: their molecular species, gene families, and functions. *Ann. Rev. Plant. Physiol. Plant. Mol. Biol.* **44:** 617–645.

Furuya, M. and Song, P.-S. 1994. Assembly and properties of holophytochrome. In Kendrick and Kronenberg, 1994 pp. 105–140

Gärtner, W. 1995. Effect of site directed mutagenesis in phytochrome (phy-A from oat) on the kinetics of the light induced transformation. In: Tetrapyrrole Photoreceptors, Freising, Germany, Abstracts, P 4.2.

Gärtner, W., Ruddat, A., P. Schmidt, P., Braslvsky, S.E., Schaffner, K., Quail, P.H. and C. Gatz 1995. Recombinant phytochromes from potato. In: European Symposium on Photomorphogenesis in Plants, Sitges, Barcelona, Spain. Abstracts, p. 28.

Gärtner, W., Schmidt, P., Worm, K., Westphal, U.H., Braslavsky, S.E., Schaffner, K. and Quail, P.H. 1995. Effect of chromophore substitution on the kinetics of phytochrome A from oat. In: European Symposium on Photomorphogenesis in Plants, Sitges, Barcelona, Spain. Abstracts, p. 79.

Gensch, T., Braslavsky, S.E. and Schaffner, K. 1996. The photoequilibrium between Pr and the intermediate I_{700} of 124 kDa-phytochrome A from oat. Efficience of the photobackreaction $I_{700} \rightarrow$ Pr. 12th International Congress on Photobiology, Vienna, Austria, Abstracts, O 19.

Hamazato, F., Shinomura, T., Hanzava, H., Chory, J. and Furuya, M. 1996. Phytochrome species differentially regulate *cab* gene expression in *Arabidopsis thaliana.* In: 12th International Congress on Photobiology, Vienna, Austria, Abstracts, P82.

Harter, K., Frohnmeyer, H., Kircher, S., Kunkel, T., Muhlbauer, S. and Schäfer, E. 1994. Light induccs rapid changcs of thc phosphorylation pattern in the cytosol of evacuolated parsley protoplasts. *Proc. Natl. Acad. Sci. USA* **91:** 5038–5042.

Heim, B., Jabben, M. and Schäfer, E. 1981. Phytochrome destruction in dark- and light-grown *Amaranthus caudatus* seedlings. *Photochem. Photobiol.* **34:** 89–93.

Heyer, A.G., Mozley, D., Landschutze, V., Thomas, B., and Gatz, C. 1995. Function of phytochrome A in potato plants as revealed through the study of transgenic plants. *Plant Physiol.* **109:** 53–61.

Hill, C., Gärtner, W., Towner, P., Braslavsky, S.E. and Schaffner, K. 1994. Expression of phytochrome apoprotein from *Avena sativa* in *Escherichia coli* and formation of photoactive chromoproteins by assembly with phycocyanobilin. *Eur. J. Biochem.* **223:** 69–77.

Hillman, W.S. 1967. The physiology of phytochrome. *Annu. Rev. Plant Physiol.* **18:** 301–324.

Holzwarth, A.R., Venuti, E., Braslavsky, S.E. and Schaffner, K. 1992. The phototransformation process in phytochrome. 1. Ultrafast fluorescence component and models for the intital Pr $\rightarrow$ Pfr transformation steps in native phytochrome. *Biochim. Biophys. Acta* **1140:** 59–68.

Hughes, J., Lamparter, T., Mittmann, F., Hartmann, E., Gärtner, W., Wilde, A. and Börner, T. 1997. A prokoryotic phytochrome *Nature* **386:** 663.

Inoue, Y., Rüdiger, W., Grimm, R. and Furuya, M. 1990. The phototransformation pathway of dimeric oat phytochrome from the red-light-absorbing form to the far-red-absorbing form at physiological temperature is composed of four intermediates. *Photochem. Photobiol.* **52:** 1077–1083.

Jackson, S.D., Heyer, A., Dietze, J. and Prat, S. 1996. Phytochrome B mediates the photoperiodic control of tuber formation in potato. *Plant Journal.* **9:** 159–166.

Johnson, E., Bradley, M., Harberd, N. P. and Whitelam, G.C. 1994. Photoresponses of lightgrown phyA mutants of *Arabidopsis*. Phytochrome A is required for the perception of daylength extensions. *Plant Physiol.* **105:** 141–149.

Jones, A.M. and Edgerton, M.D. 1994. The anatomy of phytochrome, a unique photoreceptor in plants. *Semin. Cell Biol.* **5:** 295–302.

Jordan, B.R. 1992. Cellular interactions of phytochrome and their functional significance. In: Plant Photoreceptors and Photoperception, British Photobiology Society, St John's College, Cambridge, 21 p.

Jordan, E.T., Cherry, J.R., Walker, J.M. and Vierstra, R.K. 1996. The amino-terminus of phytochrome A contains two distinct functional domains. *Plant Journal.* **9:** 243–257.

Jordan, E.T., Hatfield, P.M., Hondred, D., Talon, M., Zeevaart, J.A. and Vierstra, R.D. 1995. Phytochrome A overexpression in transgenic tobacco. Correlation of dwarf phenotype with high concentrations of phytochrome in vascular tissue and attenuated gibberellin levels. *Plant Physiol.* **107:** 797–805.

Kandori, H., Yoshihara, K. and Tokutomi, S. 1992. Primary process of phytochrome: initial step of photomorphogenesis in green plants. *J. Am. Chem. Soc.* **114:** 10958–19959.

Kendrick, R.E., Kerckhoffs, L.H.J., Van Tuinen, A. and Koornneef, M. 1997. Photomorphogenic mutants of tomato. *Plant Cell Environ.* **20:** 746–751.

Kendrick, R.E. and Kronenberg, G.H.M. (eds.) 1994. Photomorphogenesis in Plants. 2nd Edition. Kluwer Academic Publishers, Dordrecht/Boston/London.

Kendrick, R.E. and Nagatani, A. 1991. Phytochrome mutants. *Plant Journal* **1:** 133–139.

Kendrick, R.E. and Spruit, C.J.P. 1977. Phototransformations of phytochrome. *Photochem. Photobiol.* **26:** 201–214.

Kolukisaoglu, H.U., Marx, S., Wiegmann, C., Hanelt, S. and Schneider-Poetsch, H.A. 1995. Divergence of the phytochrome gene family predates angiosperm evolution and suggests that *Selaginella* and *Equisetum* arose prior to *Psilotum*. *J. Mol. Evol.* **41:** 329–337.

Koornneef, M. and Kendrick, R.E. 1994. Photomorphogenic mutants of higher plants. In Kendrick and Kronenberg, 1994 pp. 601–630.

Kunkel, T., Henning, L., Poppe, C. and Schäfer, E. 1997. Evidence for a dimer based mechanism of different physiolocal actions of phyA and phyB. European Symposium on Photomorphogenesis, July 12–18. Programme and Abstract, **1:** 15.

Kunkel, T., Speth, V., Buche, C. and Schäfer, E. 1995. *In vivo* characterization of phytochrome-phycocyanobilin adducts in yeast. *J. Biol. Chem.* **270:** 20193–20200.

Lagarias, J.C., Lagarias, D.M., Murphy, J.T., Yeh, K.-C. and Wu, S.-H. 1996. Structure-function analysis of the phytochrome photoreceptor. In: 12th International Congress on Photobiology, Vienna, Austria, Abstracts, S159.

Lamparter, T., Lutterbuse, P., Schneider-Poetsch, H.A.W. and Hertel, R. 1992. A study of membrane-asociated phytochrome: hydrophobicity test and native size determination. *Photochem. Photobiol.* **56:** 697–707.

Li, L. and Lagarias, J.C. 1994. Phytochrome assembly in living cells of the yeast *Saccharomyces cerevisiae*. *Proc. Natl. Acad. Sci. USA*. **91:** 12535–12539.

Li, L., Murphy, J.T. and Lagarias, J.C. 1995. Continuous fluorescence assay of phytochrome assembly *in vitro*. *Biochemistry* **34:** 7923–7930.

Lindemann, P., Schlamann, W., Braslavsky, S.E. and Schaffner, K. 1992. The effects of Ca^{2+} and Ca^{2+}-calmodulin on the decay of the intermediates I-700(1, 2) from native *Avena* phytochrome. *J. Photochem. Photobiol. B: Biol.* **14:** 293–310.

Lippitsch, M.E., Hermann, G., Brunner, H., Müller, E. and Aussenegg, F.R. 1993. Picosecond events in the phototransformation of phytochrome—a time-resolved absorption study. *J. Photochem. Photobiol. B: Biol.* **18:** 17–25.

Lippitsch, M.E., Riegler, H., Aussenegg, F.R., Hermann, G. and Müller, E. 1988. Picosecond absorption and fluorescence studies on large phytochrome from rye. *Biochem. Physiol. Pflanz.* **183:** 1–6.

Millar, A.J., McGrath, R.B., and Chua, N.H. 1994. Phytochrome phototransduction pathways. *Annu. Rev. Genet.* **28:** 325–349.

Nakasako, M., Wada, M., Tokutomi, S., Yamamoto, K.T., Sakai, J., Kataoka, M., Tokunaga, F. and Furuya, M. 1990. Quaternary structure of pea phytochrome I dimer studied with small-angle X-ray scattering and rotary-shadowing electron microscopy. *Photochem. Photobiol.* **52:** 3–12.

Nakazawa, M. and Manabe, K. 1996. Carboxy-terminal region of phytochrome A can be divided into physiologically functional parts, "active" and "regulatory" sites. In: 12th International Congress on Photobiology, Vienna, Austria, Abstracts, P89.

Nakazawa, M., Kosugi, K. and Manabe, K. 1996. Characterization of C-terminal deleted phytochrome A. In: 12th International Congress on Photobiology, Vienna, Austria, Abstracts, P93.

Nakazawa, M., Hayashi, H., Yoshida, Y. and Manabe, K. 1993. Indentification of surface-exposed parts of red-light- and far-red-light-absorbing forms of native pea phytochrome by limited proteolysis. *Plant Cell Physiol.* **34:** 83–91.

Parker, W., Goebel, P., Ross, C.G., Song P.-S. and Stezowski, J.J. 1994. Molecular modeling of phytochrome using constitutive C-phycocyanin from *Fremyella diplosiphon* as a putative structural template. *Bioconjugate Chem.* **5:** 21–30.

Parks, B.M., Quail, P.H. and Hangarter, R.P. 1996. Phytochrome A regulates red-light induction of phototropic enhancement in *Arabidopsis*. *Plant Physiol.* **110:** 155–162.

Partis, M.D. and Grimm, R. 1990. Computer analysis of phytochrome sequences from five species: implication for the mechanism of action. *Z. Naturforsch.* **45c:** 987–998.

Pasentsis, K., Paulo, N., Algarra, P. and Thümmler, F. 1995. Characterization and expression of two different phytochrome types in the moss *Ceratodon purpureus*. In: Tetrapyrrole Photoreceptors, Freising, Germany, Book of abstracts, P4.3.

Peters, J.L. 1992. Photomorphogenetic mutants of higher plants. Ph.D. Thesis, 117, p. Wageningen Agricultural Univ., Wageningen.

Poppe, C., Ehmann, B., Frohnmeyer, H., Furuya M., and Schäfer, E. 1994. Regulation of phytochrome A mRNA abundance in parsley seedlings and cell-suspension cultures. *Plant Mol. Biol.* **26:** 481–486.

Pratt, L.H. 1994. Distribution and localization of phytochrome within the plant. In Kendrick and Kronenberg, 1994, pp. 163–186.

Pratt, L.H., Cordonnier-Pratt, M.-M., Hauser B. and Caboche, M. 1995. Tomato contains two differentially expressed genes encoding B-type phytochromes, neither of which can be considered an ortholog of *Arabidopsis* phytochrome *B. Planta* **197:** 203–206.

Pratt, L.H., Cordonnier-Pratt, M.-M., Kelmenson, P.M., Lazarova, G.I., Kubota, T. and Alba, R.M. 1997. The phytochrome gene family in tomato (*Solanum lycopersicum* L.). *Plant Cell Environ.* **20:** 672–677.

Pratt, L.H., Cordonnier M.-M., Wang, Y.-C., Stewart, S.J. and Moyer, M. 1991. Evidence for three phytochrome in *Avena*. In: Phytochrome Properties and Biological Action (Editors B. Thomas and C.B. Johnson) NATO ASI series H: Cell Biology, Springer, Berlin. 50, pp. 39–55.

Quail, P.H. 1991. Phytochrome: a light-activated molecular switch that regulates plant gene expression. *Annu. Rev. Genet.* **25:** 389–409.

Quail, P.H. 1994. Phytochrome genes and their expression. In Kendrick and Kronenberg, 1994, pp. 71–104.

Quail, P.H. 1997. An emerging molecular map of the phytochromes, *Plant Cell Environ.,* **20:** 657–665.

Quail, P.H., Boylan, M.T., Parks, B.M., Short, T.W., Xu, Y. and Wagner, D. 1995. Phytochromes: photosensory perception and signal transduction. *Science* **268:** 675–680.

Reed, J.W., Nagpal, P. and Chory, J. 1992. Searching for phytochrome mutants. *Photochem. Photobiol.* **56:** 833–838.

Robson, P.R.H. and Smith, H. 1996. Genetic and transgenic evidence that phytochromes A and B act to modulate the gravitropic orientation of *Arabidopsis thaliana* hypocotyls. *Plant Physiol.* **110:** 211–216.

Rospendovski, B.N., Farrens, D.L., Cotton, T.M. and Song, P.-S. 1989. Surface Enhanced Resonance Raman Scattering as a probe of the structural differences between the Pr and Pfr forms of phytochrome. *FEBS Lett.* **258:** 1–4.

Rüdiger, W. 1992. Events in the phytochrome molecule after irradiation. *Photochem. Photobiol.* **56:** 803–810.

Rüdiger, W. and López–Figueroa, F. 1992. Photoreceptors in algae. *Photochem. Photobiol.* **55:** 949–954.

Rüdiger, W. and Thümmler, F. 1991. Phytochrome, the visual pigment of plants. *Angew. Chem.* Int. Ed. Eng. **30:** 1216–1228.

Rüdiger, W. and Thümmler, F. 1994. The phytochrome chromophore. In Kendrick and Kronenberg, 1994, pp. 51–70.

Sakamoto, K. and Nagatani, A. 1997. Nuclear localization activity of phytochrome B. *Plant J.* **10:** 859–868.

Schaffner, K., Braslavsky, S.E. and Holzwarth, A.R. 1990. Photophysics and photochemistry of phytochrome. *Adv. Photochem.* **15:** 229–277.

Schäfer., E., Kunkel, T. and Frohnmeyer. H. 1997. Signal trunsduction in the photocontrol of chalcone synthase gene expression. *Plant Cell Environ.* **20:** 722–727.

Schendel, R, Dörnemann, D., Rüdiger, W. and Sineshchekov, V. 1996. Comparative investigations of the effect of 5-aminolevulinate feeding on phytochrome and protochlorophyll(ide) content in dark-grown seedlings of barley, cucumber and cress. *J. Photochem. Photobiol. B: Biology.* **36:** 245–253.

Scheuerlein, R. and Braslavsky, S.E. 1985. Induction of seed germination in *Lactuca sativa* L. by nanosecond dye laser flashes. *Photochem. Photobiol.* **42:** 173–178.

Scheuerlein, R., Inoue, Y. and Furuya, M. 1988. Intermediates in the photoconversion of functional phytochrome in fern spores of *Dryopteris.* II. *In Vivo*-kinetics of the decay of I_{700} and the formation of P_{fr} as studied with a double-laser apparatus. *Photochem. Photobiol.* **48:** 519–524.

Scurlock, R.D., Braslavsky, S.E. and Schaffner, K. 1993. A phytochrome study using two-laser/two-color flash photolysis: I_{700} is a mandatory intermediate in the Pr $\rightarrow$ Pfr phototransformation. *Photochem. Photobiol.* **57:** 690–695.

Scurlock, R.D., Evans, C.H., Braslavsky, S.E. and Schaffner, K.A. 1993. A phytochrome phototransformation study using two-laser/two-color flash photolysis: analysis of the decay mechanism of I_{700}. *Photochem. Photobiol.* **58:** 106–115.

Serlin, B.S., Lew, R.R., Krasnoshtein, F., Krol, J. and Sumida, K.D. 1996. Phytochrome activation of K^+ channels and chloroplast rotation in *Mougeotia*—the escape times. *Plant Cell Physiol.* **37:** 175–179.

Siebert, F., Grimm, R., Rüdiger, W., Schmidt, G. and Scheer, H. 1990. Infrared spectroscopy of phytochrome and model pigments. *Eur. J. Biochem.* **194:** 921–928.

Shinomura, T., Nagatani, A., Hanzawa, H., Kubota, M., Watanabe, M. and Furuya, M. 1996. Action spectra for phytochrome A- and B-specific photoinduction of seed germination in *Arabidopsis thaliana. Proc. Natl. Acad. Sci. USA* **93:** 8129–8133.

Sineshchekov, V.A. 1994. Two spectroscopically and photochemically distinguishable phytochromes in etiolated seedlings of monocots and dicots. *Photochem. Photobiol.* **59:** 77–86.

Sineshchekov, V.A. 1995a. Photobiophysics and photobiochemistry of the heterogeneous phytochrome system. *Biochim. Biophys. Acta* **1228:** 125–164.

Sineshchekov, V.A. 1995b. Evidence for the existence of two phytochrome A populations. *J. Photochem. Photobiol., B: Biology* **28:** 53–55.

Sineshchekov, V.A. and Akhobadze, V.V. 1992. Phytochrome states in etiolated pea seedlings: fluorescence and primary photoreactions at low temperatures. *Photochem. Photobiol.* **56:** 743–749.

Sineshchekov, V.A. and Rüdiger, W. 1992. Fluorescence spectroscopy of stable and labile phytochrome in stems and roots of etiolated cress seedlings. *Photochem. Photobiol.* **56:** 735–742.

Sineshchekov, V.A., Frances, S. and White, M.J. 1995. Fluorescence and photochemical characterization of phytochrome in de-etiolated pea mutant *lip. J. Photochem. Photobiol., B: Biology* **28:** 47–51.

Sineshchekov, V.A., Heyer, A.G. and Gatz, C. 1996. Phytochrome states in transgenic potato plants with altered phyA levels. *J. Photochem. Photobiol., B: Biol.* **34:** 137–142.

Sineshchekov, V., Hughes, J., Hartmann, E. and Lamparter, T. 1998a. Fluorescence and photochemistry of recombinant phytochrome from the cyanobacterium *Synechocystis. Photochem. Photobiol.* **67:** 263–267.

Sineshchekov, V., Lamparter, T. and Hartmann, E. 1994. Evidence for the existence of membrane-associated phytochrome in the cell. *Photochem. Photobiol.* 60, 561–520.

Sineshchekov, V.A., Lapko, V.N., Sineshchekov, A.V., Kozhukh, G.V. and Udal'tsov, A.V. 1990. *J. Photochem. Photobiol.* **5:** 219–229.

Sineshchekov, V.A., Ogorodnikova, O.B., Devlin, P.F. and Whitelam, G.C. 1998b. Fluorescence spectroscopy and photochemistry of phytochromes A and B in wild-type, mutant and transgenic strains of *Arabidopsis thaliana. J. Photochem. Photobiol. B: Biol.* **42:** 133–142.

Sineshchekov, V.A. and Sineshchekov, A.V. 1990. Different photoactive states of the red phytochrome form in the cells of etiolated pea and oat seedlings. *J. Photochem. Photobiol. B: Biol.* **5:** 197–217.

Smith, H. 1995. The adaptive significance of the phytochromes: evolutionary implications. In: European Symposium on Photomorphogenesis in Plants, Sitges, Barcelona, Spain. Abstracts, p. 15.

Smith, H. (ed.) 1997. Photomorphogenesis (Special issue). *Plant Cell Environ.* **20:** 657–840.

Smith, H. and Fork, D.C. 1992. Direct measurement of phytochrome photoconversion intermediates at high photon fluence rates. *Photochem. Photobiol.* **56:** 599–607.

Smith, H. and Whitelam, G.C. 1990. Phytochrome, a family of photoreceptors with multiple physiological roles. *Plant Cell Environ.* **13:** 695–707.

Smith, H. and Whitelam, G.C. 1997. The shade avoidance syndrome: multiple responses mediated by multiple phytochromes. *Plant Cell Environ.* **20:** 840–844.

Somers, D.E. and Quail, P.H. 1995. Temporal and spatial expression patterns of PHYA and PHYB genes in *Arabidopsis. Plant J.* **7:** 413–427.

Sommer, D. and Song, P.-S. 1990. The chromophore topography and secondary structure of 124–kDa *Avena* phytochrome probed by Zn^{2+}-induced chromophore modification. *Biochemistry* **29:** 1943–1948.

Song, P.-S. 1985. Primary molecular events in aneural cell photoreception. In: Sensory Perception and Transduction in Aneural Organisms. (Colombetti, G., Lenci, F. and Song, P.-S., eds.) pp. 47–59. Plenum Press, London.

Song, P.–S. (1988) The molecular topography of phytochrome. *J. Photochem. Photobiol., B: Biol.* **2:** 43–57.

Song, P.-S. 1994, Phytochrome as a light switch for gene expression in plants. *The Spectrum* **7:** 2–7.

Song, P.-S., Park, M.H. and Furuya, M. 1997. Chromophore: apoprotein interactions in phytochrome A. *Plant Cell Environ.*, **20:** 707–712.

Song, P.-S., Sommer, D., Wells, T.A., Hahn, T.R., Park, H.J. and Bhoo, S.H. 1996. Light signal transduction mediated by phytochromes—preliminary studies and possible approaches. *Indian J. of Biochem. Biophys.* **33:** 1–19.

Sopory, S.K. and Munshi, M. 1998. Protein kinases and phosphatases and their role in cellular signalling in plants. *Crit. Rev. Pl. Sci.* (in press).

Starostzik, C. and Marwan, W. 1995. A photoreceptor with characteristics of phytochrome triggers sporulation in the true slime mould *Physarum polycephalum. FEBS Lett.* **370:** 146–148.

Terry, M.J., Hall, J.L. and Thomas, B. 1992. The association of type 1 phytochrome with wheat leaf plasma membranes. *J. Plant Physiol.* **140:** 691–698.

Terry, M.J., McDowell, M.T. and Lagarias, J.C. 1995. (3Z)-and (3E)- phytochromobilin are intermediates in the biosynthesis of the phytochrome chromophore. *J. Biol. Chem.* **270:** 11111–11118.

Thümmler, F., Algarra, P. and Fobo, G.M. 1995a. Sequence similarities of phytochrome to protein kinases: implication for the structure, function and evolution of the phytochrome gene family. *FEBS Lett.* **357:** 149–155.

Thümmler, F., Dufner, M., Kreisl, P., and Dittrich, P. 1992. Molecular cloning of a novel phytochrome gene of the moss *Ceratodon purpureus* which encodes a putative light-regulated protein kinase. *Plant Mol. Biol.* **20:** 1003–1017.

Thümmler, F., Herbst, R., Algarra, P. and Ullrich, A. 1995b. Analysis of the protein kinase activity of moss phytochrome expressed in fibroblast cell culture. *Planta* **197:** 592–596.

Tokutomi, S., Kandori, H. and Yoshihara, K. 1995. A novel primary photoproduct from the excited state of red-absobing form of phytochrome based on picosecond flash photolysis. In: Tetrapyrrole Photoreceptors, Freising, Germany, Abstracts, P 4.6.

Tokutomi, S. and Mimuro, M. 1989. Orientation of the chromophore transition moment in the 4-leaved shape model for pea phytochrome molecule in the red-light absorbing form and its rotation induced by the phototransformation to the far-red-light absorbing form. *FEBS Lett.* **255:** 350–353.

Tokutomi, S., Sugimoto, T. and Mimuro, M. 1992. A model for the molecular structure and orientation of the chromophore in a dimeric phytochrome molecule. *Photochem. Photobiol.* **56:** 545–552.

Tomizawa, A.K., Stockhaus, J., Chua, N.H. and Furuya, M. 1995. Spectrophotometric and molecular properties of mutated rice phytochrome. *Plant Cell Physiol.* **36:** 511–516.

VanDerWoude, W.J. (1987) Application of the dimeric model of phytochrome action to high irradiance responses. In: Phytochrome and Photoregulation in Plants (Edited by M. Furuya), pp. 249–258. Academic Press, Tokyo.

Vierstra, R.D. and Quail, P.H. 1983a. Photochemistry of 124 kilodalton *Avena* phytochrome *in vitro. Plant. Physiol.* **72:** 264–267.

Vierstra, R.D. and Quail, P.H. 1983b. Purification and initial characterization of 124 kdalton phytochrome from *Avena. Biochemistry*, **22:** 2498–2505.

Vierstra, R.D., Jordan, E.T. and Clough, R.C. 1996. Analysis of structural domains required for phytochrome function by *in vitro* mutagenesis. In: 12th International Congress on Photobiology Vienna, Austria, Abstracts, S164.

Volotovsky, I.D. 1992. Phytochrome-regulatory photoreceptor of plants. 168 p. "Navuka i Tekhnika", Minsk, Beloruss.

Wada, M., Grolig, F. and Haupt, W. 1993. Light-oriented chloroplast positioning. Contribution to progress in photobiology. *J. Photochem. Photobiol. B: Biol.* **17:** 3–25.

Wada, M., Kanegae, T., Nozue, K. and Fukuda, S. 1997. Cryptogam phytochromes. *Plant Cell Environ.* **20:** 685–690.

Wagner, D. and Quail, P.H. 1995 Mutational analysis of phytochrome B identifies a small COOH-terminal-domain region critical for regulatory activity. *Proc. Natl. Acad. Sci. USA* **92:** 8596–8600.

Wagner, D., Tepperman, J.M. and Quail, P.H. 1991. Overexpression of phytochrome B induces a short hypocotyl phenotype in transgenic *Arabidopsis. Plant Cell* **3:** 1275–1288.

Wells, T.A., Nakazawa, M., Manabe, K. and Song, P.S. 1994. A conformational change associated with the phototransformation of *Pisum* phytochrome A as probed by fluorescence quenching. *Biochemistry* **33:** 708–712.

Wester, L., Somers, D.E., Clack, T. and Sharrock, R.A. 1994. Transgenic complementation of the *hy3* phytochrome B mutation and response to PHYB gene copy number in *Arabidopsis. Plant J.* **5:** 261–272.

Whitelam, G.C. 1995. The roles of phytochromes A and B and other phytochromes. In: European Symposium on Photomorphogenesis in Plants, Sitges, Barcelona, Spain. Abstracts, p. 51.

Whitelam G.C. and Devlin, P.F. 1997. Roles of different phytochromes in *Arabidopsis* photomorphogenesis. *Plant Cell Environ.* **20:** 752–758.

Wu, S.-H. and Lagarias, J.C. 1997. The phytochrome photoreceptor in the green alga *Mesotaenium caldariorum:* implication for a conserved mechanism of phytochrome action. *Plant Cell Environ.* **20:** 691–699.

Xu, Y., Parks, B.M., Short, T.W., and Quail, P.H. 1995. Missense mutations define a restricted segment in the C-terminal domain of phytochrome A critical to its regulatory activity. *Plant Cell* **7:** 1433–1443.

Yang, Y.Y., Nagatani, A., Zhao, Y.J., Kang, B.J., Kendrick, R.E. and Kamiya, Y. 1995. Effects of gibberellins on seed germination of phytochrome-deficient mutants of *Arabidopsis thaliana* (L). *Plant Cell Physiol.* **36:** 1205–1211.

Yeh, K.-C., Wu, S.-H., Murphy, T. and Lagarias, J.C. 1997. A cyanobacterial phytochrome two-component light sensory system. *Science* **277:** 1505–1507.

Zhang, C.-F., Farrens, D.L., Björling, S.C., Song, P.-S. and Kliger, D.S. 1992. Time-resolved absorption studies of native etiolated oat phytochrome. *J. Am. Chem. Soc.* **114:** 4569–4580.

Concepts in Photobiology: Photosynthesis and Photomorphogenesis
G.S. Singhal, G. Renger, S.K. Sopory, K-D. Irrgang and Govindjee (Eds)

26. Blue Light Perception and Signal Transduction in Higher Plants

Jitendra P. Khurana[1] and Kenneth L. Poff[2]

[1]Department of Plant Molecular Biology,
University of Delhi South Campus, New Delhi 110021, India

[2]MSU-DOE Plant Research Laboratory,
Michigan State University, East Lansing, Michigan-48824, USA

Summary

Blue light regulates a diverse range of plant responses that include phototropism, stem growth inhibition, leaf expansion, chloroplast development, stomatal opening, anthocyanin accumulation and changes in expression of various genes. Some of these responses are elicited by low fluence and others by a higher fluence of blue light. These responses are regulated by the photoreceptors absorbing principally in the UV-A (320–390 nm) and blue (390–500 nm) region of the electromagnetic spectrum. Until very recently, the biochemical identity of the blue light receptors was considered a major unsolved problem. The combined application of physiological, biochemical and molecular genetic approaches has now generated new insight into not only the photoperception mechanism but also the signal transduction components that mediate the effects of blue light on plant development. Convincing evidence has now emerged for the existence of multiple blue light receptors which may harbor more than one chromophore. The long-standing controversy on whether flavins and/or carotenoids are the receptors for blue light still persists. The isolation and characterization of *CRY* genes has provided evidence for the involvement of a flavin, in addition to a pterin, in blue light-induced responses. There has been reasonable progress in deciphering some of the blue light signaling components, but the demonstration of the hierarchial role of these components in any single response pathway is presently lacking. Blue light also profoundly influences the expression of genes, and their light-responsive elements (*cis*-acting sequences) and the *trans*-acting (protein) factors are being identified. The wealth of information that has accumulated on these aspects mentioned above has been evaluated in the following text. A greater emphasis has been laid on the studies where the power of genetics has been exploited to provide answers to some of the unsolved mysteries that surround the perception of blue light and its mechanism of action.

1. Introduction

Light is one of the most critical environmental factors and profoundly influences the growth and development of higher plants. Not only is light utilized as a primary source of energy for photosynthesis, it also is perceived as an informational signal to synchronize plant development throughout the life cycle. Examples of such developmental processes which are under light control include onset of seed germination, cessation of stem growth, apical hook opening, tropic responses, rapid leaf expansion, chloroplast development, stomatal movement, flower induction and onset of senescence. In addition, of course, light does influence changes at the cellular and molecular level associated with these developmental processes. This light-dependent development of plants is called photomorphogenesis. To perceive and respond appropriately to the changing ambient conditions, higher plants have evolved several classes of highly sensitive sensory pigment systems which can measure light intensity and spectral quality, and sense its direction as well as duration. These include the red/far-red reversible phytochromes, UV-A/blue-sensitive pigments and UV-B absorbing pigments. Of these three pigment systems, phytochrome is the predominant and best characterized photosensory pigment system which is ubiquitous in higher plants (for details, see Kendrick and Kronenberg, 1994, and relevant chapters of this volume). Our understanding of the blue light receptors has also advanced significantly. However, very little is known about the UV sensing mechanism in higher plants (see Stapleton, 1992; Bharti and Khurana, 1997).

Although the red/far-red reversible phytochromes are considered to comprise the principal class of photomorphogenic receptors, blue light also plays an important regulatory role in plant development. Some of the salient features of typical blue light responses that differentiate it from responses governed by phytochromes are as follows: (1) They are induced most effectively by blue and in most situations also by UV-A. (2) They are generally stimulated by low incident energies. (3) The low-fluence blue responses are not photoreversible. (4) They can be observed against a background of red light. (5) For responses where both blue and red/far-red radiations are effective, the response kinetics to the two are usually different and the response to blue light becomes apparent more rapidly. (6) Phytochrome mutants retain normal blue sensitivity for some responses. However, some of these points may have to be retracted (or qualified) in light of the convincing evidence emerging for an interactive role of phytochromes and blue light receptors. In comparison to phytochrome, much less is known about the blue light sensory systems and the associated signalling mechanisms. Most of the earlier studies have concentrated upon phenomenological details and the properties of the probable photoreceptors and signalling components inferred from physiological experiments conducted on whole organisms. The scenario has, however, changed rapidly in the past few years with the molecular characterization of at least two blue photoreceptor pigments and elucidation of several potential components of the blue light-activated transduction chains.

In this chapter, an attempt has been made to evaluate the status of research on blue light-dependent photosensory systems. Although the major emphasis is placed on recent developments due emphasis has been given to the ground work done earlier which in fact laid the foundation for the present day research.

2. Blue Light-induced Responses

The diverse array of responses elicited by blue light include phototropism, apical hook opening, inhibition of shoot elongation of seedlings, stomatal opening, chloroplast development, leaf unfolding, tendril coiling, stimulation of flavonoid biosynthesis, and changes in expression of certain genes. Some of these responses currently receiving greater attention for understanding the mechanism of action of blue light are elaborated below.

2.1 Phototropism

Phototropism is the most extensively studied blue light response and can be defined as the development of curvature of an organ in relation to the direction of light. Directional responses like phototropism endow physiological advantages to organisms, such as more efficient utilization of light energy and protection against photodestruction. Although the phenomenon of phototropism was documented over a century ago in the monograph "The Power of Movement in Plants" (Darwin, 1880), the physiological basis of this phenomenon remains poorly understood. In nature, the directional differences in light, its quality and intensity are sensed by the plant, which results in differential growth and eventually the development of curvature. In fact, some of the early experiments on phototropism lead to the discovery of auxins, the causal agent for differential growth. By and large, the role of auxin in curvature development has stood the test of time, although there have been some challenges to it. The photophysiology of this process is quite complex as is evident by the fluence-response curve shown in Fig. 1. In a typical phototropism experiment, the young dark-grown seedlings (3-day-old) are irradiated unilaterally with blue light of known fluence and the curvature allowed to develop in dark and measured after about two hours (in the case of *Arabidopsis;* Steinitz and Poff, 1986); it has been observed that the seedlings with the position of the hook away from the light direction are more responsive and develop higher curvature (Khurana et al., 1989a). The curvature developed in response to low fluence and short irradiation times is usually small and is referred to as 'first positive' curvature. The curvature increases with increasing fluence of blue light up to a certain limit (the ascending arm of the curve) and it is followed by a decline (the descending arm of the curve). Further increase in fluence (in a specific range) does not elicit any curvature and it is called the zone of 'indifferent' response. Still higher doses of blue photons, usually for prolonged durations, result in development of 'second positive' curvature, which can be of a much higher magnitude than the 'first positive' response. In the fluence-response curve, the curvature developed in the range of ascending arm of the 'first positive' phototropism is primarily dependent on number of quanta impinging on the plant (which is a function of the fluence rate of light and time of irradiation). The physiological basis of the descending arm of the 'first positive' and the 'second positive' curvature is, however, not clear, although the duration of irradiation is the key factor for the intensity of the 'second positive' phototropism. The fluence-response relationship for phototropism of monocot and dicot seedlings is essentially similar and has been exploited in designing protocols for screening mutants of phototropism. The results of some physiological experiments, coupled with analysis of these mutants, have revealed that multiple photoreceptors may be involved in regulating phototropic response to blue light (see Konjević et al., 1989, 1992; Briggs and Liscum, 1997).

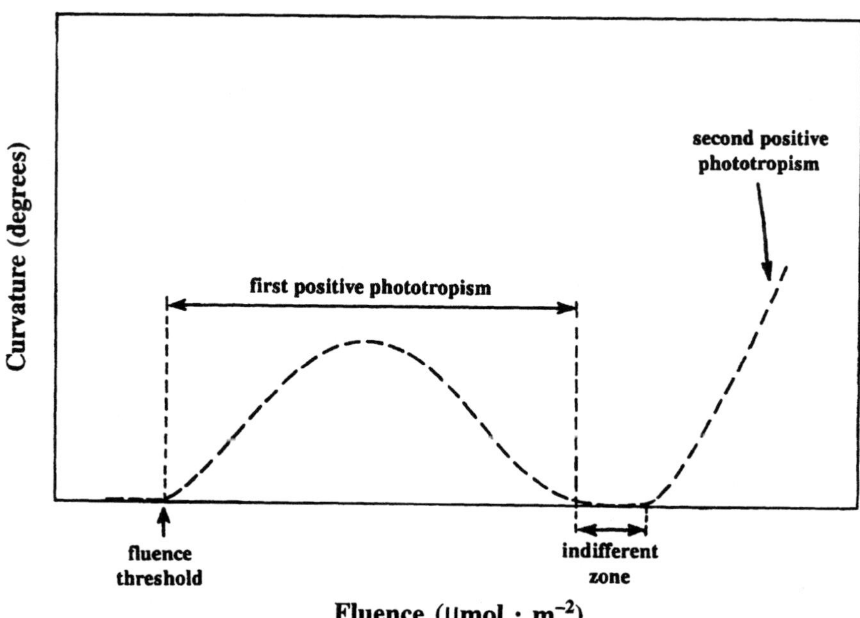

Fig. 1. A diagrammatic representation of the fluence-response relationship for phototropism of *Arabidopsis* seedling hypocotyl. Adapted from Steinitz and Poff (1986).

2.2 Stomatal Opening

When light intensity diminishes or water stress develops, guard cells shrink and stomata close. Stomatal movement is vital for modulating gas exchange between the leaf cells and the surrounding environment, and controlling water potential. In addition to photosynthesis in the guard cells that can contribute to osmotic build-up required for opening, a non-photosynthetic photosensory system sensitive to blue light also regulates movement of stomatal apparatus. The blue light response is involved in stomatal opening in early morning and stomatal responses to sunflecks. The stomatal adaptation to sun, shade and temperature are probably regulated by guard cell chloroplasts, with abscisic acid playing an important role in stomatal closure. These changes in stomatal aperture are mediated by fluctuations in turgor of the guard cells, which require active ion transport. More details on this aspect have been provided while discussing the mechanism of blue light action.

2.3 Stem Growth Inhibition

Young dark-grown seedlings acquire an etiolated habit with long hypocotyl, apical hook and unexpanded cotyledons. The transition from etiolated to de-etiolated habit results in inhibition of hypocotyl (or stem) elongation, hook opening and cotyledon expansion. Light plays an important role in this adaptive mechanism through the combined action of red, far-red and blue radiations. Blue light inhibits the elongation growth of young seedlings grown in dark (or even in red light). The effect of blue light is additive to that of red/far-red and the kinetics of the growth inhibition is also different. The response (growth inhibition) initiates within seconds of blue light irradiation and continues on with the prolonged exposure. The relationship between the initial and the long-term response remains unclear, but it is likely that the initial response occurs due to loosening of the cell wall and the long-term response involves changes in hormone flow. As will become obvious from the discussion later, these changes in the cell wall, with or without the involvement of hormones, are preceded by changes at the membrane level.

The phenomenology of some of these blue light-induced responses has been worked out in considerable detail. However, the mechanisms by which plants are able to sense blue light and respond appropriately have been largely unknown until recently. The primary reason for this has been our ignorance about the precise nature of the blue light photoreceptor pigments and the diversity of these pigments that may exist to regulate a wide array of plant responses.

3. Nature of Blue Light Photoreceptors

The identification of blue light receptor in higher plants and the nature of the chromophore(s) have been a subject of considerable debate. Although some of the earlier studies provided evidence for flavins to be the most likely candidate, evidence is now emerging not only for the existence of multiple blue light photoreceptors but also a variety of different chromophores.

3.1 Clues from Physiological Experiments

The comparison of the action spectra and the absorption spectra of various known pigments has been a standard practice for deducing the nature of the regulatory photoreceptors. However, an exact comparison is never achieved because of factors such as (a) the absorption spectrum of a pigment is greatly influenced by the surrounding molecular environment, (b) action spectra may be altered due to interference by other pigments/ compounds and (c) multiple pigments may be involved in controlling a particular response.

The detailed action spectra, indicating the relative effectiveness of different wavelengths of light in eliciting a response, have been determined for several blue light-induced responses (see Briggs and Iino, 1983). The action spectra of blue light respones typically show three absorbance bands in the blue region of the spectrum, with maxima at 425, 450 and 475 nm (Fig. 2). In addition, a long wavelength band with an absorption maximum at 370 nm is also characteristically present. Based upon the comparison of these action spectra with the absorption spectra of probable pigments, carotenoids and flavins, which display absorbance in the blue region (and also absorb in the UV range), have received greater attention as probable blue light receptors. However, studies with metabolic inhibitors have provided evidence more in favour of a flavin. For example, phenylacetic acid and potassium iodide, that quench the excitation of flavins, have been shown to inhibit blue light-mediated phototropism in etiolated maize coleoptiles (Vierstra and Poff, 1981a). On the other hand, experiments with norflurazon, a bleaching herbicide, that inhibits carotenoid biosynthesis, suggest that carotenoids may have only a secondary role in blue light-mediated phototropism (Vierstra and Poff, 1981b).

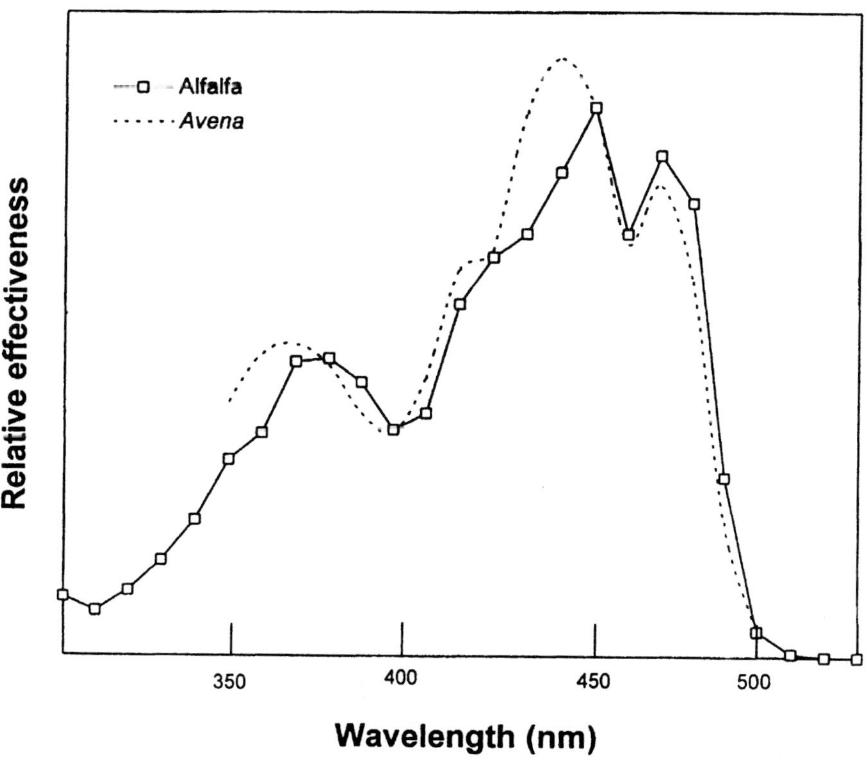

Fig. 2. Action spectra for coleoptile phototropism of *Avena* (a monocot) and seedling hypocotyl phototropism of alfalfa (a dicot). Adapted from Thimann and Curry (1960) and Baskin and Iino (1987).

Studies on blue light-induced absorbance changes (LIACs) also implicated a flavin as the probable photoreceptor. These LIACs, with action spectra similar to other blue light-mediated responses, were first recorded, *in vivo,* in *Phycomyces* and *Dictyostelium* mycelial mats (Poff and Butler, 1974) and subsequently shown to represent flavin-mediated reduction of b-type cytochrome (Muñoz and Butler, 1975). Subsequent studies, including those on maize coleoptiles, established that all components necessary for exhibiting LIACs are essentially associated with the plasma membrane (see Short and Briggs, 1994).

Although the possibility of carotenoids as probable photoreceptors for blue light was nearly given up, recently, Quiñones and Zeiger (1994) have documented several correlations between the level of zeaxanthin, a chloroplastic carotenoid of xanthophyll cycle, and phototropism in maize coleoptiles. They have also shown that not only action spectra (in the blue region) of phototropism and stomatal opening but also for other blue light responses of guard cells and coleoptile chloroplasts match the absorption spectrum of zeaxanthin (Quiñones et al., 1996). However, Briggs and coworkers have discounted this possibility, since in experiments on maize coleoptiles, they failed to observe any clear difference in the blue light-dependent phototropism and protein phosphorylation (that has been implicated as an early step in phototropism) responses of seedlings regardless of the presence or absence of carotenoids (Palmer et al., 1996).

3.2 Genetic Evidence

To complement the physiological and biochemical studies, genetic mutants have also been isolated in *Arabidopsis* in the past decade, that have indeed provided some useful clues towards the identification of putative blue light receptors.

Inhibition of hypocotyl elongation in young seedlings is influenced by red/far-red and blue/UV-A light. Mutants of *Arabidopsis* with long hypocotyls have been identified which are insensitive to either red and/or far-red light or are exclusively insensitive to blue light. The latter category is represented by *hy*4 (long hypocotyl) and *blu* (*b*lue *l*ight *u*ninhibited) mutants (Koornneef et al., 1980; Liscum and Hangarter, 1991). The focus here would be on *hy*4 since all the *blu* mutants turned out to be allelic to *hy*4 (*see* Ahmad and Cashmore, 1996; Hangarter, personal communication). Strikingly, *hy*4 shows incomplete dominance, with the heterozygote being about half as sensitive to blue light as wild-type. Earlier studies have shown a similar gene dosage dependence for phytochrome mutants *hy*3 and *hy*8, which are harboring mutations in the *PHYB* and *PHYA* genes, respectively. It thus provides a clue that *hy*4 may represent a mutation in a gene encoding a protein of the blue light-sensing mechanism, and possibly a receptor pigment. The *hy*4 mutant, that is insensitive to blue light for its hypocotyl growth inhibition, a high-irradiance blue response, was, however, found to retain normal phototropism, a low-fluence blue response.

Among the several phototropism mutants isolated (Khurana and Poff, 1989), at least strain JK224 satisfies certain criteria to be considered as a putative photoreceptor mutant. For JK224, the threshold and saturation fluence requirements for first positive curvature in young etiolated seedling shoots are shifted towards higher fluences by a factor of 20–30, when compared to wild-type parent. The second positive phototropic response is normal in JK224 and so is its gravi-sensing ability. The increased quantum requirement for first positive curvature in JK224 indicates that JK224 may be deficient in a functional blue photoreceptor pigment for phototropism. This view also finds support from the observation that JK224 lacks the normal blue light-induced protein phosphorylation response (discussed in more detail later). Another phototropism mutant, JK218, also deserves consideration here although it has generally been considered a response mutant and not a photoreceptor mutant *per se*. Strain JK218 represents a 'null' response mutant that does not respond at all to prolonged exposures of unilateral blue light (or even white light) and is impaired in both first and second positive curvature. Like JK224, this strain also responds normally to gravity stimulus. As a heterozygote with wild-type, JK218 shows intermediate response (Khurana and Poff, 1989), i.e. incomplete dominance, a characteristic of receptor mutations not only in phytochrome apoprotein (*PHYA, PHYB*) and cryptochrome (*HY*4) genes but also for ethylene sensing (i.e. *ETR*1). However, further work is necessary before JK218 can be considered a photoreceptor mutant because, unlike JK224, JK218 displays a normal protein phosphorylation response.

Recently, Liscum and Briggs (1995) have isolated several more phototropism mutants in *Arabidopsis* and named them *nph* (*n*on-*p*hototropic *h*ypocotyl). On complementation analysis, JK224 and JK218 (isolated originally by Khurana and Poff, 1989) have been reported to be allelic to *nph*1 and *nph*3, and redesignated as *nph*1–2 and *nph*3–3, respectively. However, *nph*1 represents a class of 'null' response mutants but JK224/ *nph*1–2 is only partially affected (as stated above). This photophysiological difference could be due to the point mutations generated in JK224/*nph*1–2 by EMS treatment (Khurana and Poff, 1989) and the gross chromosomal changes that may have occurred in other alleles of *nph*1 by fast neutron bombardment (Liscum and Briggs, 1995). Analysis of *nph* alleles, including *nph*1-2/JK224, has revealed that these mutants lack or have severely reduced blue light-dependent *in vitro* phosphorylation of 120 kDa target protein. Combining this observation and the photophysiological behaviour of *nph*1 alleles, and *nph*1–2/JK224 in particular, it was speculated that this 120 kDa autophosphorylating protein may have a receptor function (Reymond et al., 1992; Liscum and Briggs, 1995). The cloning and characterization of *NPH*1 gene (Huala et al., 1997) has proved to be more revealing in this regard.

In addition to the blue, UV-A and green regions of the spectrum, red light also has been shown to influence, though indirectly, the phototropism response. It has been found that the red light preirradiated seedlings usually exhibit an exaggerated phototropic curvature in response to unilateral blue light (Chon and Briggs, 1966; Janoudi and Poff, 1992). Recently, employing the phyA-deficient mutants of *Arabidopsis,* the role of *phyA* has been implicated in the enhancement of phototropic response to low-fluence red light (Parks et al., 1996). In a more elaborate study on *phy*A and *phy*B mutants and transgenic lines overexpressing these phytochrome species, Janoudi et al. (1997a, b) have indicated that the involvement of phyA is essential in the very-low- to low-fluence range for enhancement. Either phyA or phyB is required for the high-fluence enhancement by red light. The mechanism for this complex interaction in enhancement remains to be worked out.

3.3 Molecular Cloning and Characterization of *CRY*1, *CRY*2 and *NPH*1 Genes

3.3.1 CRY1 gene

The gene defined by the long hypocotyl mutant *hy*4 has now been characterized and represents a landmark discovery in the history of blue light sensory physiology. An allele of *hy*4 was identified in a collection of T-DNA tagged mutant lines and cloned by marker rescue (Ahmad and Cashmore, 1993). The *HY*4 gene has been named as CRY1 (for cryptochrome 1) by Cashmore and coworkers and encodes a protein of 681 amino acids, with the amino-terminal half of the gene product having high sequence homology to microbial DNA photolyases, a rare class of flavoproteins that on photoactivation catalyze the repair of pyrimidine dimers in DNA damaged by UV light. However, CRY1 gene product lacks the photolyase activity. In all probability, the homology between the CRY1 gene product and DNA photolyase is due to striking similarity in the chromophore-binding domains—a flavin attachment for reduced flavin at the carboxy-terminal and for deazaflavin (or a pterin) at the amino terminus. The carboxy-terminal region of *HY*4 product also harbors a tropomyosin-like domain, a feature altogether absent in DNA photolyases, and may have a functional significance. Incidentally, seven of the twenty mutant alleles of *hy*4 have lesions in this region of the gene. In others, the lesions are spread throughout the structural gene, including the chromophore-binding domains. It thus becomes obvious that both the chromophore-binding and the carboxy-terminal domains are essential for CRY1 function.

The CRY1 cDNA encodes a protein of ca. 76 kDa and is apparently devoid of any membrane-spanning domain, indicating the soluble nature of the protein. This is in contrast to the generally held belief that the blue light receptor is membrane localized. However, most of the physiological and biochemical studies favouring this hypothesis were confined to low-fluence blue responses such as phototropism and stomatal movement. Although a conclusive molecular genetic evidence in support of this is yet to emerge, but it may well be that the receptor pigment for low-fluence blue responses is membrane localized and that for high-fluence blue response is soluble in nature. It is also possible that eventually even the receptor involved in low-fluence blue responses may turn out to be of soluble nature, a situation reminiscent of the phytochromes which too were originally thought to be associated with the pelletable fraction of cell extracts but, as it later turned out, are largely soluble (cytosolic proteins); only a small fraction may be associated with the membrane and that small fraction may be necessary for its biological activity.

To decipher the precise nature of the chromophores and also to provide credibility for the assumption that the deduced amino acid sequence of CRY1 harbors chromophore-binding domains similar to microbial DNA photolyases, CRY1 protein was overexpressed in a baculovirus expression system and was found to bind FAD non-covalently (Lin et al., 1995). Interestingly, a semiquinone redox intermediate of CRY1 receptor absorbed green light (> 500 nm), justifying the broad action spectrum of the responses mediated by flavin-type cryptochromes. The existence of a second chromophore which may be responsible for absorbing principally in the blue region was identified by overexpression of the photolyase homologous region of CRY1 in *E. coli* as a fusion protein containing maltose-binding protein. In addition to binding FAD, this fusion protein was found to bind a pterin, methyltetrahydrofolate, which normally absorbs light in the UV-A region although its bound form has absorption maxima shifted to blue region (Malhotra et al., 1995). This sounds paradoxical since based upon the action spectrum and inhibitor studies, invariably, a flavin-type photoreceptor was thought to be regulating blue light responses. It is likely that different chromophores when bound to different apoproteins may cause profound shifts in the absorption spectrum. The situation may even get more complicated given the fact that multiple chromophores bind to the same apoprotein. This may result in emergence of complicated action spectra, which need to be analysed more critically to deduce the nature of the probable photoreceptors.

3.3.2 CRY2 gene

A CRY1 homologue has been identified in *Arabidopsis* and designated as CRY2 (Hoffman et al., 1996; Lin et al., 1996). The chromophore binding amino-terminal domain of CRY2 is similar to that of CRY1, whereas carboxy-terminal half bears little homology to that of CRY1. Overexpression studies of *CRY2* indicate that its function overlaps with that of CRY1 in regulating inhibition of hypocotyl elongation and cotyledon expansion (Lin et al., 1998). By overexpression of CRY1 and CRY2 fusion proteins in transgenic *Arabidopsis*, Ahmad et al. (1998a) have demonstrated that their domains are functionally interchangeable. Strikingly, unlike CRY1, the CRY2 protein is unstable in UV-A, blue and green light (that activate CRY2 receptor) and is stable in dark and red light (Lin et al., 1998), a characteristic reminiscent of the dark-abundant, light-labile form of the phytochrome, PHYA. It has also been proposed that CRY2 may have functional significance under conditions where light is limiting (Ahmad et al., 1998a).

As mentioned earlier, the long-hypocotyl mutant *hy*4, which is defective in CRY1, was found to exhibit normal phototropism, a low-fluence blue response. In a breakthrough study, however, Cashmore and coworkers have recently claimed that transgenic *Arabidopsis* seedlings overexpressing CRY1 and CRY2 genes show hypersensitivity to unilateral blue light for 'first positive' phototropism. There was virtually no sign of the 'first positive' phototropism in the *cry*1 *cry*2 double mutant seedlings (Ahmad et al., 1998b). In addition, the blue light-induced phosphorylation of 120 kDa polypeptide (linked earlier to phototropism) was absent in *cry*1 *cry*2 double mutant seedlings. Based upon these data, it has been proposed that cryptochrome is one of the photoreceptors mediating phototropism in higher plants (Ahmad et al., 1998b).

For deciphering the functional significance of CRY2, Guo et al. (1998) isolated *cry*2 mutants that did not accumulate CRY2 protein. These mutants were impaired not only in hypocotyl growth inhibition response and cotyledon expansion, they also flowered later than the wild-type. The *cry*2 mutant was found to be allelic to another late-flowering mutant, *fha*. Further analysis revealed that *cry*2/*fha* mutant plants are incompletely responsive to photoperiod and that CRY2 is a positive regulator of flowering-time gene *CONSTANS*(*CO*). A comparison of the flowering behaviour of *cry*2 and *phy*B mutants under red and blue light also indicated that flowering in *Arabidopsis* is regulated by the antagonistic action of PHYB and CRY2 (Guo et al., 1998).

3.3.3 NPH1 gene

Based upon physiological and biochemical studies, it was proposed that *nph*1/JK224 mutant represents a blue light receptor for phototropism and it might undergo autophosphorylation in response to blue light. The cloning of the NPH1 gene (Huala et al., 1997) has finally provided some credibility to this assumption. The NPH1 gene encodes a protein of 996 amino acids (112 kDa) and the coding region consists of 20 exons extending for 5.4 kb. The deduced amino acid sequence suggests that the carboxy-terminal region of NPH1 encodes a serine/threonine protein kinase. The amino-terminal region of NPH1 has two repeated domains, LOV1 and LOV2 (LOV for light, oxygen and voltage), that share similarity with diverse proteins of archaea, eubacteria and eukaryotes. The redox status of some of these proteins is known to be regulated by light, oxygen or voltage, and it probably involves a flavin moiety (see Huala et al., 1997). The elucidation of the functional significance of LOV1 and LOV2 domains in NPH1 is thus crucial for understanding how the cryptochromes (CRY1 and/or CRY2) and NPH1 interact for controlling blue light-induced phototropism in plants?

4. Transduction of Blue Light Signals

A broad consideration of the phenomenology of a whole array of blue light-induced responses in higher plants (described earlier) suggests the existence of multiple signal transduction chains. The recent reports on the existence of multiple receptors for blue light also supports this assumption. Because of this inherent complexity and the lack of biochemical identity of the blue light receptors until recently, it is not surprising that the progress in unravelling the mechanism of blue light action has been rather slow. However, the mechanism by which the blue light signal is converted into a biochemical entity and transduced to elicit a biological response has come under extensive investigation lately. Consequently, some of the signalling components have been identified. But, as would be apparent from the discussion below, efforts have been largely response based and there is thus an urgent need to develop a unified view of the blue light action mechanism.

4.1 Changes in Redox Activity

As mentioned earlier, one of the earliest studies on blue light action mechanism described blue light-induced absorbance changes (or LIACs) in the isolated plasma membrane fraction. These LIACs were ascribed to the flavin-mediated photoreduction of b-type cytochromes. The high blue light intensities required for LIACs and the difficulty in correlating it directly with a biological response has caused aspersion on its real significance as one of the early steps in blue-light triggered signalling cascade. However, a few other recent studies have examined the effect of blue light on redox activity. A flavin-mediated electron transport to ferricyanide at the plasma membrane has been implicated in blue light signalling (Rubinstein and Stern, 1991). The light-regulated translation of chloroplast mRNAs in *Chlamydomonas* has also been shown to be redox-dependent (Danon and Mayfield, 1994). Although blue light also influences plastid gene expression and the expression of nuclear genes encoding plastid proteins in higher plants, the evidence to connect redox changes with this response is lacking (see Short and Briggs, 1994).

4.2 GTP-binding Proteins

GTP-binding regulatory proteins (G-proteins) play a crucial role in transducing and amplifying the receptor signals in a variety of organisms. Their presence has been demonstrated in higher plants as well and their role implicated in transducing receptor-mediated signals including blue light.

Employing immunological probes developed for components of G-protein transducin, Warpeha et al. (1991) established the presence of components of the heterotrimeric G-protein in the plasma membrane fraction of etiolated pea apical buds. Although western blots of pea membrane proteins probed with antibodies against the transducin $\beta\gamma$ complex recognized four polypeptides, the α-subunit was identified as a 40 kDa polypeptide using antibodies against transducin α-subunit or synthetic G_α peptides. Studies using photoaffinity labelling with GTP analog, ADP-ribosylation by cholera toxin and inhibition of ADP-ribosylation by pertussis toxin, in a blue light-specific manner, also provided evidence that a 40 kDa polypeptide may be the α-subunit of a heterotrimeric G-protein involved in blue light signal transduction. Warpeha et al. (1991) also reported a transient increase in GTPase activity and GTP-γ-S binding in the plasma membrane fraction isolated from apical buds irradiated with blue light. The kinetics of GTP-γ-S binding with plasma membrane-enriched fraction in response to blue was similar to that for GTPase activity, suggesting that both responses are manifestation of the same process. In addition, the tissue-sensitivity and the fluence-response characteristics of blue light-mediated stimulation of GTPase activity and blue light-stimulated transcription of specific *CAB* genes has also been found to be correlative.

G-proteins may also play a role in regulating stomatal movement. Studies employing non-hydrolysable inhibitor (GDPβS) and activator (GTPγS), which stimulate and inhibit inward K^+ current, respectively, suggest that the activated G-protein inhibits stomatal opening (Fairley-Grenot and Assmann, 1991). It is thus likely that blue light-mediated opening of stomates does not involve a G-protein, and it may in fact counter the effect of blue light.

4.3 Changes in Electrical Potentials and Ion Fluxes

4.3.1 Hypocotyl growth inhibition response

High-irradiance blue light is known to inhibit hypocotyl elongation of the seedlings grown in dark. Although the photoreceptor pigment responsible for this response is thought to be a flavoprotein encoded by CRY1 (Ahmad and Cashmore, 1993), the signal transduction mechanism is not completely understood.

Electrophysiological studies conducted on material during the early stages of blue light-induced hypocotyl growth inhibition have demonstrated a rapid and transient depolarization of the plasma membrane of epidermal cells, which precedes the blue light-mediated cessation of *Cucumis* hypocotyl elongation and is rectified within 2 to 3 min (Spalding and Cosgrove, 1989, 1992). This depolarization is localized to the irradiated portion and is not propagated to unirradiated regions of the hypocotyl. Studies with specific ion channel blockers have shown that tetraethyl ammonium, a K^+-channel blocker, is ineffective, whereas vanadate and KCN abolish the depolarization, indicating the involvement of plasma membrane H^+-ATPase in blue light-induced depolarization. Recently, evidence for the role of Cl^- channels, commonly called anion channels, has also emerged. An anion channel blocker

NPPB [5-nitro-2-(3-phenylpropylamino)-benzoic acid] not only blocked this anion channel reversibly but also checked the blue light-induced depolarization *in vivo* and also its inhibitory effect on hypocotyl elongation (Cho and Spalding, 1996). For repolarization, role of Ca^{2+} has been advocated (see Cosgrove, 1994). However, Ca^{2+} has also been found to be essential for blue light-induced inhibition of growth in *Cucumis sativus* (Shinkle and Jones, 1988). Cytokinin also mimicks blue light in causing inhibition of hypocotyl growth of *Cucumis sativus* (Cohen et al., 1989) and *Brassica napus* (Khurana et al., unpublished), and Ca^{2+} is required obligatorily for this response; Ca^{2+} channel blockers and Ca^{2+}-chelating agents inhibit cytokinin activity, whereas A 23187, a Ca^{2+} ionophore, can retard hypocotyl growth even in the absence of a cytokinin.

4.3.2 Changes in guard cells

It has been unequivocally established that stomatal opening is primarily driven by proton gradients across the plasma membrane and activation of the ion channels in the guard cells. By measuring the acidification of the external medium or detecting outward electrical currents, employing elegant patch clamp technique, it has been shown that protons are extruded from the guard cell protoplasts immediately upon blue light irradiation (Assmann et al., 1985; Shimazaki et al., 1986). The mechanism by which the protons are extruded, however, remains controversial. The possibility of the involvement of a H^+-ATPase (Shimazaki et al., 1986) or a plasma membrane redox system (Raghavendra, 1990) has been considered but the interdependence or relative involvement of these two remains to be worked out. Subsequent steps are, however, better understood and it is known that consequent to membrane hyperpolarization, influx of K^+ takes place through voltage-gated channels, which eventually leads to increase in turgor in guard cells, causing stomatal opening. Studies with agonists and antagonists of signal transduction components have suggested a role for Ca^{2+}/calmodulin and a calmodulin-dependent myosin light chain kinase-like protein in proton extrusion from guard cell protoplasts in *Vicia faba* (Shimazaki et al., 1992). In a similar study, a role for diacylglycerol and protein kinase has been suggested in stomatal opening in *Commelina communis* and regulation of outward current in guard cell protoplasts of *Vicia faba.* Surprisingly, an elevated cytosolic Ca^{2+} level blocks inward rectifying K^+ channels of the guard cell (Schroeder and Hagiwara, 1989), leading to loss of turgor and finally stomatal closing. Thus, there may be multiple roles of Ca^{2+} in regulating stomatal conductance. The specific inhibitors of protein phosphatase 2B blocked the inhibitory effect of Ca^{2+} on K^+ channels, indicating that Ca^{2+} probably acts through dephosphorylation of some protein(s).

4.4 Regulation of Phosphorylation of Plasma Membrane Proteins

Rapid blue light-induced phosphorylation of a 120 kDa microsomal membrane protein in elongating stem tissue of dark-grown pea seedlings was first reported by Gallagher et al. (1988). Subsequently, it was demonstrated that this protein is localized in the plasma membrane, and the fluence required for threshold and saturation of blue light-induced phosphorylation is close to that for phototropism. The tissue that perceives the phototropic stimulus is the one that shows the strongest phosphorylation response to blue light (see Short and Briggs, 1994). As mentioned earlier, studies employing phototropism mutants of the *nph*1 (*n*on-*p*hototropic *h*ypocotyl) class, including JK224 (now redesignated *nph*1–2), which lack blue light-induced phosphorylation of the 120 kDa polypeptide in crude membrane preparation (Reymond et al., 1992; Liscum and Briggs, 1995), provided support for the assumption that this blue light-regulated phosphoprotein constitutes a signalling step in phototropism transduction chain. Besides pea and *Arabidopsis*, blue light-induced phosphorylation of a membrane protein in the range of 100 to 130 kDa has also been reported for several other plant species (zucchini, sunflower, tomato, barley, wheat, maize and oat), representing both monocots and dicots (see Short and Briggs, 1994). Recently, blue light-induced phosphorylation of the polypeptides of the microsomal fraction of young dark-grown wheat seedlings has been investigated in detail in our laboratory. As is evident from the autoradiogram in Fig. 3, blue light stimulates phosphorylation of at least four polypeptides of the plasma membrane fraction in the range of 110 and 70 kDa (for details, see Sharma et al., 1997). The highly sensitive tip portion of the etiolated seedlings is most responsive to blue light and the intensity of the phosphorylation response is higher in the leaf than the tubular coleoptile fraction. The kinetics of induction of this response and the presence of all the four phosphopolypeptides in both the coleoptile sheath and the enclosed leaf, suggest that phosphorylation of plasma membrane proteins may be one of the early steps not only in phototropism transduction chain but perhaps the signalling cascade of low-fluence blue responses in general.

The biochemical characterization of the 120 kDa polypeptide of pea by Briggs and coworkers suggests that the blue light receptor, the kinase activity and the substrate protein may be embodied in a single polypeptide or in a stable complex. This assumption is based upon the observations that: (a) the phosphorylated substrate contains an ATP-binding site that is essential for kinase activity, (b) phosphorylation photosensitivity is not affected by detergent treatment, and (c) the substrate fraction migrates as a large complex of 335 kDa on denaturing gel, retaining sensitivity for light-induced phosphorylation subsequent to electrophoresis (see Short and Briggs, 1994). Further analysis is, however, required to decipher whether the photosensitive 120 kDa polypeptide has integral photoreceptor and protein kinase activity or it co-migrates with the protein possessing these functions.

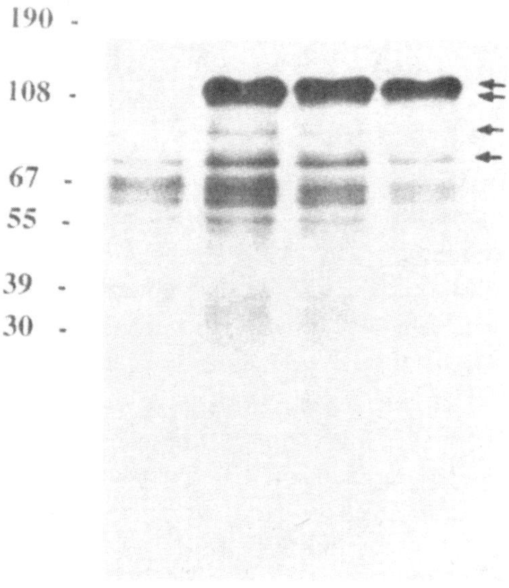

Fig. 3. Blue light-induced phosphorylation of microsomal membrane proteins of the 4-day-old dark-grown wheat seedlings. Autoradiogram shows the changes in phosphorylation of polypeptides (indicated by arrows) of the microsomal fraction irradiated with blue light for specified duration, followed by *in vitro* phosphorylation and SDS-PAGE. For details, see Sharma et al. (1997).

4.5 Genetic Analysis of the Signalling Cascade

To dissect the signalling pathway initiated following photoexcitation of blue light receptor(s), *Arabidopsis* mutants have been isolated that are aberrant in blue light-dependent inhibition of hypocotyl elongation, a high-irradiance response, or are defective in phototropism of etiolated seedling hypocotyl to unilateral low-fluence blue light.

4.5.1 Mutants of hypocotyl growth inhibition response

This class is represented by *blu* (*b*lue *l*ight *u*ninhibited) mutants. As mentioned earlier, all the three *blu* mutants described (Liscum and Hangarter, 1991) have been found allelic to *hy*4, a photoreceptor pigment mutant (Hangarter, personal communication). Therefore, caution must be exercised in extrapolations from the published data on the *blu* mutants. In an earlier study, however, these *blu* mutants also showed reduced cotyledon expansion and anthocyanin accumulation under blue light, but like *hy*4, they exhibited normal phototropism and stomatal function, two of the most extensively studied low-fluence blue responses (see Liscum and Hangarter, 1994), suggesting that low-fluence and high-irradiance blue responses operate through genetically distinct sensory systems. This is also consistent with the physiological data from *Cucumis sativus* and *Sinapis alba*, where phototropism and hypocotyl growth inhibition have been shown to involve separate pathways (see Cosgrove, 1994).

To regulate the growth of seedling hypocotyl, it appears that the blue light photoreceptor pigment and phytochrome may interact. For example, the phytochromobilin-deficient *hy*2 and *hy*6 mutants are insensitive to low-fluence blue light for hypocotyl growth inhibition. Likewise, the sensitivity of *phy*A mutants to low-fluence blue light for apical hook opening is reduced, whereas *phy*B mutants retain normal sensitivity (see Liscum and Hangarter, 1993), indicating a probable role for *phy*A in this response. However, double mutant analysis between *blu*1/*hy*4 and *hy*6 (a phytochrome-deficient mutant), which exhibits an etiolated phenotype in white, blue, green, red and far-red light, and between *blu*1/*hy*4 and *hy*8 (phyA-deficient), which has long hypocotyl only in blue and far-red light (see Liscum and Hangarter, 1994), indicates as well the independent action of blue and red/far-red light in regulating hypocotyl growth. In a more recent report, Ahmad and Cashmore (1997) have shown that in severely deficient phytochrome mutants, the CRY1-mediated blue light responses are drastically reduced, even though the levels of CRY1 *per se* are not affected. It has thus been concluded that the presence of active phytochrome is essential for full expression of CRY1 activity for mediating high irradiance responses.

4.5.2 Phototropism mutants

Seedling hypocotyls/shoots of *Arabidopsis* display positive phototropism and roots exhibit negative phototropism in response to unilateral blue light of 450 nm. More than 35 mutant strains of *A. thaliana,* cv. Estland, have been identified that either lack or display altered phototropic behaviour (Khurana and Poff, 1989; Khurana et al., 1989b). Six of these mutants, JK224/*nph*1 (non-*p*hototropic *h*ypocotyl 1), JK218/*nph*3, JK345/*pho*1, JK229, ZR8 and ZR19, have been characterized in detail. Strain JK218 represents a null allele as it lacks both first and second positive phototropic response, whereas JK229 and JK345 lack the first positive phototropism and show alteration in second positive phototropism, thus genetically delineating the two response pathways. On the other hand, ZR8 and ZR19 are reduced amplitude response mutants. Both the ZR strains and JK345 are also defective in gravitropic response and may represent components common to both tropic response pathways.

Strain JK224 has been described earlier as a probable photoreceptor mutant along with the *nph*1 class of phototropism mutants isolated by Liscum and Briggs (1995). Mutants of *nph*1 class are impaired in blue light-induced protein phosphorylation response. However, *nph*2, *nph*3 and *nph*4 alleles respond normally in this respect, indicating that these later mutants represent downstream signalling components. Further analysis of these mutants has also revealed that NPH2 and NPH3 are specifically required for phototropism, whereas NPH4 is involved in both phototropism and gravitropism, and probably functions directly in the differential growth response (see Briggs and Liscum, 1997). Interestingly, all the phototropism mutants tested are genetically distinct and retain a normal hypocotyl growth inhibition response.

In contrast to the seedling shoots that display positive phototropism, the bending response of the roots away from the light source is an example of negative phototropism. Okada and Shimura (1992) have isolated a large number of putative mutants in root phototropic response. Two of them, *rpt*1 and *rpt*2, have been characterized in detail. Both *rpt*1 and *rpt*2, though defective in root phototropism, retain normal hypocotyl phototropism as well as wild-type-like response of root and shoot gravitropism, indicating the existance of multiple pathways controlling tropic responses.

4.5.3 Other blue light response mutants

Some of the blue light-induced responses, such as hypocotyl growth inhibition, cotyledon expansion and anthocyanin production, are also induced by photoactivation of phytochrome. It is thus not surprising that some of the phytochrome response mutants (described elsewhere in this volume) display aberration in these responses. A long hypocotyl mutant, *hy*5, has been found to be partially impaired in both blue and red light-mediated hypocotyl growth inhibition (Koornneef et al., 1980; Liscum and Hangarter, 1993). Two of the constitutive photomorphogenic mutants of *Arabidopsis, det*1 and *det*2, have been found to be epistatic to *hy*2 and *hy*6 (phytochromobilin-deficient mutants) and also to *hy*4, the cryptochrome mutant, indicating that *DET* products also function in blue response pathway. Likewise, another set of constitutively photomorphogenic mutants, designated *pho* (for plumular hook open phenotype, in dark), also lack normal tropic response to both light and gravity (see Khurana et al., 1996, 1998). One of these mutants, *pho*1/JK345 was originally isolated as a phototropism mutant and is impaired in both first and second positive response. Epistatic analysis of JK345/*pho*1–1, vis-à-vis other constitutive photomorphogenic mutants and the photoreceptor mutants, and its molecular characterization will be useful in understanding the interactive effects of red, far-red and blue light.

5. Blue Light-Mediated Changes in Gene Expression

The perception of blue light by specific photoreceptors activates a signalling mechanism that transduces the signal to gene expression machinery, leading eventually to regulation of more overt morphogenic responses. In comparison to phytochrome, very little is known about the precise regulation of expression of genes exclusively by blue light. Such studies have probably been hampered because phytochrome can also respond to blue light, particularly in case of long-term high irradiance responses. Although an exhaustive coverage has been given to this aspect in other chapters of this volume, some select examples have been considered here for a direct correlation of these studies which form a part of the mechanism of blue light action.

The steady-state transcript levels of several nuclear- and plastid-encoded genes were found to change when tobacco and *Chenopodium* cell suspension cultures were irradiated with blue light for prolonged duration (Richter, 1984; Richter et al., 1987). The plastid differentiation and changes in gene expression were primarily induced by blue light. Although red light alone only had a marginal effect, it strongly inhibited the blue light-stimulated increase in *RBCS* and *RBCL* transcripts (Richter and Wessel, 1985). In a more detailed analysis, blue light-specific regulation of individual genes and gene families encoding proteins of the photosynthetic apparatus has been examined. Blue light altered the accumulation of transcripts in nine of the thirteen gene families examined in apical buds of the pea plants grown normally under red light (see Thompson et al., 1985). Using gene-specific probes, Chua and coworkers provided evidence that relative effectiveness of blue and red light in enhancing abundance of *RBCS*-3A and –3C transcripts is development-dependent (Fluhr and Chua, 1986; Fluhr et al., 1986). Blue light did not cause any significant change in levels of either *RBCS*-3A or –3C transcripts of etiolated pea leaves, whereas their transcript abundance was regulated by phytochrome in a red/far-red reversible manner. However, when dark-adapted mature plant leaves were irradiated with monochromatic light for an extended duration (8 h), blue light was found to be at least five-fold more effective in elevating levels of *RBCS*-3A and –3C transcripts.

The analysis of transcript abundance of *CAB* genes and three unidentified cDNA clones, pEA25, pEA27 and pEA215, in red light-grown pea seedlings exposed to blue radiation indicated the existence of a low-fluence and a high-fluence blue response (Warpeha and Kaufman, 1990). The transcript levels of *CAB* and pEA215 increased at low fluences and it was followed by a decline at higher fluences, whereas that of pEA25 and pEA207 were negatively regulated. In dark-grown plants, however, exposure to high-fluence blue light enhanced the accumulation of *CAB* and pEA215 transcripts. The changes in transcript abundance detected in response to blue light were cycloheximide insensitive (Marrs and Kaufman, 1989), suggesting that no intervening regulatory genes or the translation of pre-existing transcripts is involved in this response. High fluence rates of blue light are also known to specifically induce accumulation of mRNA of early light inducible proteins (ELIPs) in mature pea plants (Adamska et al., 1992); these proteins are related to the *CAB* gene family.

Besides regulating the expression of genes associated with plastid development, blue light has also been shown to influence the expression of genes encoding enzymes of the phenylpropanoid pathway (see Jenkins, 1997). Most of the work has been confined to *CHS*, encoding chalcone synthase, the key enzyme in flavonoid biosynthesis. Earlier studies have shown that *CHS* gene expression is coordinately regulated by red, blue and UV-B radiations in cultured parsley cells (Ohl et al., 1989). In *Sinapis alba*, *CHS* expression is regulated by monochromatic lights in a development-dependent manner, with red light being most effective in cotyledons of young seedlings and blue/UV more effective at later stages of development (Batschauer et al., 1991). In *Arabidopsis* too, expression of *CHS* (and also of other flavonoid genes) is regulated by multiple photoreceptors in a temporal and spatial manner (see Jenkins, 1997; Bharti and Khurana, 1997). The light induction experiments, together with analysis of mutant and transgenic lines, have shown that a blue light receptor is involved in the regulation of flavonoid gene expression (Kubasek et al., 1992; Ahmad et al., 1995); however, UV-B was found to be more effective than blue light for the induction process. Employing phytochrome A- and phytochrome B- deficient mutants, Batschauer et al. (1996) have demonstrated that inductive effect of blue (and UVA) light on *CHS* expression is independent of at least these phytochrome species. However, in systems like parsley and mustard, phytochrome is also thought to be involved in the induction of flavonoid genes (see Beggs and Wellmann, 1994).

It is apparent from these studies that multiple receptors are involved in the regulation of flavonoid gene expression, with UV and blue light playing a predominant role. A detailed analysis for delineating the specific effects exerted by various sensory systems and understanding their interaction, however, needs to be performed, and mutants defective in specific photoreceptor functions may turn out to be quite useful for such studies.

To identify the regulatory sequences in the upstream promoter regions of the genes that confer sensitivity to blue light and the corresponding *trans*-acting (protein) factors which transduce the light signal, the promoter region of the *CHS* gene has been analysed in some detail. Feinbaum et al. (1991) studied the expression of several chimeric constructs of variable/truncated 5' upstream promoter regions of *CHS* fused to β-glucuronidase (GUS) coding sequence in transgenic *Arabidopsis*. This analysis revealed that sequences between –528 and –186 bp of the promoter region are essential for responsiveness to blue light. With regard to the identification of possible *trans*-acting factors involved in transducing the blue light signal, no significant development has taken place and this area of research remains wide open for exploration.

6. Conclusions and Future Prospects

Our understanding of the blue light physiology of higher plants has surely advanced in the past few years. The ignorance that prevailed over the decades for the identification of the receptors for blue light has been overcome.

There is now unequivocal evidence for the existence of multiple blue light receptors that govern specific subset of responses. Molecular and physiological analysis of distinct genetic mutants has revealed that cryptochromes may be the predominant blue light photoreceptors in plants. Apparently, there is considerable overlap in function between at least two members of the CRY family. Both CRY1 and CRY2 affect high-fluence blue responses, such as inhibition of hypocotyl elongation and stimulation of anthocyanin accumulation, and the low-fluence response, phototropism. In addition, CRY2 also plays a major role in controlling flowering time. In contrast, the putative blue light receptor for phototropism, NPH1, which encodes an auto-phosphorylating protein kinase with a redox-sensing domain, does not significantly affect other cryptochrome-regulated responses. Why and how these multiple blue light absorbing species interact and contribute to blue light-induced responses like phototropism, would be interesting to decipher. Unravelling the signal transduction pathways for different photoreceptors also continues to be the hottest area of research. The application of genetic and biochemical approaches has been useful in providing new insights into the signalling events that may transduce the blue light signal across the membrane barriers. However, most of the information obtained is fragmentory and has been generated while investigating diverse plant responses. Synthesis of this fragmentory knowledge and elucidation of signal transduction chain(s) which have relevance to a wide range of blue light-induced responses would be of immense significance. To identify the protein immediately next in hierarchy to the sensory receptors, and also the down-stream components, exploiting the potential of yeast two-hybrid system may prove to be revolutionary. How different photosensory signalling pathways are integrated to control complex plant developmental processes also deserves serious consideration and probably will be the focus of research in coming years. The characterization of *cis*-acting regulatory sequences in the promoter region and the *trans*-acting (protein) factors that interact with these gene sequences in blue light-specific manner is another area that needs greater attention for understanding the molecular mechanism of blue light action. However, the involvement of multiple photoreceptors in regulating the expression of light-regulated genes makes it a daunting task. This limitation can probably be circumvented by employing mutants defective in specific regulatory photoreceptors.

Acknowledgements

The research work in the first author's laboratory is financially supported by the University Grants Commission, the Department of Science and Technology and the Department of Biotechnology of the Government of India. The help rendered by Ms Anju Kochhar in preparation of illustrations is gratefully acknowledged.

References

Adamska, I., Ohad, I. and Kloppstech, K. 1992. Synthesis of the early light-inducible protein is controlled by blue light and related to light stress. *Proc. Natl. Acad. Sci. USA* **89**: 2610–2613.

Ahmad, M. and Cashmore, A.R. 1993. *HY4* gene of *A. thaliana* encodes a protein with characteristics of a blue-light photoreceptor. *Nature* **366**: 162–166.

Ahmad, M. and Cashmore, A.R. 1996. Seeing blue: the discovery of cryptochrome. *Plant Mol. Biol.* **30:** 851–861.

Ahmad, M. and Cashmore, A.R. 1997. The blue-light receptor cryptochrome 1 shows functional dependence on phytochrome A or phytochrome B in *Arabidopsis thaliana. Plant J.* **11:** 421–427.

Ahmad, M., Jarillo, J.A. and Cashmore, A.R. 1998a. Chimeric proteins between CRY1 and CRY2 *Arabidopsis* blue light photoreceptors indicate overlapping functions and varying protein stability. *Plant Cell* **10:** 197–207.

Ahmad, M., Jarillo, J.A., Smirnova, O. and Cashmore, A.R. 1998b. Cryptochrome blue-light photoreceptors of *Arabidopsis* implicated in phototropism. *Nature* **392:** 720–723.

Ahmad, M., Lin, C. and Cashmore, A.R. 1995. Mutations throughout an *Arabidopsis* blue-light photoreceptor impair blue-light-responsive anthocyanin accumulation and inhibition of hypocotyl elongation. *Plant J.* **8:** 653–658.

Assmann, S.M., Simoncini, L. and Schroeder, J.I. 1985. Blue light activates electrogenic ion pumping in guard cell protoplasts of *Vicia faba. Nature* **318:** 285–287.

Baskin, T.I. and Iino, M. 1987. An action spectrum in the blue and ultraviolet for phototropism in alfalfa. *Photochem. Photobiol.* **46:** 127–136.

Batschauer, A., Ehmann, B. and Schäfer, E. 1991. Cloning and characterization of a chalcone synthase gene from mustard and its light-dependent expression. *Plant Mol. Biol.* **16:** 175–185.

Batschauer, A., Rocholl, M., Kaiser, T., Nagatani, A., Furuya, M. and Schäfer, E. 1996. Blue and UV-A light-regulated *CHS* expression in *Arabidopsis* independent of phytochrome A and phytochrome B. *Plant J.* **9:** 63–69.

Beggs, C.J. and Wellmann, E. 1994. Photocontrol of flavonoid biosynthesis. *In*: Kendrick, R.E. and Kronenberg, G.H.M. (eds), Photomorphogenesis in Plants, pp. 733–752. Kluwer Academic Publishers, Dordrecht.

Bharti, A.K. and Khurana, J.P. 1997. Mutants of *Arabidopsis* as tools to understand the regulation of phenylpropanoid pathway and UVB protection mechanisms. *Photochem. Photobiol.* **65:** 765–776.

Briggs, W.R. and Iino, M. 1983. Blue-light-absorbing photoreceptors in plants. *Phil. Trans. R. Soc. London* **B303:** 347–359.

Briggs, W.R. and Liscum, E. 1997. The role of mutants in search for the photoreceptor for phototropism in higher plants. *Plant Cell Environ.* **20:** 768–772.

Cho, M.H. and Spalding, E.P. 1996. An anion channel in *Arabidopsis* hypocotyls activated by blue light. *Proc. Natl. Acad. Sci. USA* **93:** 8134–8138.

Chon, H.P. and Briggs, W.R. 1966. Effect of red light on the phototropic sensitivity of corn coleoptiles. *Plant Physiol.* **41:** 1715–1724.

Cohen, L., Gepstein, S. and Horwitz, B.A. 1991. Similarity between cytokinin and blue light inhibition of cucumber hypocotyl elongation. *Plant Physiol.* **95:** 77–81.

Cosgrove, D.J. 1994. Photomodulation of growth. *In:* Kendrick, R.E. and Kronenberg, G.H.M. (eds), Photomorphogenesis in Plants, pp. 631–658. Kluwer Academic Publishers, Dordrecht.

Danon, A. and Mayfield, S. 1994. Light-regulated translation of chloroplast messenger RNAs through redox potential. *Science* **266**: 1717–1719.

Darwin, C. 1880. The Power of Movement in Plants. John Murray, London.

Fairley-Grenot, K.G. and Assmann, S.M. 1991. Evidence for G-protein regulation of inward K^+ channel current in guard cells of fava bean. *Plant Cell* **3**: 1037–1044.

Feinbaum, R.L., Storz, G. and Ausubel, F.M. 1991. High intensity and blue light regulated expression of chimeric chalcone synthase genes in transgenic *Arabidopsis thaliana* plants. *Mol. Gen. Genet.* **226:** 449–456.

Fluhr, R. and Chua, N.-H. 1986. Developmental regulation of two genes encoding ribulose-bisphosphate carboxylase small subunit in pea and transgenic petunia plants: phytochrome response and blue-light induction. *Proc. Natl. Acad. Sci. USA* **83:** 2358–2362.

Fluhr, R., Moses, P., Morelli, G., Coruzzi, G. and Chua, N.-H. 1986. Expression dynamics of the pea *rbcS* multigene family and organ distribution of the transcripts. *EMBO J.* **5:** 2063–2071.

Gallagher, S., Short, T.W., Ray, P.M., Pratt, L.H. and Briggs, W.R. 1988. Light-mediated changes in two proteins found associated with plasma membrane fractions from pea stem sections. *Proc. Natl. Acad. Sci. USA* **85:** 8003–8007.

Guo, H., Yang, H., Mockler, T.C. and Lin, C. 1998. Regulation of flowering time by *Arabidopsis* photoreceptors. *Science* **279:** 1360–1363.

Hoffman, P.D., Batschauer, A. and Hays, J.B. 1996. *PHH*1, a novel gene from *Arabidopsis thaliana* that encodes a protein similar to plant blue light photoreceptors and microbial photolyases. *Mol. Gen. Genet.* **253:** 259–265.

Huala, E., Oeller, P.W., Liscum, E., Han, I.-S., Larsen, E. and Briggs, W.R. 1997. *Arabidopsis* NPH1: a protein kinase with a putative redox-sensing domain. *Science* **278:** 2120–2123.

Janoudi, A.-K., Gordon, W.R., Wagner, D., Quail, P. and Poff, K.L. 1997a. Multiple phytochromes are involved in red-light-induced enhancement of first-positive phototropism in *Arabidopsis thaliana. Plant Physiol.* **113:** 975–979.

Janoudi, A.-K., Konjević, R., Whitelam, G. Gordon, W. and Poff, K.L. 1997b. Both phytochrome A and phytochrome B are required for the normal expression of phototropism in *Arabidopsis thaliana* seedlings. *Physiol. Plant.* **101:** 278–282.

Janoudi, A.-K. and Poff, K.L. 1992. Action spectrum for enhancement of phototropism by *Arabidopsis thaliana* seedlings. *Photochem. Photobiol.* **56:** 655–659.

Jenkins, G.I. 1997. UV and blue light signal transduction in *Arabidopsis. Plant Cell Environ.* **20:** 773–778.

Kendrick, R.E. and Kronenberg, G.H.M. (eds) 1994. Photomorphogenesis in Plants. Kluwer Academic Publishers, Dordrecht.

Khurana, J.P., Best, T. and Poff, K.L. 1989a. Influence of hook position on phototropic and gravitropic curvature by hypocotyls of *Arabidopsis thaliana. Plant Physiol.* **90:** 376–379.

Khurana, J.P., Kochhar, A. and Jain, P.K. 1996. Genetic and molecular analysis of light-regulated plant development. *Genetica* **97:** 349–361.

Khurana, J.P., Kochhar, A. and Tyagi, A.K. 1998. Photosensory perception and signal transduction in higher plants—molecular genetic analysis. *Crit. Rev. Pl. Sci.* (In Press).

Khurana, J.P. and Poff, K.L. 1989. Mutants of *Arabidopsis thaliana* with altered phototropism. *Planta* **178:** 400–406.

Khurana, J.P., Ren, Z., Steinitz, B., Parks, B., Best, T. and Poff, K.L. 1989b. Mutants of *Arabidopsis thaliana* with decreased amplitude in their phototropic response. *Plant Physiol.* **91:** 685–689.

Konjević, R., Khurana, J.P. and Poff, K.L. 1992. Analysis of multiple photoreceptor pigments for phototropism in a mutant of *Arabidopsis thaliana. Photochem. Photobiol.* **55:** 789–792.

Konjević, R., Steinitz, B. and Poff, K.L. 1989. Dependence of the phototropic response of *Arabidopsis thaliana* on fluence rate and wavelength. *Proc. Natl. Acad. Sci. USA* **86:** 9876–9880.

Koornneef, M., Rolff, E. and Spruit, C.J.P 1980. Genetic control of light-inhibited hypocotyl elongation in *Arabidopsis thaliana* (L.) Heynh. *Z. Pflanzenphysiol.* **100:** 147–160.

Kubasek, W.L., Shirley, B.W., McKillop, A., Goodman, H.M., Briggs, W.R. and Ausubel, F.M. 1992. Regulation of flavonoid biosynthetic genes in germinating *Arabidopsis* seedlings. *Plant Cell* **4:** 1229–1236.

Lin, C., Ahmad, M., Chan, J. and Cashmore, A.R. 1996. CRY2: a second member of the *Arabidopsis* cryptochrome gene family. *Plant Physiol.* **110:** 1047.

Lin, C., Robertson, D.E., Ahmad, M., Raibekas, A.A., Jorns, M.S., Dutton, P.L. and Cashmore, A.R. 1995. Association of flavin adenine dinucleotide with the *Arabidopsis* blue-light receptor CRY1. *Science* **269:** 968–970.

Lin, C., Yang, H., Guo, H., Mockler, T., Chen, J. and Cashmore, A.R. 1998. Enhancement of the blue-light sensitivity of the *Arabidopsis* young seedlings by a blue-light receptor CRY1. *Proc. Natl. Acad. Sci. USA* **95:** 2686–2690.

Liscum, E. and Briggs, W.R. 1995. Mutations in the NPH1 locus of *Arabidopsis* disrupt the perception of phototropic stimuli. *Plant Cell* **7:** 473–485.

Liscum, E. and Hangarter, R.P. 1991. *Arabidopsis* mutants lacking blue light-dependent inhibition of hypocotyl elongation. *Plant Cell* **3:** 685–694.

Liscum, E. and Hangarter, R.P. 1993. Photomorphogenic mutants of *Arabidopsis thaliana* reveal activities of multiple photosensory systems during light-stimulated apical-hook opening. *Planta* **191:** 214–221.

Liscum, E. and Hangarter, R.P. 1994. Mutational analysis of blue-light sensing in *Arabidopsis. Plant Cell Environ.* **17:** 639–648.

Malhotra, K., Kim, S.T., Batschauer, A., Dawut, L. and Sancar, A. 1995. Putative blue-light photoreceptors from *Arabidopsis thaliana* and *Sinapis alba* with a high degree of sequence homology to DNA photolyase contain the two photolyase cofactors but lack DNA repair activity. *Biochemistry* **34:** 6892–6899.

Marrs, K.A. and Kaufman, L.S. 1989. Blue light regulation of transcription for nuclear genes in pea. *Proc. Natl. Acad. Sci. USA* **86:** 4492–4495.

Muñoz, V. and Butler, W.L. 1975. Photoreceptor pigment for blue light in *Neurospora crassa. Plant Physiol.* **55:** 421–426.

Ohl, S., Hahlbrock, K. and Schäfer, E. 1989. A stable blue-light derived signal modulates ultraviolet-light-induced activation of the chalcone-synthase gene in culture of parsley cells. *Planta* **177:** 228–236.

Okada, K. and Shimura, Y. 1992. Mutational analysis of root gravitropism and phototropism of *Arabidopsis thaliana* seedlings. *Aust. J. Plant Physiol.* **19:** 439–448.

Palmer, J.M., Warpeha, K.M.F. and Briggs, W.R. 1996. Evidence that zeaxanthin is not the photoreceptor for phototropism in maize coleoptiles. *Plant Physiol.* **110:** 1323–1328.

Parks, B.M., Quail, P.H. and Hangarter, R.P. 1996. Phytochrome A regulates red-light induction of phototropic enhancement in *Arabidopsis. Plant Physiol.* **110:** 155–162.

Poff, K.L. and Butler, W.L. 1974. Absorbance changes induced by blue light in *Phycomyces blakesleeanus* and *Dictyostelium discoideum. Nature* **248:** 799–801.

Quiñones, M.A., Lu, Z. and Zeiger, E. 1996. Close correspondence between the action spectra for the blue light responses of the guard cell and coleoptile chloroplasts, and the spectra for blue light-dependent stomatal opening and coleoptile phototropism. *Proc. Natl. Acad. Sci. USA* **93:** 2224–2228.

Quiñones, M.A. and Zeiger, E. 1994. A putative role of the xanthophyll, zeaxanthin, in blue-light photoreception of corn coleoptiles. *Science* **264:** 558–561.

Raghavendra, A.S. 1990. Blue light effects on stomata are mediated by the guard cell plasma membrane redox system distinct from the proton translocating ATPase. *Plant Cell Environ.* **13:** 105–110.

Reymond, P., Short, T.W., Briggs, W.R. and Poff, K.L. 1992. Light-induced phosphorylation of a membrane protein plays an early role in signal transduction for phototropism in *Arabidopsis thaliana. Proc. Natl. Acad. Sci. USA* **89:** 4718–4721.

Richter, G. 1984. Blue light control of the level of two plastid mRNAs in cultured plant cells. *Plant Mol. Biol.* **3:** 271–276.

Richter, G., Dudel, A., Einspanier, R., Danhauer, I. and Husemann, W. 1987. Blue-light control of mRNA level and transcription during chloroplast differentiation in photomixotrophic and photoautotrophic cell cultures (*Chenopodium rubrum* L.). *Planta* **172:** 79–87.

Richter, G. and Wessel, K. 1985. Red light inhibits blue light-induced chloroplast development in cultured plant cells at the mRNA level. *Plant Mol. Biol.* **5:** 175–182

Rubinstein, B. and Stern, A.I. 1991. The role of plasma membrane redox activity in light effects in plants. *J. Bioenerg. Biomem.* **23:** 393–408.

Schroeder, J.I. and Hagiwara, S. 1989. Cytosolic calcium regulates ion channels in the plasma membrane of *Vicia faba* guard cells. *Nature* **338:** 427–430.

Sharma, V.K., Jain, P.K., Maheshwari, S.C. and Khurana, J.P. 1997. Rapid blue-light-induced phosphorylation of plasma-membrane-associated proteins in wheat. *Phytochemistry* **44:** 775–780.

Shimazaki, K., Iino, M. and Zeiger, E. 1986. Blue light-dependent proton extrusion by guard-cell protoplasts of *Vicia faba. Nature* **319:** 324–326.

Shimazaki, K., Kinoshita, T. and Nishimura, M. 1992. Involvement of calmodulin and calmodulin-dependent myosin light chain kinase in blue light-dependent H^+ pumping by guard cell protoplasts from *Vicia faba* L. *Plant Physiol.* **99:** 1416–1421.

Shinkle, J.R. and Jones, R.L. 1988. Inhibition of stem elongation in *Cucumis* seedlings by blue light requires calcium. *Plant Physiol.* **86:** 960–966.

Short, T.W. and Briggs, W.R. 1994. The transduction of blue light signals in higher plants. *Ann. Rev. Plant Physiol. Plant Mol. Biol.* **45:** 143–171.

Spalding, E.P. and Cosgrove, D.J. 1989. Large plasma-membrane depolarization precedes rapid blue-light-induced growth inhibition in cucumber. *Planta* **178:** 407–410.

Spalding, E.P. and Cosgrove, D.J. 1992. Mechanism of blue-light-induced plasma-membrane depolarization in etiolated cucumber hypocotyls. *Planta* **188:** 199–205.

Stapleton, A.E. 1992. Ultraviolet radiation and plants: burning questions. *Plant Cell* **4:** 1353–1358.

Steinitz, B. and Poff, K.L. 1986. A single positive phototropic response induced with pulsed light in hypocotyls of *Arabidopsis thaliana* seedlings. *Planta* **168:** 305–315.

Thimann, K.V. and Curry, G.M. 1960. Phototropism and phototaxis. *In:* Florkin, M. and Mason, H. (eds), Comparative Biochemistry. Vol. 1, pp. 243–309. Academic Press, New York.

Thompson, W.F., Kaufman, L.S. and Watson, J.C. 1985. Induction of plant gene expression by light. *BioEssays* **3:** 153–159.

Vierstra, R.D. and Poff, K.L. 1981a. Mechanism of specific inhibition of phototropism by phenylacetic acid in corn seedlings. *Plant Physiol.* **67:** 1011–1015.

Vierstra, R.D. and Poff, K.L. 1981b. Role of carotenoids in the phototropic response of corn seedlings. *Plant Physiol.* **68:** 798–801.

Warpeha, K.M.F., Hamm, H.E., Rasenick, M.M. and Kaufman, L.S. 1991. A blue-light-activated GTP-binding protein in the plasma membranes of etiolated peas. *Proc. Natl. Acad. Sci. USA* **88:** 8925–8929.

Warpeha, K.M.F. and Kaufman, L.S. 1990. Two distinct blue-light responses regulate the levels of transcripts of specific nuclear-coded genes in pea. *Planta* **182:** 553–558.

Concepts in Photobiology: Photosynthesis and Photomorphogenesis
G.S. Singhal, G. Renger, S.K. Sopory, K-D. Irrgang and Govindjee (Eds)

27. UV-B Effects: Receptors and Targets

L.O. Björn

Department of Plant Physiology, University of Lund
Box 117, S-221 00 Lund, Sweden

Summary

Evidence is presented that there is a signalling pathway for UV-B (ultraviolet-B radiation) effects on gene activity that differs from that of the UV-A/blue light signalling pathway. Evidence is also presented for a specific UV-B receptor pigment, with peak sensitivity near 290 nm, which is active in growth regulation and induction of UV-protecting pigmentation. A possible chromophore in this photoreceptor is a tetrahydropterin derivative. There are many interactions between this UV-B photoreceptor and other photoreceptors in plants, and some of the effects mediated by the UV-B receptor may not take place via gene regulation. UV-B regulation of genes may also proceed via photon absorption in DNA. Various types of radiation-induced changes in DNA are discussed, and the conclusion drawn that the most important one is formation of (6–4) photoproducts of pyrimidine bases in DNA. Calculations show that stratospheric ozone depletion of realistic magnitude does not cause significant increase in excitation of the 290 nm UV-B photoreceptor, in contrast to photon absorption in DNA.

1. Introduction

Effects of ultraviolet-B radiation in plants can be broadly classified into two categories: damaging (stress, inhibitory effects) and regulation (photomorphogenesis). The present chapter is a complement to Chapter 19 of this volume and will focus on regulatory aspects, and on the primary events, photoreceptors and other radiation targets, and signal transduction. Ecological, agricultural and economic aspects of UV-B impact on plants have been extensively treated in a number of recent reviews (e.g., Caldwell et al., 1995; Björn, 1996; Rozema et al., 1997).

2. Evidence for a Specific UV-B Signalling Pathway

When it comes to mechanisms of ultraviolet-B specific regulatory effects, most work has been done with (a) processes related to the formation of flavonoids and other protective (UV-shielding) pigments, and (b) inhibition of stem extension. In *Arabidopsis thaliana*, the enzyme chalcone synthase (CHS, controlling flavonoid synthesis) can be induced by either UV-B radiation or longer wavelength (UV-A or blue) radiation, but the signalling pathway seems to be different in the two cases. Using cell suspension culture, Christie and Jenkins (1996) compared the effects of different inhibitors for the short and the long wavelength induction. The calcium channel blocker nifedipine was found to inhibit both UV-B and UV-A/blue induction of CHS mRNA formation. Lanthanum ions had no effect, from which it was concluded that the calcium channels involved are not in the plasma membrane, which is accessable to externally added lanthanum ions. On the other hand, ruthenium red, known to block calcium channels in mitochondria and endoplasmic reticulum, inhibited both UV-B and UV-A/blue light effects, and the effect was specific; there was no general inhibition of transcription.

However, although these experiments indicate that elevation of the cytosolic calcium ion concentration is one step in both UV-B and UV-A/blue signal transduction, elevation of calcium ion concentration alone can not replace either photosignal. This was shown by applying the ionophore A23187 in combination with 10 mM calcium ion in the medium.

Both UV-B and UV-A/blue response were inhibited by the protein kinase inhibitors staurosporine and K252, implying protein phosphorylation as a step in the transduction of both signals. It is to note that these compounds are specific for serine and threonine phosphorylation, and that the tyrosine/histidine kinase inhibitor genistein, which is a potent inhibitor of phytochrome effects, was inactive. Thus different protein phosphorylations are involved in the phytochrome responses and in the blue and UV responses. Other effects of UV-B radiation on the phosphorylation pattern of plant proteins have been directly demonstrated by Yu and Björn (1997).

All these results point to similarities between the transduction channels for UV-B signals on one hand and UV-A/blue signals on the other. However, a difference was found when the effect of the calmodulin agonist W-7 was tested. This had very little effect on the UV-A/blue response in concentrations up to 200 μM, while the UV-B response was decreased to a small fraction of the control already at a concentration of 25 μM. This indicates that the calmodulin is involved in the UV-B pathway, but not in the UV-A/blue pathway.

3. Action Spectra and Attempts to Define a Specific UV-B Photoreceptor

If there are differences in the UV-B and UV-A/blue signalling pathways, the photoreceptors must also differ. The UV-A/blue photoreceptors (cryptochromes) are now starting to emerge from the fog in which they have so long been hiding, and characterization of one of them, CRY1, is now well in progress (see Ahmad and Cashmore, 1993 and review by Jenkins et al., 1995). The UV-B receptor(s) still remain more elusive.

Action spectroscopy is a useful technique for characterizing molecular radiation targets. Action spectroscopy is the determination of effect per incident photon as a function of wavelength; the result of such a determination is called an action spectrum. A full investigation of this kind involves the determination of effects at a number of photon fluences ("radiation doses") for a number of wavelengths, and then, for each wavelength interpolate the photon fluence necessary to produce a certain standard effect. The action spectrum should resemble the absorption spectrum of the molecular target. Sometimes it is necessary to correct for filtering effects by shielding pigments in the tissue before a comparison of action and absorption spectra is meaningful.

Wellmann (1983) makes a comparison of action spectra for damaging processes and regulations. We shall later return to the damaging processes, peaking in the ultraviolet-C region (below 280 nm). Wellmann exemplifies regulatory processes by induction of flavonoid synthesis in parsley (Wellmann 1975) and of anthocyanin synthesis in maize coleoptiles. In both the latter processes induction of CHS, the enzyme mentioned in the previous section, is probably the bottleneck; another key enzyme is phenylalanine ammonia lyase (PAL). Both processes also have very similar action spectra, peaking at about 290 nm. Yatsuhashi et al. (1982) and Hashimoto et al. (1991) studied the effect of different irradiations on anthocyanin accumulation in *Sorghum*. It was found that irradiations that would create the far-red absorbing form of phytochrome could induce anthocyanin accumulation, and that this effect could be reversed by far-red light. However, there was also a phytochrome-independent induction, with an action spectrum confined to the ultraviolet region and peaking at 290 nm. In the same plant material, a separate damaging effect of ultraviolet radiation, with an action spectrum rising almost monotonically to below 260 nm, was seen.

As mentioned above, that damaging ultraviolet radiation effects usually have action peaks at shorter wavelengths than 280 nm. This is indicative of photon absorption in DNA. However, photon absorption in DNA can also result in the accumulation of protective pigments. In *Phaseolus vulgaris* leaves, accumulation of a protecting pigment (in this case the isoflavonoid coumestrol) has an action spectrum rising steeply and monotonically from 300 to 254 nm (Beggs et al., 1985).

It is probably this kind of response that is mediated by the *cis*-regulatory elements in another leguminous plant (soybean) studied by Wingender (1990).

Up- or downregulation of various genes in response to radiation damage to DNA is known from animal systems and probably common also in plants, and therefore it is natural that a section on ultraviolet radiation changes in DNA (see below) is included also in this chapter dealing mainly with regulation. Because so little spectral information is available, it is in most cases not known whether an observed effect occurs via photon absorption in DNA, in the 290 nm-photoreceptor, or in some other target. Examples of gene regulations in which absorption in DNA is a likely mechanism are the activations of genes coding for stress proteins, and the down-regulations of genes coding for proteins involved in photosynthesis (Strid et al., 1994).

Inhibition of stem extension by ultraviolet-B radiation, which has proven to be of great importance in the competition between plants (Barnes et al. 1988, 1990) also clearly is a photomorphogenetic response rather than sign of damage. Evidence for this is that it is not necessarily accompanied by decreased dry matter production (Ballaré et al., 1991), and that administration of gibberellin can reverse the inhibition (Ballaré et al., 1991). It is not certain that, in the initial phase, the effect proceeds via gene regulation: it is more rapid (noticeable within about 3 h) than the regulation of pigment production (Ballaré et al., 1995b). Perhaps the situation is analogous with that for phytochrome, which regulates genes, but also has an immediate effect on stem elongation which does not proceed via gene regulation, and the hormone auxin, which also has such a dual action mechanism.

The rapid UV-B specific inhibition of stem elongation has an action spectrum, peaking at about 295 nm (Ballaré et al., 1995a), which is in agreement with what has been found for pigment induction (Wellmann, 1975; Yatsuhashi et al., 1982; Beggs and Wellmann, 1985; Hashimoto et al., 1991; Beggs and Wellmann, 1994). Thus there are now action spectra available for various processes and plants pointing to a common type of UV-B photoreceptor. In Fig. 1 I have collected action spectra from the literature, which can be interpreted as signatures of a specific UV-B photoreceptor. On the whole this series of spectra indicates a rather uniform type of receptor with peak efficiency at about 290 nm. Other action spectra of a similar appearance are those obtained by Hsiao and Björn (1982) for carotenogenesis in the fungus *Verticillium agaricinum,* by Kumagai (1983) for induction of sporulation in the ascomycetes *Alternaria tomato* and *Helminthosporium oryzae*, and by Ng et al. (1964) for anthocyanin formation in *Spirodela oligorhiza.* The latter action spectrum has a peak also in the red region, which could indicate that in this case phytochrome could mediate the same effect as the UV-B receptor.

4. The Relation Between the UV-B Receptor and Phytochrome

As mentioned above, UV-B responses are in some cases similar to phytochrome mediated responses; this is true for instance for inhibition of stem extension and induction of anthocyanin formation. In other cases it is necessary that the far-red absorbing form of phytochrome (Pfr) is produced, before the effect of UV-B irradiation can be manifested. Under natural conditions this is not evident, since Pfr is always produced under the conditions when a plant is exposed to ultraviolet-B radiation. But under experimental conditions it can be shown that for some responses it is necessary that both the UV-B receptor and phytochrome are activated (e.g., Wellmann, 1971). In some cases it is even necessary that three different photoreceptors (UV-B photoreceptor, UV-A/blue photoreceptor, and phytochrome) become activated before a response can be manifested. This "coaction" has been reviewed in detail by Mohr (1994).

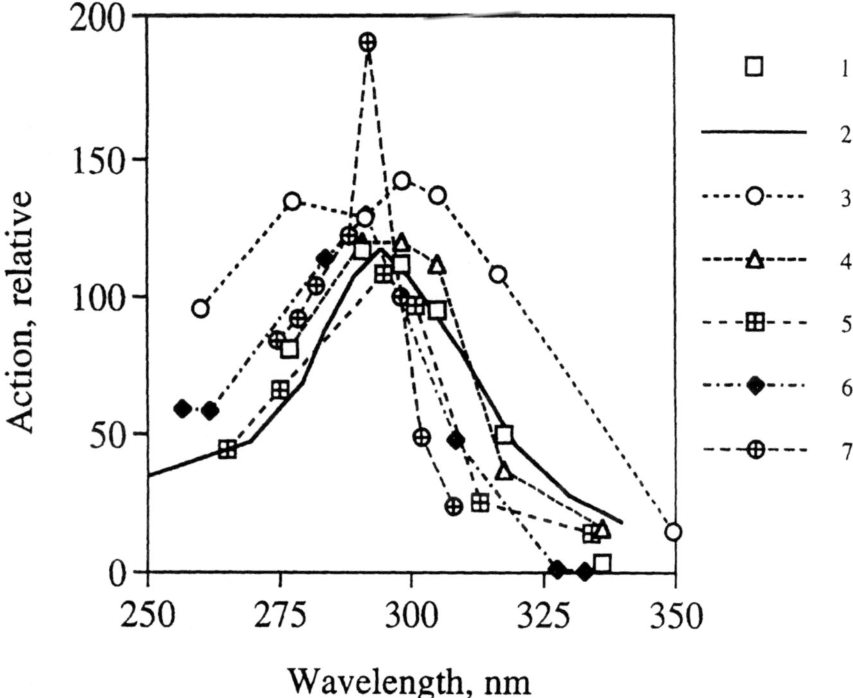

Fig. 1. **Action spectra for UV-B specific regulation in plants replotted from various sources: 1. Induction of synthesis of flavonoid glycosides in cell cultures of parsley (Wellmann, 1975 as cited by Wellmann, 1983). 2. The same as 1 but according to Beggs and Wellmann (1994). 3. Induction of anthocyanin synthesis in maize coleoptiles (Beggs and Wellmann, 1985). 4. The same as 3 but according to Wellmann (1983). 5. Inhibition of hypocotyl elongation in tomato (Balleré et al., 1995). 6. Induction of anthocyanin formation in *Sorghum bicolor* in a background of red light (Yatsuhashi et al., 1982). 7. The same 6 but according to Hashimoto et al. (1991).**

In other cases it has been shown that the UV-B effect is developed without the participation of phytochrome; even mutants lacking one type of phytochrome or almost completely lacking all phytochromes can react in a normal way to UV-B or even be more sensitive than the wild type (Lercari, et al., 1990; Ballaré et al., 1991; Brandt et al., 1995), or the UV-B effect can be induced on a far-red background. Some effects of UV-B on the photosynthetic system resemble effects of decreased red/far-red ratio, i.e. a shade response by phytochrome action (Middleton and Teramura, 1994).

5. UV-B Action and Flavin Photoreceptors

The absorption of cryptochrome type photoreceptors containing flavin extends into the UV-B region, and these photoreceptors can therefore be activated by UV-B. Several authors have proposed that "the" UV-B photoreceptor contains flavin. Iodide ion, which is a quencher of the triplet state of flavins, has been shown to cancel the effect of UV-B on a specific UV-B photoreceptor, as have some other less specific inhibitors (Khare and Guruprasad, 1993; Ballaré et al., 1995a, b). The only that can be said from this is, however, as the latter authors cautiously put it, that it suggests an "involvement of a flavin chromophore in the elicitation". That the UV-B specific photoreceptor itself would contain flavin is in contradiction to the action spectra and the UV-B specificity of some responses.

A possible explanation for the iodide inhibition of UV-B specific responses is that they require "coaction" between a UV-B photoreceptor and a flavin (UV-A/blue) photoreceptor as described by Mohr (1994) and others. In the experiments by Ballaré et al. (1995a, b) any flavin photoreceptor would have been excited by the white background against which the action spectrum was measured. That the flavin is rather tightly coupled in some way to the action of UV-B in this case is evident from the fact that iodide applied after the UV-B had no effect.

Hsiao and Björn (1984) were probably among the first to suggest a role of pterins in biological photoperception. Of photoreceptor chromophores so far discussed in the literature, tetrahydro-6, 7-dimethyl pterin or a related compound is a likely candidate for the 290 nm photoreceptor (Galland and Senger, 1988). The absorption spectrum by Fuller et al. (1971) is compared in Fig. 2 to two of the UV-B action spectra.

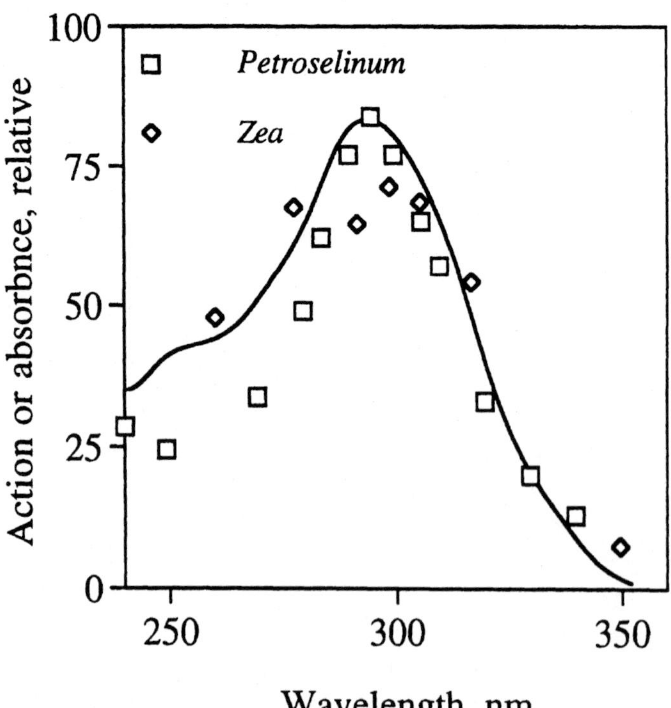

Fig. 2. Comparison of the action spectra for flavonoid formation in parsley, *Petroselinum hortense* (Beggs and Wellmann, 1994) and anthocyanin synthesis in maize, *Zea mays* (Beggs and Wellmann, 1985) with the absorption spectrum of tetrahydro 6, 7-dimethyl pterin (Fuller et al., 1971).

6. Diversity of UV-B Induced Morphogenetic Responses

Although inhibition of shoot extension and induction of pigment accumulation are the two most studied types of UV-B regulation, there are many other responses specifically induced by UV-B, which should be characterized as photoregulatory. UV-B down-regulates many enzymes involved in photosynthetic carbon dioxide assimilation (reviewed by Strid et al., 1994). It can cause cotyledon curling (Wilson and Greenberg, 1993). Very commonly, it causes thickening of leaves, but it can also cause leaves to become thinner (Johanson et al., 1995). UV-B also affects the distribution of assimilates and growth between root and shoot. Such allocation changes and photomorphogenetic effects may have surprising effects on the harvest index (Deckmyn and Impens, 1995).

7. Diversity of UV-B Targets

It is impossible to state at present which of the UV-B effects enumerated in the previous section that are mediated by the specific UV-B receptor with a peak at 290 nm. The best-known plant photoreceptors, phytochrome and the UV-A/blue receptors, both absorb in the UV-B region, but probably this has no importance under daylight conditions, since the fluence rate is so much higher in the UV-A/blue and red/far-red spectral bands than in the UV-B band. Damage by UV-B to Photosystem II, DNA, membranes and other targets may have surprising secondary effects which may in some cases be difficult to separate from photomorphogenetic, photoreceptor mediated, effects. This is no place for a thorough treatment of damage by UV-B, but I shall say a few words about a special aspect of UV-B damage to DNA.

8. DNA as a Target for Ultraviolet Radiation

We shall now return to the damaging effects briefly mentioned above. They can often be demonstrated in the same plant organs and tissues as the inductive effects in which the 290 nm UV-B photoreceptor mediates inductive effects, but their action spectra, as already mentioned, rise steeply monotonically towards shorter wavelengths to 260 nm. Already the action spectra point to DNA as a possible target. In many of the cases further proof of DNA being a target is provided by photoreactivation experiments.

Photoreactivation means that the damaging effect of the ultraviolet radiation can be reversed by visible light or UV-A radiation. In addition to what is described below, it should be noted that DNA can be indirectly acted upon (without being the photon absorber) by free radicals or active oxygen species created by the radiation.

Ultraviolet-B radiation can alter DNA and thereby impair its function in several ways. There are also a number of different processes by which the cells can repair these lesions. The two most frequent types of lesions produced by ultraviolet radiation are the cyclobutane type dimers of pyrimidine bases, and the pyrimidine (6–4) pyrimidine type dimers, commonly referred to as (6–4) photoproducts. Both these lesions (and, as far as is known, no others) can be repaired by light-dependent enzymes called photolyases (a process called photoreactivation), and in strong daylight this type of repair is much more rapid than other repair mechanisms. There is one type of photolyase repairing cyclobutane type lesions, and a separate one repairing (6–4) type lesions.

Photoreactivation was first discovered as prevention by subsequent visible light of the browning of banana skin when the cells had been exposed to ultraviolet radiation (Hausser and v. Oehmcke, 1933). Both in the banana skin (unpublished observations by the present author) and in many other cases, the damaging effect of ultraviolet radiation can be *completely* reversed by visible light as long as the ultraviolet fluence is not too large. This shows that lesions other that the two kinds of photoreactivable damage (i.e. other than the two kinds of DNA dimer formation) are of minor physiological importance, but it does not show which of the two dimer types that is most important physiologically, neither can this be deduced from the relative frequency of the two types of lesion determined by molecular techniques. Cyclobutane type dimers have in the past often been assumed to be the most important UV-B induced type of DNA lesion. I shall try to show here that this view is highly questionable. Some reasons for this statement have already been given by Björn (1996), but I shall add some further arguments here.

When judging the relative importance of different types of DNA lesions in plants, we may draw upon a huge treasure of information collected for mammals and microorganisms. Most likely the basic damage processes are the same, although the rates of rapair processes may differ.

DNA damage may be important for three different reasons: (1) it may decrease replication and transcription and thereby decrease protein synthesis, (2) it may cause mutations, and (3) it may up- or downregulate certain genes. The deleterious effects of somatic mutations are probably less important in plants than in man, where some of them may cause cancer.

For human fibroblasts it has been shown (Petit-Frère et al., 1996) that inhibition of replication and transcription is dominated by (6–4) photoproduct formation rather than by cyclobutane dimer formation. In this investigation the different types of pyrimidine base combinations (TT, TC, CT or CC) were not considered.

As for mutagenesis in *Escherichia coli,* Horsefall and Lawrence (1994) have shown that the TC (6–4) adduct is less mutagenic that the TT (6–4) adduct, but still more mutagenic that the cyclobutane dimers. The TC Dewar photoisomer, which is specifically formed under UV-B irradiation rather than UV-C had enhanced mutagenicity. The TC (6–4) adduct is the most abundant of the (6–4) lesions.

I suspect that some of the earlier investigations have erronously mistaken (6–4) photolesions for cyclobutane dimers, as they have used photoreactivation as an indicator of cyclobutane dimers, not realizing that (6–4) photolesions can also be photoreactivated. This has given a false impression of cyclobutane dimer importance.

Furthermore, earlier estimates of the relative frequencies of cyclobutane dimers and (6–4) photoproducts seem to be in error for other reasons, too (Cadet et al., 1992). It is likely that both types of lesion are about equally frequent, but (6–4) photoproduct formation more deleterious. A contributing reason for this is that (6–4) photoproduct formation is confined to actively transcribed regions, while cyclobutane type dimers are formed all over the genome (Mitchell et al., 1993).

The only investigation on plants relevant to this question is that of Britt et al. (1993), who found that an *Arabidopsis* mutant deficient in repair of (6–4) photoproducts exhibited increased sensitivity to ultraviolet radiation.

One has to consider the whole radiation environment to judge the relative importance of different lesions. After (6–4) photoproducts have been photoconverted to Dewar isomers, they can no longer be photorepaired. The action spectrum for this photoconversion peaks at 320 nm, and has a tail extending beyond 340 nm (Mitchell and Rosenstein, 1987), and under daylight conditions the fluence rate in the spectral band causing (6–4) photoproduct isomerisation is therefore many times higher than that causing initial DNA lesions. This fact increases the suspicion that laboratory experiments have been biased in the sense that importance of cyclobutane dimers has been exaggerated.

My conclusion is that more attention should be paid to the formation of (6–4) photoproducts, and less to cyclobutane type dimers.

9. Influence of Stratospheric Ozone Depletion on UV-B Action

There has been much concern about what impact increased UV-B due to stratospheric ozone depletion might have on biological processes. To which extent the effects of ultraviolet radiation will be affected by stratospheric ozone depletion and the concomitant change of the spectral fluence rate of daylight depends on the action spectra for the various processes. We shall make here a comparison between a process mediated by the 290 nm-photoreceptor and DNA related processes. As an example of the former, we pick the UV-induction of flavonoid accumulation so ably studied by Wellmann and coworkers. We convolute the spectrum for this process by the daylight spectrum, either under 300 dobson units of ozone, or under an ozone layer with a 10% depletion, i.e. under 270 dobson units. The relative areas under these convolution graphs correspond to the impact of the radiation in the two cases. The outcome is that 10% ozone depletion results in 4.2% more action on the photoprocess. This process then has what we call a radiation amplification factor of 0.42. This value was obtained using the more recent values of Beggs and Wellmann (1994). If we use the older values for the same process reported by Wellmann (1975) we obtain the radiation amplification factor 0.58. In both cases the calculations were carried out for summer daylight in Lund, southern Sweden (15 June at noon, 55.7°N, no clouds, solar elevation 57.6° above the horizon). The calculation was carried out using a modification of the ultraviolet daylight program described by Björn and Murphy (1985). The slight discrepancy in radiation amplification factor using two sets of data for the same process points to a common difficulty: The long wavelength tails of action spectra are in many cases not known with sufficient accuracy for reliable factor calculations. Because the convoluted spectra do not approach zero within the UV-B range, the true radiation amplification factor is even lower than 0.42, which was obtained by terminating the integration at 340 nm.

In any case, it is certain that the radiation amplification factor for processes mediated by the 290 nm UV-B photoreceptors is very low compared to processes dependent on photon absorption in DNA. For "naked" DNA the radiation amplification factor is close to 2 (Madronich et al. 1991), and it can be somewhat lower *in vivo* due to screening pigments. It can therefore be concluded that stratospheric ozone depletion has a much larger importance for processes dependent on DNA photochemistry, and very little importance for processes mediated by the 290 nm UV-B photoreceptor.

References

Ahmad, M. and Cashmore, A.R. 1993. *HY4* gene of *A. thaliana* encodes a protein with the characteristics of a blue-light photoreceptor. *Nature* **366:** 162–166.

Ballaré, C.L., Barnes, P.W. and Kendrick, R.D. 1991. Photomorphogenic effects of UV-B radiation on hypocotyl elongation in wild type and stable-phytochrome-deficient mutant seedlings of cucumber. *Physiol. Plant.* **83:** 652–658.

Ballaré, C.L., Barnes, P.W. and Flint, S.D. 1995a. Inhibition of hypocotyl elongation by ultraviolet-B radiation in de-etiolating tomato seedlings. I. The photoreceptor. *Physiol. Plant.* **93:** 584–592.

Ballaré, C.L., Barnes, P.W., Flint, S.D. and Price, S. 1995b. Inhibition of hypocotyl elongation by ultraviolet-B radiation in de-etiolating tomato seedlings. II. Time-course, comparison with flavonoid responses and adaptive significance. *Physiol. Plant.* **93:** 593–601.

Barnes, P.W., Jordan, P.W., Gold, W.G., Flint, S.D. and Caldwell, M.M. 1988. Competition, morphology, and canopy structure in wheat (*Triticum aestivum*) and wild oat (*Avena fatua*) exposed to enhanced ultraviolet-B radiation. *Func. Ecology* **2:** 319–330.

Barnes, P.W., Flint, S.D. and Caldwell, M.M. 1990. Morphological responses of crop and weed species of different growth forms to ultraviolet-B radiation. *Am. J. Bot.* **77:** 1354–1360.

Beggs, C.J. and Wellmann, E. 1985. Analysis of light-controlled anthocyanin formation in coleoptiles of *Zea mays* L.: The role of UV-B, blue, red and far-red light. *Photochem. Photobiol.* **41:** 481–486

Beggs, C.J. and Wellmann, E. 1994. Photocontrol of flavonoid synthesis. In

Photomorphogenesis in plants (Kendrick, R.E. and Kronenburg, G.H.M., eds), 2nd ed., pp. 733–751. Kluwer Academic Publishers, Dordrecht.

Beggs, C.J., Stolzer-Jehle, A. and Wellmann, E. 1985. Isoflavonoid formation as an indicator of UV-stress in bean (*Phaseolus vulgaris* L.) leaves. The significance of photorepair in assessing potential damage by increased solar UV-B radiation. *Plant Physiol.* **79:** 630–634.

Björn, L.O. 1996. Effects of ozone depletion and increased UV-B on terrestrial ecosystems. *Int. J. Environ. Stud.* **51/3(A):** 217–243.

Björn, L.O. and Murphy, T.M. 1985. Computer calculation of solar ultraviolet radiation at ground level. *Physiol. Vég.* **23:** 555–561.

Brandt, K., Gianni, A. and Lercari, B. 1995. Photomorphogenic responses to UV radiation III: A comparative study of UVB effects on anthocyanin and flavonoid accumulation in wild-type and aurea mutant of tomato (*Lycopersicon esculentum* Mill.). *Photochem. Photobiol.* **62:** 1081–1087.

Britt, A.B., Chen, J.J., Wykoff, D. and Mitchell, D.L. 1993. A UV-sensitive mutant of *Arabidopsis* defective in the repair of pyrimidine pyrimidone (6–4) dimers. *Science* **261:** 1571–1574.

Cadet, J., Anselmino, C., Douki, T. and Voituriez, L. 1992. Photochemistry of nucleic acids in cells. *J. Photochem. Photobiol. B.: Biol.* **15:** 277–298.

Caldwell, M.M., Teramura, A.H., Tevini, M., Bornman, J.F., Björn, L.O. and Kulandaivelu, G. 1995. Effects of increased solar ultraviolet radiation on terrestrial plants. *Ambio* **24:** 166–173.

Christie, J.M. and Jenkins, G.I. 1996. Distinct UV-B and UV-A/blue light signal transduction pathways induce chalcone synthase gene expression in *Arabidopsis* cells. *Plant Cell* **8:** 1555–1567.

Deckmyn, G. and Impens, I. 1995. UV-B increases the harvest index of bean (*Phaseolus vulgaris* L.). *Plant Cell Environ.* **18:** 1426–1433.

Fuller, R.C., Kidder, D.W., Nugent, N.A., Dewey, V.C. and Rigopoulos, N. 1971. The association and activities of pteridines in photosynthetic systems. *Photochem. Photobiol.* **14:** 359–371.

Galland, P. and Senger, H. 1988. The role of pterins in the photoreception and metabolism of plants. *Photochem. Photobiol.* **48:** 811–820.

Hashimoto, T., Shichijo, C. and Yatsuhashi, H. 1991. Ultraviolet action spectra for the induction and inhibition of anthocyanin synthesis in broom sorghum seedlings. *J. Photochem. Photobiol. B: Biol.* **11:** 353–363.

Hausser, K.W. and v Oehmcke, H.V. 1933. Lichtbräunung an Fruchtschalen. *Strahlentherapie* **48:** 223–229.

Horsefall, M. and Lawrence, C.W. 1994. Accuracy of replication past the T-C (6–4) adduct. *J. Mol. Biol.* **235:** 465–471.

Hsiao, K.C. and Björn, L.O. 1982. Aspects of photoinduction of carotenogenesis is the fungus *Verticillium agaricinum*. *Physiol. Plant.* **54:** 235–238.

Hsiao, K.C. and Björn, G.S. 1984. A possible photoreceptor pigment for light-induced carotenogenesis in *Verticillinum agaricium*. *Photochem. Photobiol.* 39 (suppl.) 16S

Jenkins, G.I., Christie, J.M., Fuglevand, G., Long, J.C. and Jackson, J.A. 1995. Plant responses to UV and blue light: biochemical and genetic approaches. *Plant Science* **112:** 117–138.

Johanson, U., Gehrke, C., Björn, L.O. and Callaghan, T.V. 1995. The effects of enhanced UV-B radiation on the growth of dwarf shrubs in a subarctic heathland. *Funct. Ecology* **9:** 713–719.

Khare, M. and Guruprasad, K.N. 1993. UV-B-induced anthocyanin synthesis in maize regulated by FMN and inhibition of FMN photoreactions. *Plant Sci.* **91:** 1–5.

Kumagai, T. 1983. Action spectra for blue and near ultraviolet reversible photoreaction in the induction of fungal conidiation. *Physiol. Plant.* **57:** 468–471.

Lercari, B., Sodi, F. and Lipucci di Paola, M. 1990. Photomorphogenic responses to UV radiation: involvement of phytochrome and UV photoreceptors in the control of hypocotyl elongation in *Lycopersicon esculentum. Physiol. Plant.* **79:** 668–672.

Madronich, S., Björn, L.O., Ilyas, M. and Caldwell, M.M. 1991: Changes in biologically active ultraviolet radiation reaching the earth's surface. In Environmental effects panel report, 1991 update (van der Leun, J. and Tevini, M., eds), Chapter 1. United Nations Environment Programme, Nairobi 1991.

Middleton, E.M. and Teramura, A.H. 1994. Understanding photosynthesis, pigment and growth responses induced by UV-B and UV-A irradiances. *Photochem. Photobiol.* **60:** 38–45.

Mitchell, D.L. and Rosenstein, B.S. 1987. The use of specific radioimmunoassays to determine action spectra for the photolysis of (6–4) photoproducts. *Photochem. Photobiol.* **45:** 781–786.

Mitchell, D.L., Pfeifer, G.P., Taylor, J.-S., Zdzienicka, M.Z. and Nikaido, O. 1993. Biological role of (6–4) photoproducts and cyclobutane dimers. In Frontiers of Photobiology, Proc. 11th Int. Congr. Photobiology (Shima, A., Ichahashi, M., Fujiwara, Y. and Takebe, H., eds), pp. 337–344. Elsevier, Amsterdam.

Mohr, H. 1994. Coaction between pigment systems. *In* Photomorphogenesis in Plants, 2nd ed. (Kendrick, R.E. and Kronenberg, G.H.M., eds), pp. 353–373. Kluwer Academic Publ., Dordrecht.

Ng, Y.L., Thimann, K.V. and Gordon, S.A. 1964. The biogenesis of anthocyanin. X. The action spectrum for anthocyanin formation in *Spirodela oligorhiza. Arch. Biochem. Biophys.* **107:** 550–558.

Petit-Frère, C., Clingen, P.H., Arlett, C.F. and Green, M.H.L. 1996. Inhibition of RNA and DNA synthesis in UV-irradiated normal human fibroblasts is correlated with pyrimidine (6–4) pyrimidone photoproduct formation. *Mutation Res.* **354:** 87–94.

Rozema, J., van de Staaij, J., Björn, L.O. and Caldwell, M.M. 1997. UV-B as an environmental factor in plant life: stress and regulation. *TREE* **12:** 22–28.

Strid, A., Chow, W.S. and Anderson, J.M. 1994. UV-B damage and protection at the molecular level in plants. *Photosynth. Res.* **39:** 475–489.

Wellmann, E. 1971. Phytochrome mediated flavone glycoside synthesis in cell suspension cultures of *Petroselinum hortense* D after pre-irradiation with UV light. *Planta* **101:** 283–286.

Wellmann, E. 1975. Der Einfluss physiologischer UV-Dosen auf Wachstum und Pigmentierung von Umbelliferenkeimlingen. In Industrieller Pflanzenbau (Bancher, E., ed.), pp. 229–239. Tech. Univ. Wien Selbstverlag.

Wellmann, E. 1983. UV radiation in Photomorphogenesis. *In* Shropshire Jr, W. and Mohr, H. (eds), *Enc. Plant Physiol.,* New Series **16B:** 745–756. Springer Verlag.

Wilson, M.I. and Greenberg, B.M. 1993. Specificity and photomorphogenetic nature of ultraviolet-B induced cotyledon curling in *Brassica napus L. Plant Physiol.* **102:** 671–677.

Wingender, E. 1990. Transcription regulating proteins and their recognition sequences. *Crit. Rev. Eukaryotic Gene Expression* **1:** 11–48.

Yatsuhashi, H., Hashimoto, T. and Shimizu, S. 1982. Ultraviolet action spectrum for anthocyanin formation in broom *Sorghum* first internodes. *Plant Physiol.* **70:** 735–741.

Yu, S.G. and Björn, L.O. 1997. Effects of UVB radiation on light-dependent and light-independent protein phosphorylation in thylakoid proteins. *J. Photochem. Photobiol. B:* Biology (in press).

PART C

Gene Regulation and Photomorphogenesis

Section VIII: Photoresponses and Molecular Mechanism

Concepts in Photobiology: Photosynthesis and Photomorphogenesis
G.S. Singhal, G. Renger, S.K. Sopory, K-D. Irrgang and Govindjee (Eds)

28. Photomorphogenesis in Lower Plants

Jon Hughes and Elmar Hartmann
Pflanzenphysiologie, Freie Universität Berlin,
Königin-Luise-Str. 12–16, D-14195 Berlin, Germany

Summary

Lower plants and recently even cyanobacteria offer unique experimental possibilities and insights into photomorphogenesis at the genetic, biochemical, physiological and structural levels. Here we offer an overview of light-regulated development in ferns and mosses, touching on the photoresponses of other lower organisms, including prokaryotes. Additionally we discuss three related topics of special interest in photomorphogenesis—membrane-associated phytochrome, light-regulated axis and polarity development, and primitive phytochromes.

1. Introduction

In this chapter we describe and discuss some of the photomorphogenic concepts stemming from the study of phytochrome in lower organisms, rather than presenting an all-encompassing review of the effects of light on growth and development (see Hartmann and Jenkins, 1984; Wada and Kadota, 1989; Senger and Schmidt, 1994; Wada and Sugai, 1994, for such overviews). In keeping with this we also ignore some of the restrictions deriving from arbitrary definitions and terminology. For example we include ferns, fungi, slime-moulds and even some prokaryotes alongside the mosses and algae with which we are principally concerned—this motley assemblage of groups could never serve to define "the lower plant". Similarly we include within "photomorphogenesis" (correctly the light-regulated development of form) also readily-reversible intracellular movements such as those shown by chloroplasts, as well as biochemical and molecular biological insights. We do not, on the other hand, discuss the phototaxes in *Chlamydomonas* and *Euglena*. This selective approach helps us bring related ideas together.

We have chosen to emphasise phytochrome-regulated processes rather than those associated with blue light receptor systems. Although the latter are important and widespread amongst lower plants, the recent cloning of the CRY photoreceptor genes in *Arabidopsis* in Cashmore's laboratory (Ahmad and Cashmore, 1993, 1996) has defined the current paradigm here within the realm of the higher plant: exactly the opposite is the case with phytochrome, where lower organisms, especially cyanobacteria, are making revolutionary contributions to the field (Kehoe and Grossman, 1996; Hughes et al., 1996, 1997; Lamparter et al., 1997; Yeh et al., 1997).

2. Structures, Lifecycles and Photobiology

The basic biology of lower plants is less familiar to most people than that of the more advanced seed-bearing plants. This is unfortunate, as especially the mosses contributed greatly to our understanding of genetics and morphogenesis during the first half of the twentieth century, the work of von Wettstein and his colleagues (for example, von Wettstein, 1932) standing alongside that of Dobshansky on the fruit-fly *Drosophila*. In recent years interest in morphogenesis lower plant systems has re-emerged, however, and is the prime focus of this review. A framework for discussing the photomorphogenesis in various groups is thus given below—any appropriate textbook will provide deeper and broader insights into the fascinating biology of these organisms.

2.1 Ferns, Mosses and their Allies

Mosses and liverworts (bryophytes) and ferns are dispersed by spores rather than seeds, and their lifecycles are characterised by the alternation of distinct gametophytic and sporophytic generations. The gametophyte form germinates from the spore as a microscopic green filament (protonema) which gives rise to plantlets or prothalli (gametophores), eventually developing to produce egg and sperm cells (gametes). The sporophyte comprises the form that develops from the fertilised egg, eventually to produce spores, completing the lifecycle. In ferns (as in higher plants) the leafy plant we generally recognise is the sporophyte, while the gametophyte comprises a rather simple, inconspicuous prothallus. Moss plantlets, on the other hand, are gametophytic. Like the prothallus, they develop from protonemata: here the sporophyte is represented only by the spore-bearing structures living more or less parasitically on the gametophyte plantlet.

Not surprisingly, plant photobiology has concentrated on angiosperms, at least ostensibly in the context of agriculture. This is difficult work for a variety of reasons including the complexity and impenetrability of the tissues that make up the body of seed plants. Plant scientists in various disciplines have attempted to avoid these problems by investigating the far simpler gametophyte of lower plants. The fern prothallus and moss leaflet consist in general of a single layer of rather similar, undifferentiated cells, whereas the protonema of many species consists of a linear—although often branched—filament of cells, growing exclusively from the tip cell. Clearly, such easily-accessible material makes life a lot easier for the investigator. Some moss species have further advantages, however. Protonemata (Fig. 1) can be grown indefinitely in sterile culture on nutrient media, even in darkness. The cell walls can be digested away to release huge numbers of naked protoplasts, most of which can regenerate to form normal filaments within a few days (Fig. 2). This is particularly useful for molecular-genetic approaches, as the protoplasts can be persuaded to take up foreign DNA from the medium, leading to transgenic clonal filaments in a small fraction of the time required for the same manipulation of a higher plant (Schaefer et al., 1991; Knight, 1994; Zeidler et al., 1996; Schaefer and Zrÿd, 1997). Regeneration from moss protoplasts has also been introduced as a model system for basic studies of plant morphogenesis (3.2).

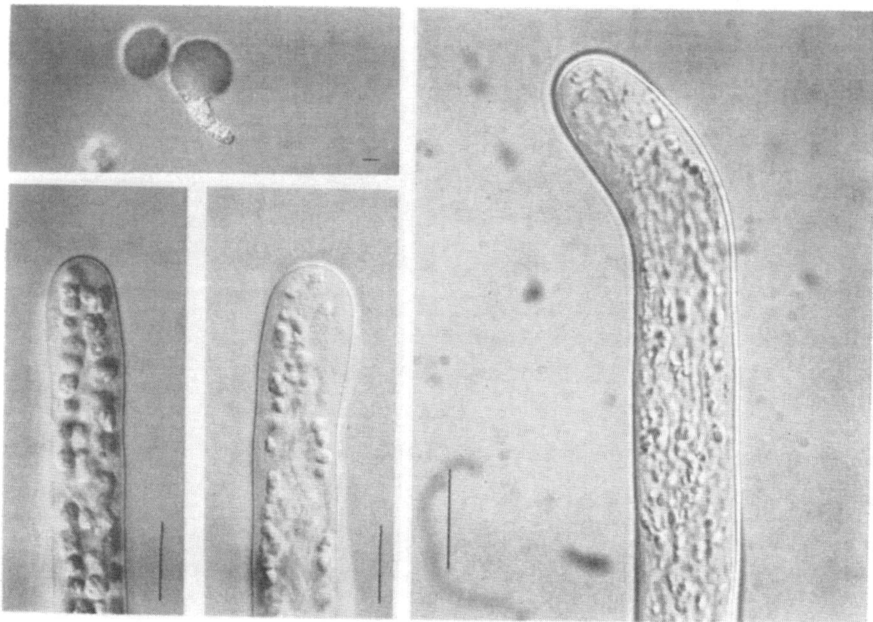

Fig. 1. Spores and protonemata of *Ceratodon purpureus*. Upper left, spore germination. Lower left, chloronemal tip cell rich in chloroplasts. Lower middle, caulonemal tip cell with fewer, smaller chloroplasts. Right, phototropism in dark-grown caulonemal tip cell (irradiated for 1 hr with 10 μmol $m^{-2}s^{-1}$ R from left) (V. Meske, unpublished). Bars: 10 μm.

A wide range of photobiological phenomena have been reported in moss and fern gametophytes. The field has been reviewed extensively by Hartmann and Jenkins (1984), Wada and Kadota (1989) and Wada and Sugai (1994). Spore germination usually requires light. At least in some species, a brief pulse of red light (R) can induce germination, whereas this is inhibited by a subsequent pulse of far-red light (FR), the control point seeming to be an initial cell division within the spore prior to emergence (Furuya et al., 1997). This photoreversibility is clear evidence for involvement of phytochrome, and is analogous to R/FR reversible germination in many higher plant species. R photons are absorbed and integrated as physiologically-active Pfr, forming an exquisitely sensitive light detection system for the spore which otherwise remains dormant in darkness. Clearly it is not sensible

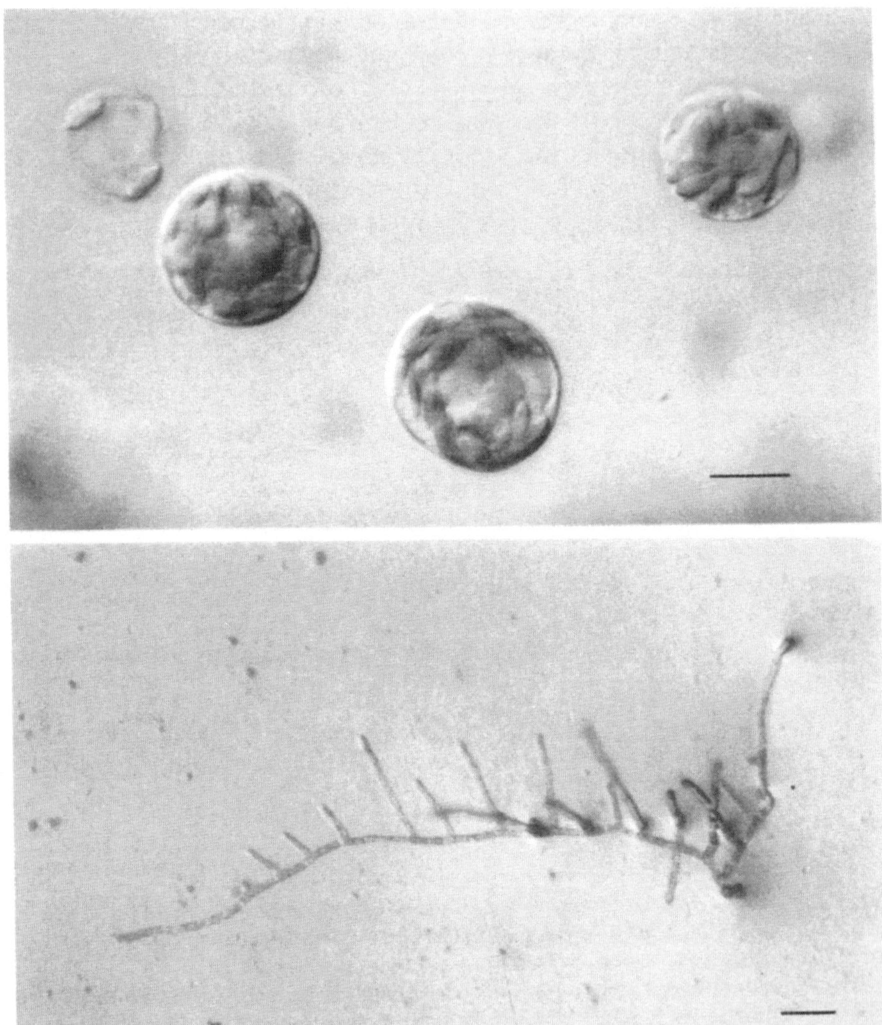

Fig. 2. Protoplast regeneration in *Ceratodon*.
Above, protoplasts released from chloronemal cells by cellulase digestion and allowed to regenerate for 24 hours. The cell near the middle has already begun to regenerate, as apparent from its asymmetry. Bar: 10 μm. Below, protonemata growing from a regenerated protoplast. Bar: 100 μm.

for a photosynthetic plant to attempt to grow unless light is available. Less clear is the ecological significance of FR inhibition of spore germination. In the case of seeds it is thought that R/FR antagonism is connected with the detection of chlorophyll-bearing leaves and hence competitors in the immediate environment (see Smith, 1994). Although this explanation seems difficult to reconcile with the lifestyle of ferns and mosses commonly growing in shady places, it might serve to encourage germination in early Spring prior to budbreak in woodland, whilst suppressing germination later.

Conversely, not all lower plants grow in shade; for example, the common mosses *Ceratodon* and *Funaria* are adapted to particularly exposed environments.

In bryophytes and many ferns, the emerging protonemal filament consists of chloroplast-rich cells, all derived from an apical cell. The intercalary cells do not elongate, growth of the filament occurring exclusively by tip growth of the apical cell. In some ferns, notably *Ceratopteris*, however, the emerging protonema is multiseriate, consisting of several parallel files of cells. Having germinated, it is necessary that protonemal growth be guided towards light, and indeed a particularly interesting set of physiological responses ensure that this is achieved: one of these is described in some detail below (3.1.1). In many species a thin rhizoid cell emerges from the spore soon after the protonema: interestingly, the rhizoid is hardly sensitive to light, tending instead to grow downwards (positive gravitropism).

In ferns such as *Adiantum* (see Wada and Sugai, 1994), the protonema is ephemeral, growing linearly for a few days before differentiating to form the prothallus. In R cell division is inhibited, the apical cell growing to a considerable length (up to 1 mm) on appropriate media, whereas in darkness or blue light (B) normal mitosis occurs leading to the formation of a linear (one-dimensional) protonema. If R is given from a particular direction, the tip cell orientates its growth towards the light—this phenomenon is called positive phototropism and is associated with a reorganisation of the actin microfilament cytoskeleton. Phototropic bending can be induced by a brief pulse of R: a subsequent FR pulse cancels the effect, implying that phytochrome is the photoreceptor involved. Longitudinal cell divisions soon take place in B, however, giving rise to the prothallus. This small (ca. 0.5 cm), often heart-shaped disc comprises a single cell layer (two dimensional) with a small number of rhizoids which grow downwards into the medium beneath, absorbing water and minerals. Its development shows interesting absorbing responses to polarised light, analogous to the dichroic effects described in 3.1.2. Furthermore, the simple structure of the prothallus makes it an ideal organ for microscopic investigation and has been extensively studied in relation to intracellular chloroplast movement (3.1.3).

Protonemata of many moss species can, on the other hand, be cultivated indefinitely. The filament emerging from the spore is termed a chloronema on account of the numerous chloroplasts each cell contains (Fig. 1). The chloronemal cells themselves are only moderately elongated, about three to five times as long as broad, and have cross-walls at right angles to the filament axis. Usually after a few divisions the apical cell either stops growing or gives rise to a caulonema in which the intercalary cells typically contain fewer chloroplasts, are strongly elongated, have diagonal cross-walls, and branch readily (Fig. 3). They are often considered "foraging" organs as they seem to be produced especially in response to low-nutrient conditions and usually show positive phototropism: when nitrogen levels

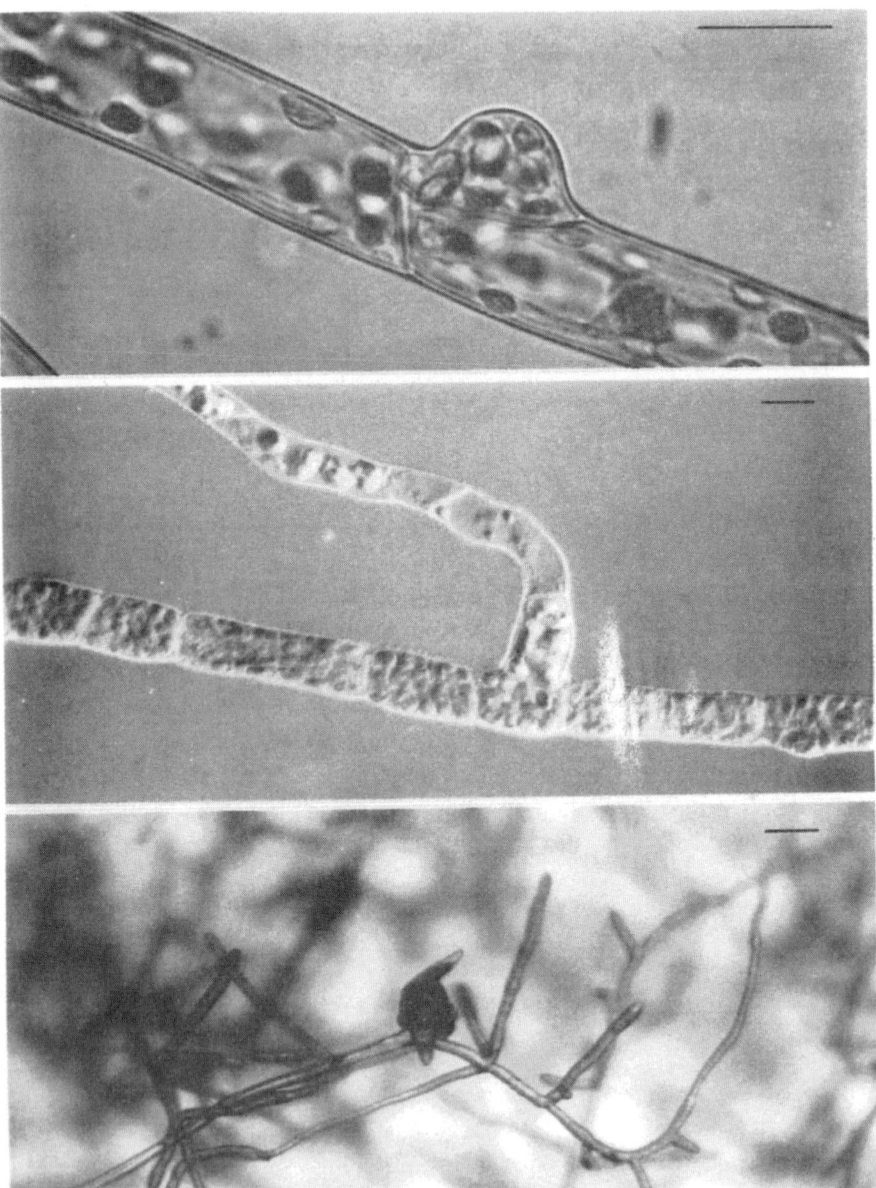

Fig. 3. Bud development in *Physcomitrella patens*.
Above, Caulonemal side branch initial; Middle, Caulonema development from chloronema. Bar: 10 μm. Below, Bud development from caulonemal filament. Bar: 100 μm.

are raised they often revert to (secondary) chloronemata. Also consistent with this role, caulonemata grow particularly rapidly in darkness although they very seldom branch under these conditions: by contrast, chloronemata do not grow at all in darkness, seeming to require light for cell division.

In *Ceratodon,* where the cytological distinction between chloro-and caulonemata is in any case rather unclear, chloronemata can at least complete the change-over to caulonemata when placed in darkness. The physiological differences between caulonema and chloronema are an interesting model system for cellular differentiation in plants (Schumaker and Dietrich, 1997; Cove et al., 1978).

The ease with which *Ceratodon* produces caulonemata in darkness is a useful feature for photobiological studies such as phototropism (Hartmann et al., 1983; 3.1.1). Dark-adapted filaments show a strong tendency to grow upwards, away from the direction of gravity (negative gravitropism), analogous to etiolation in angiosperm seedlings placed in darkness, and doubtless with the same aim of helping the buried plant find its way upwards towards daylight. We and others have isolated *Ceratodon* mutants in which this response is exactly reversed, the filaments growing downwards in darkness—hence the name *wwr* for *wrong-way response* (Wagner et al., 1997): the *gtrC* mutants in *Physcomitrella* behave similarly (Knight et al., 1991). Interestingly, the *wwr* phenotype in darkness resembles the effect of long-term B on protonemata in this species (Lamparter et al., 1998a). Dark-grown caulonema tip-cells show almost no phototropism in B but R elicits a strong bending response within minutes of irradiation (Figs. 1 and 6A), correlated with rearrangement of the actin microfilaments, in turn probably related to depolarisation of the plasma membrane and redistribution of free Ca^{2+} (Meske and Hartmann, 1995; Meske et al., 1996; Ermolayeva et al., 1996). Long-term R switches off gravitropism. Both R responses (phototropism and the effect on gravitropism) are mediated by phytochrome, as shown by reversibility studies. This is also illustrated by *ptr* mutants which contain little or no active phytochrome: they show neither response (Lamparter et al., 1996; Esch and Lamparter, 1998). Thus gravitropism in moss filaments provides a particularly nice example of how one stimulus (gravitation) can evoke dramatically different responses (positive, negative or neutral gravitropic bending) in subtly different cell types (rhizoids or protonemata) or genetic backgrounds (wildtype, *ptr* or *wwr*) or under different environmental conditions (B, R or darkness).

The moss plantlet arises from a caulonemal side-branch initial which, instead of growing as a one-dimensional daughter filament, undergoes organised transverse divisions to form a three-dimensional bud (Fig. 3). Bud induction requires light and cytokinin. Under normal conditions, buds develop into green, leafy plantlets typical of the species involved. If transferred to darkness, however, the plantlets develop small, scale-like leaves, becoming pale, spindly, and negatively gravitropic, a classical etiolation response. Liverwort development has been little studied but in general the protonema rapidly differentiates in to a prostrate prothallus. In *Marchantia* at least this too shows phytochrome-mediated etiolation in darkness, the chlorophyll content declining and the prostrate habit being replaced by a strong negative gravitropism.

Fern prothalli show an interesting response to polarized light (see Wada and Sugai, 1994) and have received attention in relation to intracellular chloroplast movements (3.1.3). Growth in darkness leads to an etiolated phenotype in the few species that have been investigated. In *Ceratopteris*, mutants have been reported in which development in total darkness is similar to that of the wildtype in light, analogous to the *cop* and *det* mutants of *Arabidopsis,* some of which involve the phytochrome signal transduction chain (Cooke et al., 1995).

Mature gametophytes produce reproductive archegonia and antheridia, the female and male sexual organs. Once the gametes have fused, the fertilized egg cell (zygote) divides and differentiates into the sporophyte. Whereas in ferns this goes on to form the dominant plant body, in bryophytes the sporophyte remains small, largely dependent on the parent gametophyte plantlet. In both cases, however, it is the sporophyte that finally gives rise to and releases the spores, completing the lifecycle.

The distinction between gametophyte and sporophyte is associated with an important change at the genetic level. The sporophyte derives from a fertilised egg produced by the fusion of two gametophytic cells and thus containing a double set of chromosomes. This diploid condition is typical of higher organisms in general. Spores, on the other hand, are produced by a special meiotic cell division in which the chromosomes are once again separated into single sets following a reshuffle. The spores and the cells of the gametophyte to which they give rise are thus haploid, each cell having only a single copy of each gene. This is convenient for the investigator it allows the effects of single-gene mutations to be observed directly, in contrast to the situation in diploid organisms where such events are generally masked by the continued presence of an unmodified (wildtype) second copy. Hence the haploid moss gametophyte is a powerful instrument for investigating the molecular genetics of photomorphogenesis (3.1) and other basic aspects of plant biology.

2.2 Algae, Fungi and their Allies

The lifecycles and morphologies of these diverse groups are generally simpler than those of the bryophytes outlined above. While they sometimes have great advantages for experimental work, the general conclusions that can be drawn from their study is weakened by their evolutionary distance from agriculturally-important plants (or indeed from animals). Conversely, especially the fungi have attracted great interest on account of their importance in fermentation and biotechnology.

Particularly two species of green alga, *Mougeotia* and *Mesotaenium* (Fig. 4), have attracted the attention of photobiologists (see Kraml, 1994; Haupt and Häder, 1994). Both species are unusual in containing a single, enormous, flattened chloroplast, the orientation of which is actively adjusted according to the light conditions (semantically, such reversible movements are not

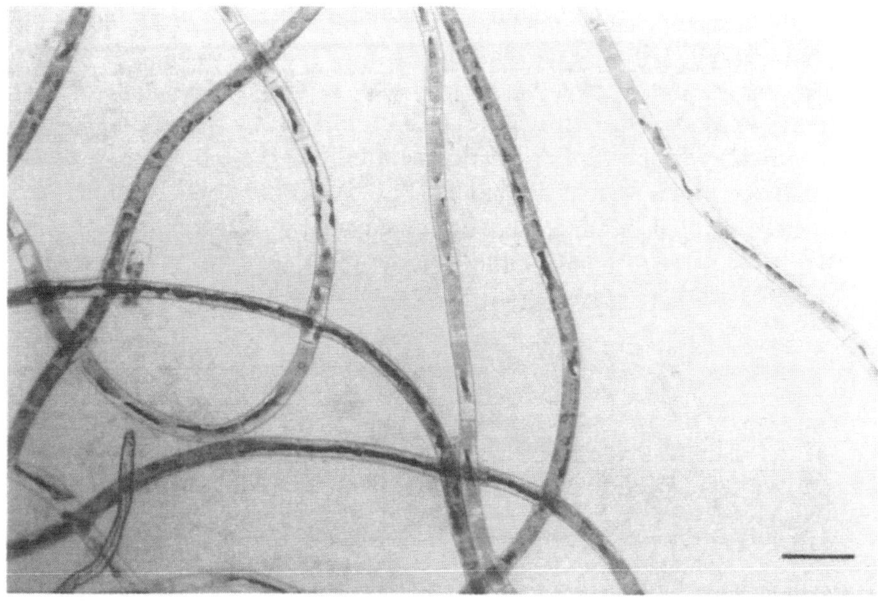

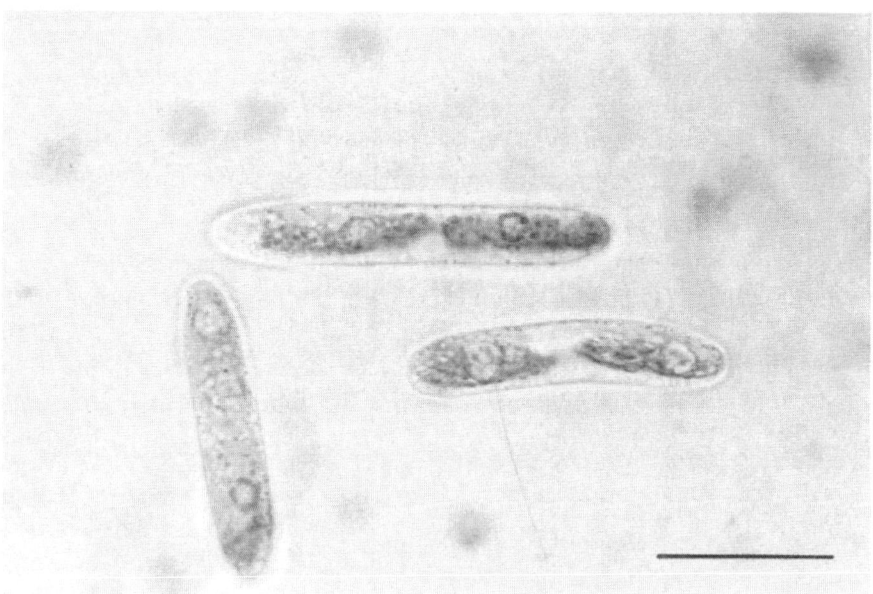

**Fig. 4. Top, *Mougeotia scalaris*. Bar: 100 μm.
Bottom, *Mesotaenium caldarium*. Bar: 10 μm.**

photomorphogenic but photonastic). Actin microfilaments are thought to bring about these movements, somehow under the control of phytochrome. The system has been investigated intensively and lead to paradoxical conclusions in relation to generally accepted ideas regarding the nature of the photoreceptor. Whereas these phenomena certainly represent a "special case", the apparent paradox might well obscure a crucial part of the mechanism of phytochrome action (3.1.3 and 3.1.4).

The marine brown alga *Fucus* has been extensively studied for quite different reasons. The fertilized egg cells in this species are apparently symmetrical, yet their further development follows the setting up of an axis and polarity which is light dependent (Fig. 5). As the eggs are rather large and easily manipulated they have been the center of research into this most fundamental level of plant morphogenesis (3.2).

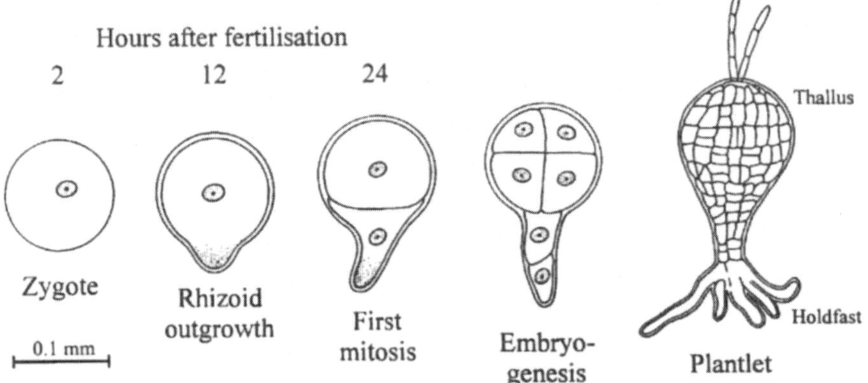

Fig. 5. ***Fucus*** **embryogenesis (modified from Goodner and Quatrano (1993), The Plant Cell 5: 1472, with permission. Copyright American Sociey of Plant Physiologists).**

Fungi are not photosynthetic, but their photoreceptive systems often parallel those known in plants. This is particularly so in relation to B, fruiting in *Coprinus* (see Manachère, 1994) and phototropism of the sporangiophore of *Phycomyces* being cases in point. The latter is one of the most light-sensitive responses known in Nature, reacting to fluence rates well below the limit of human vision. The system has attracted particular interest on account of its simple optics and the availability of mutants. Fungal responses to R are uncommon. Nevertheless, Mooney and Yager (1990) found that conidiospore induction in *Aspergillus nidulans* is light dependent, with R being most effective and its action being reversed by FR in a phytochrome-like manner. The response would have been discovered much earlier had not most laboratories unwittingly been using a mutant strain in which conidia are produced even in darkness (again possibly analogous to the *det/cop* mutations in *Arabidopsis*). The photoreceptor involved has not yet been characterized but, intriguingly, the *brlA* gene thought to be associated with the system has been cloned and the product shown to have similarities to phytochrome (Griffith et al., 1994).

Although most researchers have their pet species for study, perhaps the most congenial for cell biological investigation are the slime-moulds. *Physarum* has been studied in great detail. This strange organism lives as single-celled amoebae or as a relatively huge plasmodium consisting of a naked protoplasmic mass containing many thousands of nuclei. A seemingly phytochrome-like photoreceptor controls the differentiation of the plasmodium into a sporangiophore (Starostzik and Marwan, 1995; 3.3). Slime-moulds permit elegant experiments that would be unthinkable in any other type of organism. For instance, the interactive effects of two experimental treatments and their transduction pathways can be investigated simply by stirring together the pieces of the plasmodia at different time points following the treatments. Furthermore, like the fern and moss gametophytes, *Physarum* amoebae are haploid, allowing easy mutational genetic analysis. Thus *Physarum* might well become important in future phytochrome studies.

2.3 Prokaryotes

Although it can hardly be argued that any prokaryote is a plant, the notion that chloroplasts derive from endosymbiotic cyanobacteria has now become so widely accepted that plant biology has to take aboard this ancient group and learn from the wealth of information provided by microbiology. Particularly significant here is the relatively small genome of prokaryotes: that of *Synechocystis* has already be sequenced in its entirety at the Kazusa Research Centre in Japan (Kaneko et al., 1996) and is accessible via the internet (http://www.kazusa.or.jp).

Cyanobacteria show a variety of responses to light. Species like *Nostoc* and *Anabena* show active phototactic gliding reactions towards light of moderate fluence rates: at high light levels, however, they move away from the light source. Presumably these reactions are adaptive in that the organisms are brought into regions where light levels are optimal for photosynthesis and the danger of photooxidation is minimized. It is generally thought that they are mediated by the photosynthetic system itself, but the wavelengths most effective are very different in different species. In *Phormidium*, for example, the blue-green region is most effective, whereas in *Cylindrospermum* blue and especially red are predominant. Although this might reflect the involvement of radically different photoreceptors, the different screening effects of bulk pigments in different species can be misleading. On the other hand, the action maxima for positive and negative phototaxis in *Anabena* are different, presumably reflecting different primary photoprocesses.

In some species such as *Fremyella* and *Calothrix* the spectral distribution of the light environment leads to quite profound changes in the photosynthetic accessory pigment composition. The levels of phycoerythrin (absorbs blue-green, appears red) increase if the cells are grown in green light, whereas the levels of phycocyanin (absorbs red, appears blue-green) increase under red light. Thus light is absorbed best at wavelengths at which it is most abundant. This phenomenon known as complementary chromatic adaptation (CCA). Although the photoreceptor involved is still not known, a gene-*rcaE*-obviously involved in the adaptation response was recently cloned by genetic rescue in *Fremyella* and shown to code for a protein with similarities to phytochrome (Kehoe and Grossman, 1996). A related gene-*plpA*-was also found in *Synechocystis*, although this species does not usually show CCA (Wilde et al., 1997). The complete genome sequence of *Synechocystis* revealed a handful of other genes with sequence similarities to phytochrome, including one in which the chromophore-binding site, highly conserved in all known phytochromes, was clearly recognizable. In fact, the gene product has turned out to be a *bona fide* R/FR reversible phytochrome, with the added bonus that at least one of the receiver molecules with which it interacts could be identified by a simple inspection of the chromosome map (Hughes et al., 1997; Yeh et al., 1997; Lamparter et al., 1997). The subject of cyanobacterial phytochrome is discussed in 3.3.

Synechocystis is the first photosynthetic organism in which the coding sequences of all components of the photoperception and transduction pathways are known—one has merely to identify them! From the photobiological point of view, however, the fact that *Synechocystis* PCC6803 was chosen for the mammoth sequencing project is rather ironic, as this strain like many other unicellular cyanobacteria displays few of the responses to changes in the light environment known elsewhere in the group. Whereas one of the principle advantages of lower organisms for basic research is their relative simplicity, this reaches the stage in *Synechocystis* PCC6803 that the repertoire of responses to light is so restricted as to be a source of frustration rather than wonder for the photobiologist. This observation reflects our current ignorance of the biology of these organisms, however, rather than implying that they are blind. The photoresponses are there, we just don't yet know where to look. Indeed *Synechocystis* PCC6701 shows beautiful CCA!

The archeon *Halobacterium* has also recently provided fresh insight the context of signal transduction from photoreceptors. This species uses rotating polar flagella to swim in response both to light via photoreceptors and to dissolved amino acids via a chemoreceptor. The systems involved have been studied intensively by Oesterhelt's group at the Max-Planck-Institute in Martinsried. The simplest, bacteriorhodopsin is a light-driven proton pump whose action can interrupt flagella motion when the organism swims out of the light (Bibikov et al., 1993). Two further sensory rhodopsins contribute to photoperception (Krah et al., 1994). One transduces photorepulsion from strong light, whereas the other shows a UV/A/R photoreversibility analogous to that shown by phytochrome in R and FR: UV/A repels, R attracts. This photoreceptor has various structural features similar to those of the visual pigments in the eyes of animals, including humans, whereas its transduction pathway to the flagellar motor might have parallels to that of phytochrome, as discussed in 3.3.

3. Case Studies

As explained above, lower organisms have experimental advantages over higher plants for experimental work, in addition to their inherent interest. In line with this, certain systems have been studied in great detail and have yielded information very relevant to higher plants in which such approaches are impossible. Several such photomorphogenic and related responses are described in case studies below. Perhaps at this point a note of caution is appropriate however. The evolutionary relationship between the currently living descendants of "primitive" types and "derived" higher plants, while permitting that a particular function be conserved, does not prove that it has been: by analogy, modern mosses resemble in some respects higher plants and in many more respects their common moss-like ancestor, but both modern species are the products of hundreds of millions of years of evolutionary divergence, and are not identical.

3.1 Tropisms and Chloroplast Movements: The Paradox of Membrane-associated Phytochrome

Remarkably, despite enormous research effort over nearly half a century, the primary mechanism of plant phytochrome action is still unknown. Almost all biochemical studies imply that phytochromes are cytosolic proteins (but see Lamparter et al., 1992; Sineshchekov et al., 1994). Similarly, antibody localisation of phytochrome *in situ* implies an even distribution in the cytoplasm (see Pratt, 1994). It is important to bear in mind, however, that almost all the data refers to PHYA in etiolated angiosperm tissues, whereas the situation for other members of the phytochrome family, in green tissue, and in non-angiosperms is less well known. Phytochrome sequence data, on the other hand, is much more extensive—none of the phytochromes shows any trace of the hydrophobic structures typical of membrane proteins. Thus phytochrome is photoconverted to its active Pfr form by R, and Pfr converted back to inactive Pr by FR, all in the cytosol. This blissfully harmonious picture is ruined, however, by studies of chloroplast movements in algae and the tropic behaviour of fern and moss protonemata. The data seems to show conclusively that phytochrome is associated with the plasma membrane or structures close to it, and, worse still, that active phytochrome is not simply Pfr.

3.1.1 Phototropism in Ceratodon

Conveniently, caulonemal filaments of the moss *Ceratodon* grow readily in the dark on nutrient media, showing a strong negative gravitropism. Placed on a vertical surface in darkness, thousands of filaments develop within a day of so, and grow uniformly vertically up to 1 mm per day. If R is now given from the side, parallel to the surface on which the filaments are growing, growth is halted almost immediately, and the tip swells noticeably. After a few minutes, growth resumes, but now from the side of the bulged tip towards the source of irradiation. Within an hour, the filament shows a sharp phototropic bend associated with the changed direction of growth

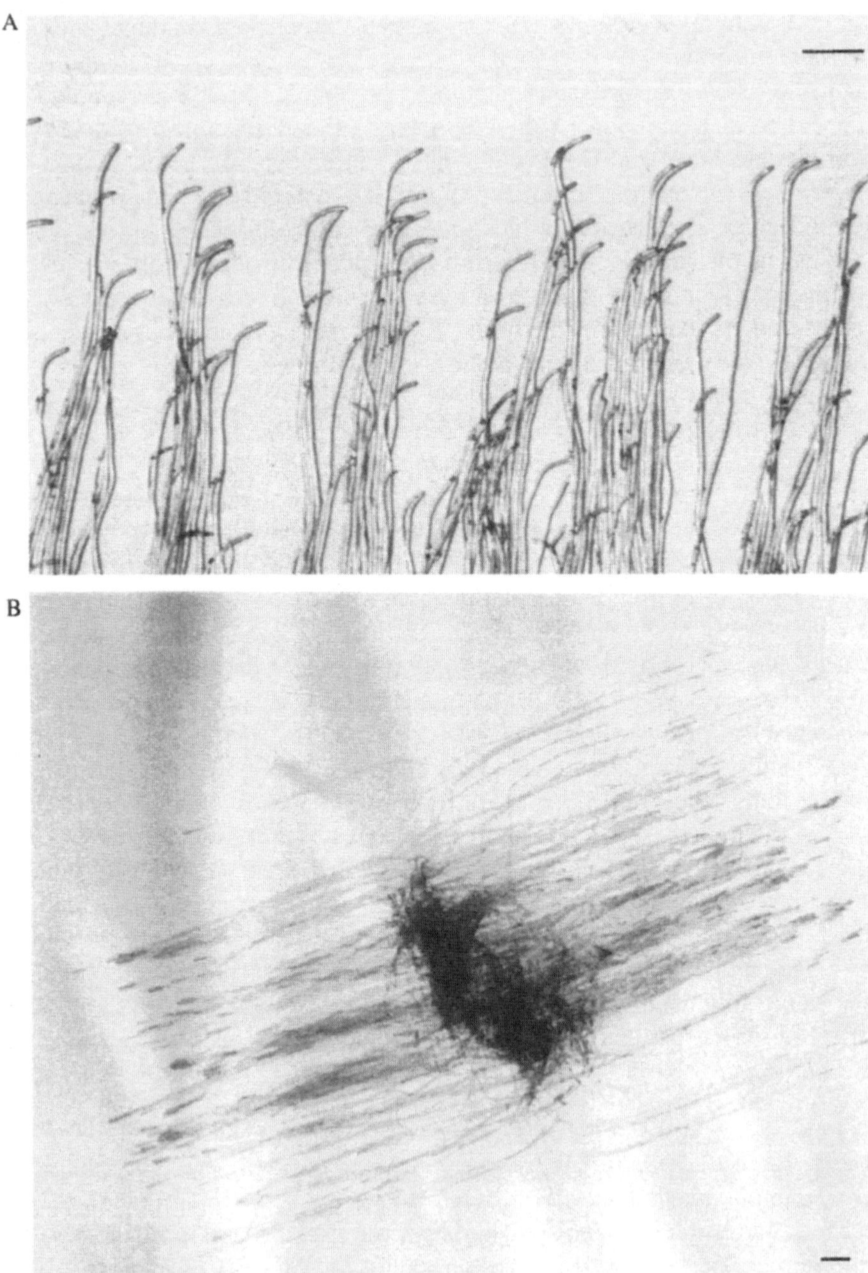

Fig. 6. Tropisms in *Ceratodon* caulonemata. A: Phototropism. Filaments were grown in darkness for 4 days on vertically-orientated agar medium then irradiated unilaterally for 2 hours with 10 μmol m^{-2}s^{-1} R from right. B: Polarotropism. Filaments were grown horizontally for two weeks on agar with polarized 10 μmol m^{-2}s^{-1} R from above. (T. Lamparter, FU Berlin; unpublished). Bars: 100 μm.

(Figs. 1, 6A and 9). Low fluence rates lead to a small degree of bending whereas bright R can induce almost horizontal growth. Continuous R irradiation can be replaced by R pulses whose action is nullified by subsequent FR: such R/FR reversibility implies phytochrome involvement (Hartmann et al., 1983). Further evidence for this comes from mutant strains isolated following UV irradiation of the haploid filaments. Aphototropic mutants have been isolated in our laboratory—in contrast to the wildtype, they contain no spectrophotometrically-detectable phytochrome. If, however, they are fed with appropriate bilin precursors, almost normal phototropism is seen, implying a defect in synthesis of the phytochrome chromophore (Lamparter et al., 1996).

Straightforward as this phenomenon might seem, it is by no means easy to explain.

- As there are no cellular structures present that might act as an eye to detect the direction of illumination, some sort of photodynamic gradient across the tip must be established in order to elicit a directional phototropic response. For our purposes the direction of this gradient is unimportant: light might be attenuated by scattering or absorption on its way through the apex, or the cylindrical cell might act as a lens, focusing light on the far side of the cell (see Fig. 7; Kraml, 1994). As it happens, microbeam irradiation of one side of the protonemal apex induces bending in that direction, showing that phototropism here follows a positive photodynamic gradient.

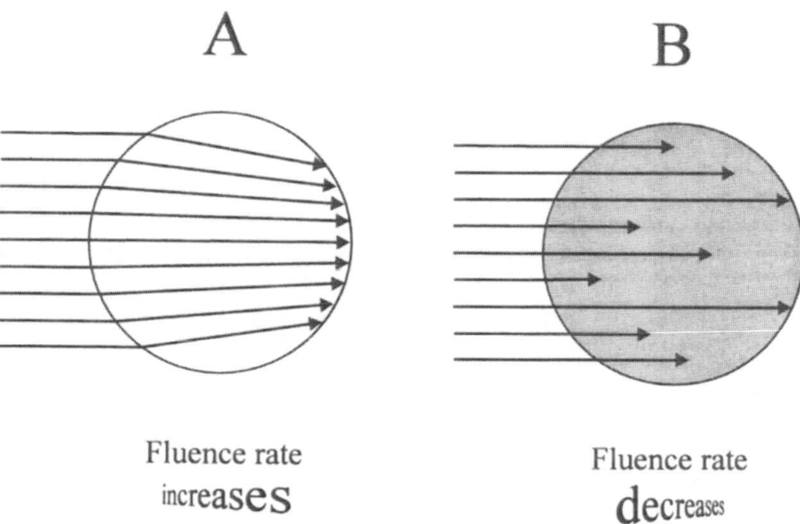

Fig. 7. Fluence rate gradients within a hypothetical cell.
If the refractive index of the cell is higher than that of the surrounding medium, light will be concentrated towards the distal side (A, lens effect), whereas if the cell contains pigments or scattering material, light will be attenuated (B).

By way of contrast, optical studies of the analogous B-induced phototropism in the *Phycomyces* sporangiophore imply a lens effect.

- The central question is, how can phytochrome detect *any* gradient? Photoconversion of Pr to Pfr is indeed driven by R, but unlike most other pigments in which the activated molecule rapidly falls back to its ground state, studies of purified phytochrome show that Pfr is rather stable thermodynamically. Hence Pfr rapidly accumulates to a maximum level (photoequilibrium) at which R photoconverts it back to Pr at the same rate as Pfr is formed. In this system, the fluence rate of R merely defines how quickly photoequilibrium is reached, not the final amount of Pfr (see Fig. 8). Thus we reach the

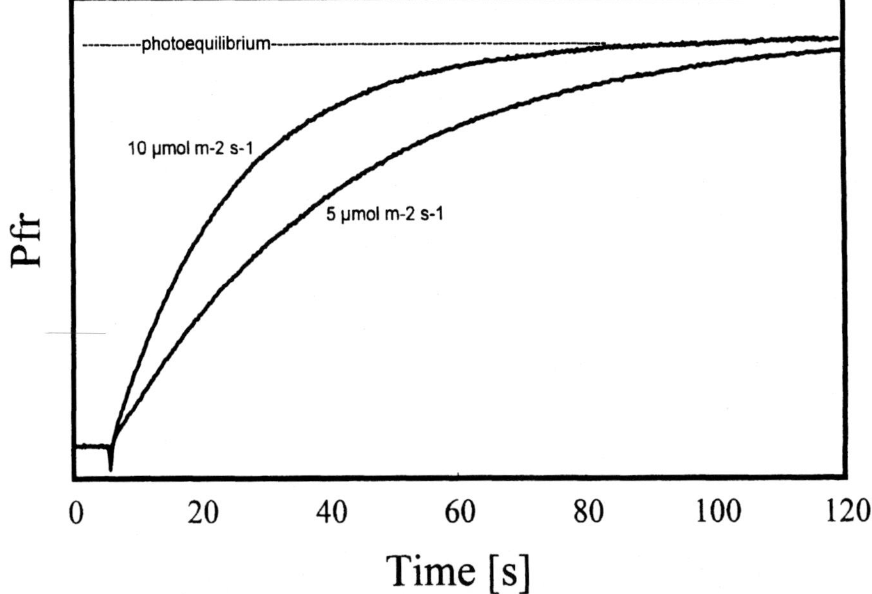

Fig. 8. Kinetics of photoconversion of *Ceratodon* phytochrome. Recombinant cDNA for the B-type phytochrome from *Ceratodon* was overexpressed in yeast, autoassembled with phycocyanobilin, and affinity purified. The amount of Pfr present was monitored via A 730 nm during irradiation of the cuvette with R (660 nm, 5 or 10 μmol $m^{-2}s^{-1}$) following saturating FR. The same photoequilibrium was reached at both fluence rates.

unfortunate conclusion that phytochrome cannot perceive the stimulus for a classical phytochrome-mediated (R/FR reversible) response.

- Unfortunately, the situation is worse still, as even if phytochrome could in some way detect a fluence gradient, how could the gradient of activated phytochrome be maintained between R pulses if it is able to move freely through the cytoplasm (by diffusion or cytoplasmic streaming)? The data indicates that, on the contrary, it is attached to something fairly immobile.

3.1.2 Polarotropism in protonemata

Perhaps surprisingly, compelling evidence that phytochrome is indeed attached at or near the plasma membrane comes from studies using polarized light, an approach introduced by Jaffe in 1958 in relation to *Fucus* zygote development (see 3.2). Photon absorption by a pigment molecule is optimal when its transition dipole moment (TDM) is parallel to the electrical vector of the radiation. Thus polarized light is only absorbed by appropriately-oriented molecules. This has no particular significance if the molecules are free to rotate in three dimensions, but if they are attached to a fixed structure the distribution of absorption is defined by the geometry of that structure and the interaction (Fig. 9).

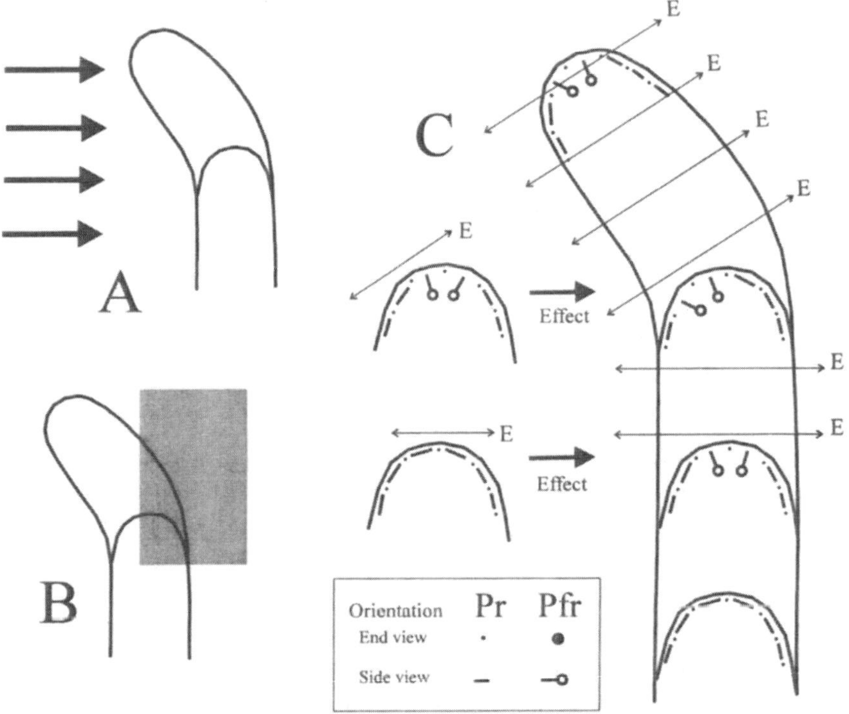

Fig. 9. Photomorphogenic responses of *Ceratodon* caulonemata to light. The tip cell grows towards the source of unilateral light (A and Fig. 6A) and away from the shaded side if partially irradiated (B). In polarized light (C and Fig. 6B), the tip cell grows perpendicularly to the electrical vector (E), probably because the Pr TDM's lie parallel to the tip surface, whereas those of Pfr are perpendicular to it. Growth occurs from the point of highest Pfr concentration. TDM's for Pr and Pfr are shown. See 3.1.1 and 3.1.2 for details.

And indeed, if moss or fern protonemata growing on a flat surface are irradiated with dim polarized R from above, their direction of growth becomes oriented within minutes to become exactly perpendicular to the electrical vector (Fig. 6B): R pulses are also effective and are FR reversible, implying phytochrome involvement. This dichroic effect is known as polarotropism and has been studied extensively in the ferns *Dryopteris* and *Adiantum* and in the mosses *Ceratodon* and *Physcomitrella* (Etzold, 1965; Jenkins and Cove, 1983; Wadas and Kadota, 1989; Kraml, 1994). The data can easily be explained on the basis of the Jaffe model with Pr being attached to the plasma membrane or something close to it, its TDM parallel to the surface of the cell tip (Fig. 9). Polarotropism might alternatively be explained by the cell wall acting as a polarizing filter and allowing only light with a certain polarisation to pass through. This is unlikely as microscopy with polarizing optics shows the cell to be dichroically inactive. Moreover, re-orientation of the filament following a polarized R pulse is most effectively prevented by FR with a polarization parallel to the filament axis (Fig. 9). This implies not only that the TDM for Pfr is shifted by 90° in relation to that of Pr, but also that it is fixed between pulses. Hence the photoreceptor must be attached to something rigid. Who would have thought it?

3.1.3 Chloroplast movement in Mougeotia and Mesotaeneum

A third line of investigation also implies that phytochrome is membrane-associated. The work of Haupt and Kraml at Erlangen (see Kraml, 1994; Haupt and Häder, 1994) on chloroplast movements in *Mougeotia* and *Mesotaenium* seems clearly to lead to this same paradoxical conclusion.

Cells of the filamentous alga *Mougeotia* are unusual in containing one large flattened chloroplast (Fig. 4A). Chloroplasts generally show photodynesis—that is, they move according to the light conditions at any time in order to optimize photosynthesis, exposing the maximum surface area for absorption in dim light, turning sideways and seeking intracellular shade to avoid photooxidative damage under strong light. Actually, in most plants these movements are in response to B, but in *Mougeotia* under appropriate conditions the phytochrome system predominates. When irradiated from above, the chloroplast rotates and flattens to face the direction of R irradiation, even if this is only given as a brief pulse, whereas the response is prevented if a FR pulse is given immediately afterwards, clear evidence of phytochrome involvement (Haupt, 1959).

As in the protonemal systems, elegant experiments with microbeams and polarized light have established seemingly beyond doubt that the phytochrome mediating these movements is membrane-associated. Generally, if the electrical vector of R is perpendicular to the axis of the filament, the chloroplast rotates to face the direction of irradiation just as in unpolarized light, whereas little movement is seen if the vector is parallel to the filament axis. This phenomenon is shown in Fig. 10. It might be argued that the

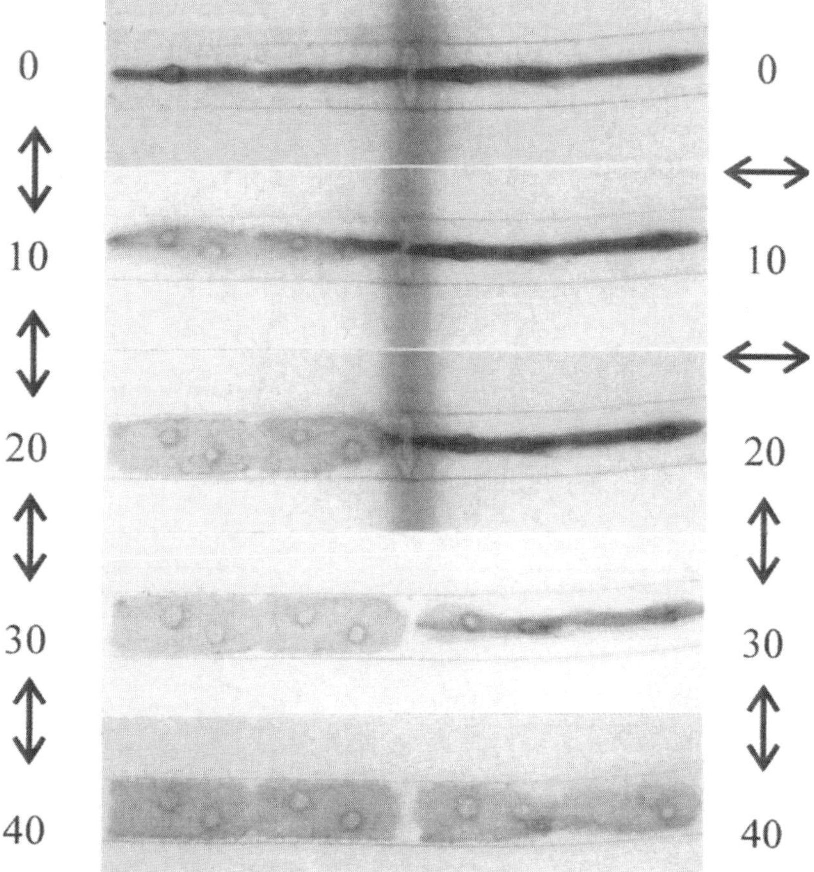

Fig. 10. ***Mougeotia*** **chloroplast movement in polarized light. The filament was irradiated with R to bring the chloroplasts to a vertical position, then with FR to remove Pfr. At t = 0 irradiation with polarised R (660 nm, 150 μmol m^{-2}s^{-1}) from the microscope's condensor was initiated. The cells on the left and right received perpendicular and parallel-vector light, respectively. Only the left cell responded, chloroplast rotation being complete within about 20 min. At this point the polarizing filter was moved to provide both cells with perpendicular-vector light, leading to chloroplast rotation in the right cell too. See 3.1.3 and Fig. 11 for details.**

light is not being absorbed for some reason. However, if perpendicular-vector R is given as a pulse followed by parallel-vector R, the induction of rotation is nullified. Thus R is absorbed in both cases, but only perpendicular-vector irradiation induces chloroplast movement. This strange behaviour too can easily be explained on the basis of the Jaffe model (Fig. 11). The edges of the chloroplast are close to the plasma membrane and seem to be repelled by Pfr, the movement being brought about via actin microfilaments in the cytoplasmic cortex (Haupt, 1960). Furthermore, partial irradiations with microbeam pulses of polarized R and FR have also clearly shown that

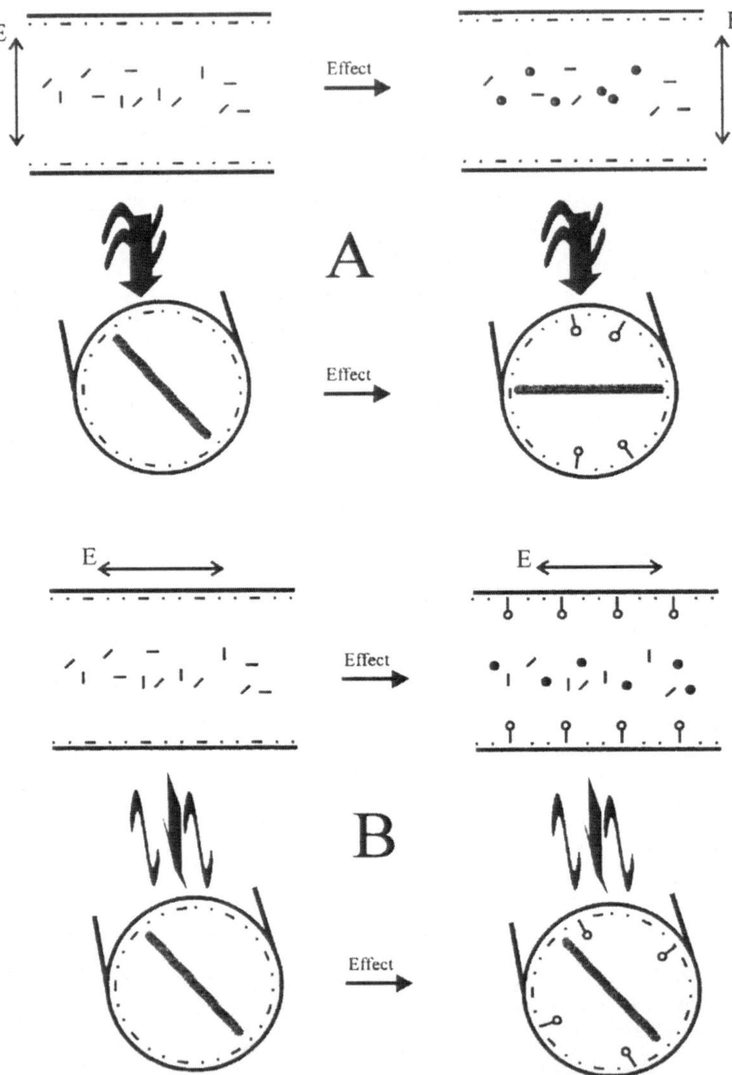

Fig. 11. Explanation of Fig. 10. The giant chloroplast is rotated to face the light, but only if the E-vector is perpendicular to the filament axis (A); parallel-vector light does not lead to rotation (B). As for polarotropism (Figs. 6B and 9), this can be explained by non-random distribution and TDM orientation of Pr and Pfr, the edges of the chloroplast being pulled from the regions of highest Pfr concentration. See 3.1.3 for details.

the TDM for R is parallel to the cell surface and perpendicular to the surface for FR reversal.

Given that the photoreceptor regulating chloroplast movement in *Mougeotia* is attached near the plasma membrane, the uniform cytosolic distribution of phytochrome in the same species implied by immunohistological work of Hanstein et al. (1992) should come as rather a shock, reflecting as it does the results of similar studies with PHYA in dark-adapted higher plant tissues (Pratt, 1994). This is indeed what would be expected on the basis of the peptide sequence derived from the *Mougeotia PHY* cDNA sequence reported by Winands and Wagner (1996): there are no hydrophobic areas typical of membrane-associated proteins and there is precious little sign of any other cytologically-relevant motifs. So, the phytochrome is cytosolic, and so it should be, whatever the photobiologists say.

Chloroplast movements of this kind are not unique to *Mougeotia*. Analogous behaviour is seen with the giant chloroplast in the unicellular alga *Mesotaenium* (Fig. 4B) (Haupt and Thiele, 1961). Phytochrome from this species has been studied in depth biochemically and molecular-genetically by Lagarias's group at UC Davis; it too behaves and looks like a typical cytosolic protein (Kidd and Lagarias, 1990; Lagarias et al., 1995). Wada and co-workers have made analogous studies in the fern *Adiantum*. For example, microbeam experiments with prothalli (Kagawa and Wada, 1996) have shown that chloroplasts move towards R within a single cell. The rate of movement is independent of fluence rate and R pulses are just as effective as continuous light. Moreover, the action of such pulses can be inhibited by subsequent FR. Thus the response is mediated by phytochrome. Using polarized light pulses with protonemata of the same species, chloroplasts move to regions of high Pfr according of the Jaffe model. Moreover, reversion by FR shows a dichroic maximum shifted by 90° as in the cases described above (Kagawa et al., 1994). Although phytochromes in *Adiantum* have not been biochemically studied in detail, sequence data for several *PHY* genes implies that they are similar to B-type higher plant phytochromes (Wada et al., 1997). Note that here, in contrast to the data for *Mesotaenium* and *Mougeotia*, the chloroplasts are attracted rather than repelled by Pfr. It should also be born in mind that whereas phytochrome mediates these movements in R, B is usually also effective and shows a quite different phenomenology.

3.1.4 Does phytochrome have a mechanism of action?

As with any set of observations, any number of general models for phytochrome action consistent with all known data can be constructed, given, enough *ad hoc* assumptions and hypothetical variables. At the other end of the scale, we do not know if all the observations are being interpreted correctly in defining such a model—experimental artefacts are just as common as models in the phytochrome field. The problem remains, how does phytochrome *really* function?

The phenomena described above occur in cells that allow irradiations under precise optical conditions and permit exacting microscopic observation. These conditions are not met in the intensely-scattering tissues of higher plants, so it is hardly surprising that such phenomena are unknown there. The underlying mechanism might nevertheless be universal. They might on the other hand be special cases, unusual responses comprising peculiar features we have yet to discover or prefer to ignore. An example of the latter is seen in *Physcomitrella* chloronemata, whereby in both photo- and polarotropism, FR potentiates the response to R, and high R fluences lead to a 90° shift in the direction of the response relative to that at low fluence rates (Jenkins and Cove, 1983). The phytochromes involved might themselves be different (see 3.3) or they might have special, perhaps membrane-associated receivers for these functions. The phytochrome receiver is known only in cyanobacteria (3.3), so it would be premature to rule out exceptions. And finally, we should perhaps be more careful with words: "cortical" or "peripheral" could be used in place of "membrane-associated"—after all, pretty much everything in the cell is associated with membranes.

In fact a membrane-associated receiver for phytochrome in *all* plant systems is perfectly feasible and even an attractive hypothesis. In the first place, it would remove any need for phytochrome to show hydrophobic structures appropriate for direct membrane attachment. Then, secondly, if phytochrome were to be present in a large excess relative to and/or if it were to interact loosely with the receiver (this would accord with two-component signalling in prokaryotes, see 3.3), phytochrome could *appear* to be cytosolic while *acting* as if fixed close to the plasma membrane. Indeed, the idea of bulk and active phytochrome pools was introduced by Siegelman and Butler (1965) more than thiry years ago. So far, so good. Now, taking the dichroic effects described above for lower plants at face value, the model predicts thirdly that photoconversion is only effective physiologically when phytochrome is attached to its receiver. The 90° shift in the TDM between Pr and Pfr is stable, thus forthly the active photoreceptors must be fairly rigidly held: the shift might result from a change within the phytochrome itself or, alternatively, in the receiver geometry. Experiments with polarised microbeams have indicated that the TDM's of Pr are not random relative to the filament axis, but arranged spirally. Thus a fifth character of the model would be that the receiver molecules are not just attached to the plasma membrane, but coherently arranged along the cell axis. While attractive, the model is of course not supported by any hard evidence. Disquietingly, such models are thereby all the more seductive.

3.2 Axis Formation and Polarity

The development of form in any organism seems hierarchical. Organs, tissues and specialized cells tend to differentiate from pre-existing, more generalized structures. As sexual reproduction leads to the formation of a zygote which gives rise to the entire body of the progeny, pattern formation begins with this cell. Nearly all organisms show a very marked degree of asymmetry in the arrangement of their body parts, which thus derives from prior developmental decisions. The earliest of these involves the setting-up of an axis and a polarity in the zygote prior to the first cell division. In animals this is sometimes predefined by the point of the entry of the sperm nucleus, whereas in most flowering plants the structure of the ovary seems to predefine the axis and polarity and, moreover, renders the zygote inaccessible to manipulation. In contrast to this, in many algae radially-symmetrical egg cells are released from the parent and develop autonomously, the polarity of development deriving from environmental cues, in particular light direction. The large (*ca.* 100 μm) zygotes of the marine brown algae *Fucus* and *Pelvetia* are especially convenient for experimental study (Fig. 5; see Kropf, 1997; Quatrano and Shaw, 1997; Goodner and Quatrano, 1993). An alternative model system for the study of these processes has recently been developed, based on the regeneration of moss protonemal protoplasts (Fig. 2; Cove et al., 1996). Both systems have contributed to our knowledge of this most fundamental level of plant morphogenesis. As in many other aspects of plant morphogenesis and in contrast to the rule in animals, the setting-up of an axis and of polarity is determined primarily by environmental factors, primarily gravity and light, as described below.

Axis and polarity definition by light in fucoid zygotes becomes possible about 4 hours following fertilization, after they have adhered to the substrate. If the cells are irradiated with B or UV/A from a particular direction around this time, a chain of cytological events is initiated culminating in the outgrowth of the rhizoid from the distal side about 7 hours later. If only part of the cell is irradiated, outgrowth occurs in the shaded region, indicating that the directional signal is perceived through attenuation rather than a lens effect (Fig. 7). Furthermore, it is not necessary to provide light continuously: the eventual point of outgrowth can be pre-programmed by a pulse of directional light given early on. The development can also the orientated by polarized light, outgrowth occurring parallel to the electrical vector (Fig. 12; Jaffe, 1958). As in the case of phytochrome-mediated polarotropism (3.1.2), this action dichroism can be explained by assuming that the photoreceptors lie close to the plasma membrane with their TDM's parallel to it. Thus light is preferentially absorbed most efficiently where the E-vector is tangential to the plasma membrane, inhibiting outgrowth at these flanks.

One of the earliest events following irradiation is a marked influx of protons through the plasma membrane at the presumptive rhizoid pole and an accompanying efflux from the opposite end. This is associated with rearrangements of the actin microfilaments, as treatment with cytochalasin B, an inhibitor of F-actin polymerisation, also interrupts this current flow as well as further morphogenesis, the zygote dividing without a rhizoid outgrowth. Similar effects result from brefeldin A, and inhibitor of Golgi-apparatus function. Presumptive Ca^{2+}-ion channels, associated with binding of dihydropyridine, also appear at this time (6–8 hours following fertilization), although the role of free Ca^{2+} gradients in determining polarity is not clear, as such gradients are notoriously difficult to measure. Nevertheless, a wide range of experimental treatments to disturb any Ca^{2+} gradient do affect

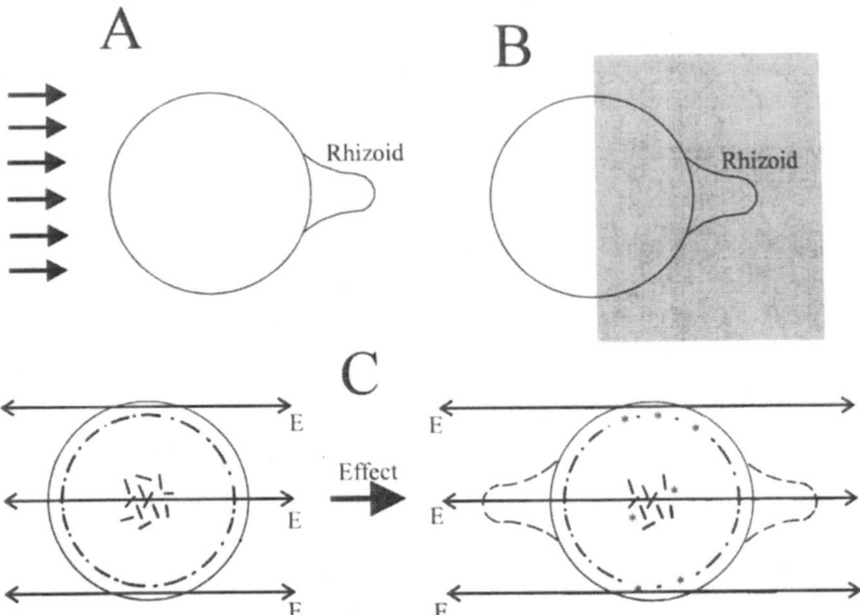

Fig. 12. Photomorphogenic responses of *Fucus* zygotes to light. The rhizoid emerges from the distal side in unilateral light (A) and from the shaded side if partially irradiated (B). In polarized light (C), rhizoids emerge parallel to the electrical vector (E), probably because the TDM's lie parallel to the cell surface. Outgrowth occurs at the points of lowest excitation. Photoreceptors in ground state,—TDM seen from side, • TDM seen end-on. Excited photoreceptors, *. See 3.2 and 3.1.2 for details.

rhizoid development. It remains possible to realign the axis by a second B irradiation up to about 8 hours following fertilization. Fixation of polarity occurs at about this time. Fixation can be suspended by treatment with cytochalasin but not, for example, the microtubule inhibitor colchicine. Two hours later, about 10–12 hours after fertilization, the asymmetry in the zygote becomes visible, followed by the outpushing of the rhizoid associated with dramatic changes in the actin microfilamental cytoskeleton and the first cell division. Although mitosis can be uncoupled from outgrowth by cytochalasin, it does not seem possible to separate the light-effect on outgrowth direction from that on the plane of division. This implies that the much earlier $P34^{cdc2}$-mediated fixation of the future cell plate perimeter is intimately associated with the spatial fixation of outgrowth. Following the first (transverse) cell division, the rhizoid attaches itself firmly to the substrate and goes on to develop into the holdfast of the adult plant. By contrast, the rest of the zygote cell containing most of the chloroplasts undergoes lateral and transverse divisions to form the embryo, finally giving rise to the thallus (Fig. 5).

In fucoid zygotes then, rhizoid development follows the establishment of polarity via environmental factors, in particular light. At least in principle, however, even this apparently fundamental morphogenic event can be dissected further. The development of polarity (front and rear) presupposes an axis which need not itself be polar. Whereas from the *Fucus* work it appears that axis and polarity development go hand in hand, recent work with regenerating *Ceratodon* protoplasts implies that it is possible to separate them physiologically both in time and according to the photoreceptor involved.

The protoplasts must be handled rather differently from zygotes. Protoplasts do not adhere to the substrate as do zygotes and thus must be immobilised in some way. Furthermore, as protoplasts they have no cell wall and hence must be provided with an osmotic buffer. Such immobilized protoplasts are spherical initially, but within about 20–60 hours a regeneration outgrowth appears, leading to the formation of a new filament (Fig. 2). If the protoplasts are irradiated unilaterally with white light or B, most regenerate roughly towards the light source. In unilateral R, however, the response is rather different. Regeneration is more closely correlated with the direction of illumination, but, whereas most protoplasts regenerate towards the light, about 30% regenerate in exactly the opposite direction; very few are intermediate. By changing the direction of irradiation at different times using R or B it is possible to dissect an early predominantly R-dependent definition of the axis from a subsequent predominantly B-dependent definition of polarity (Cove et al., 1996). Although such studies with moss protoplasts are still in their infancy, the convenience of the *Ceratodon* protoplast system makes it an attractive one for further photomorphogenic and cytological investigations. It will be interesting to see whether axis and polarity determination can be dissected in the *Fucus* system too: an important difference between the two systems is the cell wall, known to be essential for polarity fixation in zygotes. Unfortunately, studying this is more than a little complicated by the fact that, in marked contrast to animal cells, a cell wall is a prerequisite for morphogenesis in plants.

One of the key advantages in studying photoresponses is that the stimulus (light) is so easily modulated and is, effectively, non-invasive. Thus the simple photomorphogenesis systems described above are well suited to experimental manipulation. While, on the other hand, genetic approaches are limited in *Fucus*, the moss system with its small, manipulable genome would appear to be ideal for molecular-genetic approaches (Schaefer et al., 1991; Knight, 1994; Zeidler et al., 1996; Schaefer and Zrÿd, 1997; Lamparter et al., 1998b).

3.3 Origins of Phytochrome

An explanation of the seemingly paradoxical action of phytochrome revealed by studies in lower plants (see 3.1) might provide a crucial insight into the still quite unknown mechanism(s) of plant phytochrome action in general. Indeed, lower organisms have often provided useful evolutionary perspectives to systems in more advanced groups. Alternatively, some lower plant phytochromes might themselves be different and hence function differently from those in angiosperms.

In the last few years phytochrome-encoding genes and cDNAs have indeed been sequenced in various lower plant species, including *Adiantum, Selaginella, Physcomitrella, Ceratodon, Mesotaenium* and *Mougeotia* (see Wada et al., 1997; Schneider-Poetsch et al., 1998). The conceptual gene products are clearly similar to their homologues in higher plants. This wonderfully circular arguement supports the notion that they function analogously. Indeed, we have found that the dominant phytochrome in *Ceratodon* shows biochemical and optical properties rather like those of B-type phytochromes (Lamparter et al., 1995; Hughes et al., 1996; Zeidler et al., 1998). Interestingly however, one of the two phytochrome gene homologues in *Ceratodon* has very different features. Although the N-terminus of the conceptual product is phytochrome-typical, the C-domain is quite different, showing instead weak homologies to various eukaryotic kinases (Thümmler et al., 1992; Schneider-Poetsch et al., 1998). Whether this gene actually gives rise to a functional photoreceptor is still in doubt (Lamparter et al., 1995), but it does raise the question, how similar to "conventional" phytochromes does a sequence have to be for it to be recognized as a phytochrome at all? Numerous aphototropic *Ceratodon* mutants isolated in our laboratory have been screened (Lamparter et al., 1996; Esch and Lamparter, 1998): dozens shows defective chromophore biosynthesis but screening with anti-phytochrome antibodies shows than all produce normal levels of phytochrome apoprotein. Does this imply that our antibodies see the wrong phytochrome? By definition phytochrome must show R/FR reversible action, whereas the immunochemical/molecular-genetic approach currently in vogue requires phytochromes to show sequence similarity to oat PHYA, the first phytochrome to have been cloned.

In fact, the crucial contribution made in the field of lower plant phytochromes have come from an unexpected quarter. Schneider-Poetsch and co-workers were first to clone a lower plant phytochrome (Hanelt et al., 1992). Sequence comparisons revealed not only clear homologies to higher plant phytochromes, but also weaker C-terminal homologies to sensory histidine kinases, a rather diverse class of usually membrane-bound enzymes involved in environmental perception in prokaryotes. Indeed, closer inspection of higher plant phytochrome sequences revealed structural similarities to these kinases which had previously been missed (Schneider-Poetsch, 1992). Greeted with scepticism at the time, the observation was followed by the discovery of a cyanobacterial sensory histidine kinase with both N-terminal as well as C-terminal homologies to phytochrome (Kehoe and Grossman, 1996). This kinase, rcaE, was cloned in *Fremyella* where it is involved in CCA: further phytochrome-like genes were identified soon afterwards in the genome sequence of *Synechocystis*, but it remained an open question whether they were themselves photoreceptors. The overall level of similarity to plant phytochromes is low: even allowing for insertions and deletions, only about 30% of the amino acids encoded are conserved in plant phytochromes, and only one, slr0473, shows the otherwise highly-conserved chromophore-binding site (Hughes et al., 1996). Indeed, it has now been shown in our laboratory that this gene product is a true phytochrome, through its ability autocatalytically to attach bilin chromophores and thereafter show classical R/FR photoreversibility (Hughes et al., 1997; Yeh et al., 1997; Lamparter et al., 1997; Remberg et al., 1997; Sineshchekov et al., 1998).

The recombinant *Synechocystis* phytochrome offers the possibility of obtaining details of the molecular structure of this type of photoreceptor through crystallization and X-ray diffraction analysis, but its discovery also has most important implications regarding signal transduction. Sensory histidine kinases comprise one part of the so-called "two-component" signal transduction mechanism seemingly universal in prokaryotes (see Fig. 13; Hoch and Silhavy, 1995). *In vivo* the kinase exists as a dimer, each subunit consisting of two modular domains: the N-terminal input module is very variable and is responsible for the specific sensory function involved, while the C-terminal transmitter module is more conserved, bearing several almost invariate subdomains responsible for kinase function. According to the state of the input module, a particular histidine residue in the transmitter module becomes phosphorylated by the subunit partner. This phosphate then becomes transferred to an aspartate residue of a third module, the receiver, usually on a separate peptide termed the response regulator. In some cases this then acts directly (often as a DNA-binding activator of gene transcription) to bring about a response, in others further proteins are involved in a more extensive phosphorelay. The *Synechocystis* phytochrome response regulator was obvious from the genome map, lying immediately downstream of the phytochrome gene in an operon-like configuration, typical of two-component systems. This graphic illustration of the power of whole-genome analysis and the Internet was completed when Lagarias and co-workers (Yeh et al., 1997) recently demonstrated the expected R/FR-reversible phosphorelay function. Conversely, it was not at all expected that the active kinase is Pr, not Pfr. The latter is generally considered to be the "physiologically active" form in plants: Smith (1995) has recently summarised reasons for doubting this assumption. Additionally, in the phytochrome-like photoregulation of sporulation in the slime-mould *Physarum,* Pr is clearly the physiologically active species (Starostzik and Marwan, 1995).

Halobacterium provides an independent example of two-component phototransduction in prokaryotes. Here the photochromic state of the sensory rhodopsin is transduced by a further rhodopsin-like protein which then interacts with the chemotaxis receptor, a sensory histidine kinase similar to CheA in *E. coli,* thereby integrating the two signals. A phosphotransfer to a CheY-like response regulator then takes place. Phosphorylated CheY then

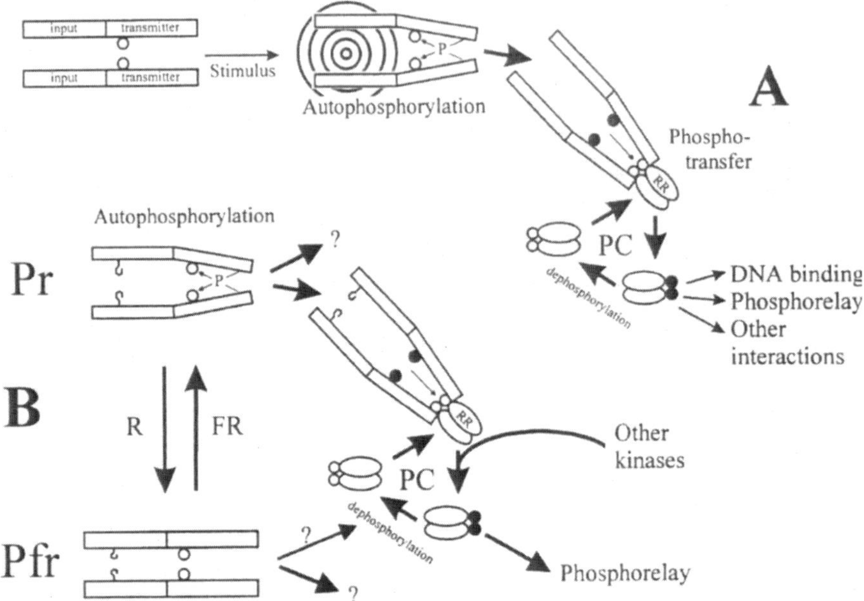

Fig. 13. Two-component signal transduction.
A. General scheme. The first component is a sensory histidine kinase, a dimer, each subunit comprising two modules or domains. The input module responds to a specific environmental signal and interacts with the transmitter module bearing highly conserved histidine kinase motifs. Once stimulated, the input module induces a mutual phosphorylation of a specific histidine residue on each subunit transmitter module. The phosphate (P) is then transferred to a specific aspartate residue the second component, the response regulator (RR). The phosphate can be transferred to further components in a phosphorelay or the regulator can act directly, for example as a transcriptional activator. PC, phosphorylation cycle. B. Scheme for phytochrome in *Synechocystis*. Pr is the active kinase leading to phosphotransfer to the response regulator which is likely to be tragetted by other kinases and, on account of its small size act as an intermediate in a further phosphorelay. The action of Pfr is unknown.

interacts directly with the flagella motor (Rudolf et al., 1995). However, the sensory rhodopsin possesses a second light-dependent activity, whereby fumarate is released as a secondary messenger and also interacts with the motor (Marwan et al., 1990). Perhaps this halobacterial system also has parallels in the case of phytochrome: it is clear that a photoreceptor can have more than one mode of action.

The work of Quail and co-workers indicates that caution is required in extrapolation from prokaryote systems to that or those of plant phytochromes, however. Firstly, a large, functionally-important domain present in plant phytochromes is missing from the *Synechocystis* phytochrome. Secondly and very significantly, several residues usually essential for two-component phosphorelay are neither conserved nor functional in plant phytochromes

(Lamparter et al., 1997; Schneider-Poetsch et al., 1998; Quail et al., 1995; Boylan and Quail, 1996). Perhaps another function of cyanobacterial phytochrome has been retained in plants, while the Pr-mediated phosphorelay has been lost.

There is certainly no doubting the existence in plants of signal transduction components related to those seen in prokaryotes. The receptors both for ethylene and for cytokinin have now been cloned and clearly show relics of two-component systems (Chang et al., 1993; Kakimoto, 1996). Sequences with two-component-like features have also recently been found in *Saccharomyces cerevisae, Neurospora* and in the cellular slime-mould *Dictyostelium.* Although no such relics have yet been found in animals, Stock and Lukat (1991) drew attention to the structural similarities between the RAS superfamily of mammalian GTPases involved in intracellar signalling and the prokaryotic CheY response regulator. This is intriguing as both RAS-like and heterotrimic GTPases have been implicated in phytochrome signal transduction in plants. The CTR1 component of the ethylene response transduction chain in plants is homologous to the RAF signalling kinase in animals; RAF associates with the plasma membrane upon phosphorylation by RAS and subsequently activates transcription factors via a phosphorelay.

The concept of phytochrome as a photoreceptor in a prokaryotic two-component system seems already to have lead to a paradigm shift in the field. They are certainly important similarities and differences between plant and cyanobacterial systems. On these bases our understanding of fundamental aspects of plant photomorphogenesis looks likely to improve rapidly in the near future. Simultaneously, work with cyanobacteria will continue: the function of the phytochrome response regulator is not yet known—it is not a DNA-binding protein—nor indeed is the response itself known. Finally, X-ray crystallography of the recombinant *Synechocystis* phytochrome might provide molecular structural data valid for phytochromes in general, contributing enormously to our understanding of these fascinating photoreceptors.

References

Ahmad, M. and Cashmore, A.R. 1993. *HY4* gene of *A. thaliana* encodes a protein with characteristics of a blue-light photoreceptor. *Nature* **366:** 162–166.

Ahmad, M. and Cashmore, A.R. 1996. Seeing blue: the discovery of cryptochrome. *Plant Mol. Biol.* **30:** 851–861.

Bibikov, S.I., Grishanin, R.N., Kaulen, A.D., Marwan, W., Oesterhelt, D. and Skulachev, V.P. 1993. Bacteriorhodopsin is involved in halobacterial photoperception. *Proc. Natl. Acad. Sci. USA* **90:** 9446–9450.

Boylan, M.T. and Quail, P.H. 1996. Are phytochromes protein kinases? *Protoplasma* **195:** 12–17.

Chang, C., Kwok, S.F., Bleecker, A.B. and Meyerowitz, E.M. 1993. *Arabidopsis* ethylene-response gene *ETR1:* similarity of product to two-component regulators. *Science* **262:** 539–544.

Cooke, T.J., Hickok, L.G. and Sugai, M. 1995. The fern *Ceratopteris richardii* as a lower plant model system for studying the genetic regulation of plant morphogenesis. *Int. J. Plant Sci.* **156:** 367–373.

Cove, D.J., Schild, A., Ashton, N.W. and Hartmann, E. 1978. Genetic and physiological studies of the effect of light on the development of the moss, *Physcomitrella patens. Photochem. Photobiol.* **27:** 249–254.

Cove, D.J., Quatrano, R.S. and Hartmann, E. 1996. The alignment of the axis of asymmetry in regenerating protoplasts of the moss *Ceratodon purpureus* is determined independently of axis polarity. *Development* **122:** 371–379.

Ermolayeva, E., Hohmeyer, H., Johannes, E. and Sanders, D. 1996. Calcium-dependent membrane depolarisation activated by phytochrome in the moss *Physcomitrella patens. Planta* **199:** 352–358.

Esch, H. and Lamparter, T. 1998. Light regulation of phytochrome content in wildtype and aphototropic mutants of the moss *Ceratodon purpureus. Photochem. Photobiol.* **67:** 450–455.

Etzold, H. 1965. Der Polarotropismus um Phototropismus der Chloronemen von *Dryopteris filix-mass* (L.) Schott. *Planta* **64:** 254–280.

Furuya, M., Kanno, M., Okamoto, H., Fukuda, S. and Wada, M. 1997. Control of mitosis by phytochrome and a blue-light receptor in fern spores. *Plant Physiol.* **113:** 677–683.

Goodner, B. and Quatrano, R.S. 1993. *Fucus* embryogenesis: a model to study the establishment of polarity. *Plant Cell* **5:** 1471–1481.

Griffith, G.W., Jenkins, G.I., Milner-White, E.J. and Clutterbuck, A.J. 1994. Homology at the amino acid level between plant phytochromes and a regulator of sexual sporulation in *Emericella (= Aspergillus) nidulans. Photochem. Photobiol.* **59:** 252–256.

Hanelt, S., Braun, B., Marx, S., Schneider-Poetsch, H.A.W. 1992. Phytochrome evolution; a phylogenetic tree with the first complete sequence of phytochrome from a cryptogamic plant (*Selaginella mertensii,* Spring). *Photochem. Photobiol.* **56:** 751–758.

Hanstein, C., Grolig, F. and Wagner, G. 1992. Immunolocalisation of cytosolic phytochrome in the green alga *Mougeotia. Bot. Acta* **105:** 55–62.

Hartmann, E. and Jenkins, G. 1984. Photomorphogenesis of mosses and liverworts. In "The experimental biology of bryophytes" Dyer A.F. and Ducket, J.G., (Eds.) Academic Press London. pp 203–228.

Hartmann, E., Klingenberg, B. and Bauer, L. 1983. Phytochrome mediated phototropism in protonemata of the moss *Ceratadon purpureus. Photochem. Photobiol.* **38:** 599–603.

Haupt, W. 1959. Die Chloroplastendrehung bei *Mougeotia.* I: Über den quantitativen und qualitativen Lichtbedarf der Schwachlichtbewegung. *Planta* **53:** 484–501.

Haupt, W. 1960. Die Chloroplastendrehung bei *Mougeotia.* II: Die Induktion der Schwachlichtbewegung durch linear polarisiertes Licht. *Planta* **55:** 465–479.

Haupt, W. and Häder, D.-P. 1994. Photomovement. In "Photomorphogenesis in plants" 2nd. Ed, Kendrick, R.E. and Kronenberg, G.H.M., (Eds.) Kluwer, Dordrecht. pp 707–732.

Haupt, W. and Thiele, R. 1961. Chloroplastenbewegung bei *Mesotaenium. Planta* **56:** 388–401.

Hoch, J.A. and Silhavy, T.J. 1995. Two-component signal transduction. 1–55581–089–6: ASM Press Washington.

Hughes, J., Lamparter, T. and Mittmann, F. 1996. Cerpu; PHY0;2, a "normal" phytochrome in *Ceratodon* (Accession No. U56698) (PGR 96–067). *Plant Physiol.* **112:** 446.

Hughes, J., Lamparter, T., Mittmann, F., Hartmann, E., Gärtner, W., Wilde, A. and Börner, T. 1997. A prokaryotic phytochrome. *Nature* **386:** 663.

Jaffe, L.F. 1958. Tropistic responses of zygotes of the Fucaceae to polarized light. *J. Exp. Res.* **15:** 282–299.

Jenkins, G.I. and Cove, D.J. 1983. Phototropism and polarotropism of primary chloronemata of the moss *Physcomitrella patens*: responses of the wild-type. *Planta* **158:** 357–364.

Kagawa, T., Kadota, A. and Wada, M. 1994. Phytochrome mediated photoorientation of cloroplasts in protonemal cells of the fern *Adiantum* can be induced by brief irradiation with red light. *Plant and Cell Physiol.* **35:** 371–377.

Kagawa, T. and Wada, M. 1996. Phytochrome and blue-light-absorbing pigment-mediated directional movement of chloroplasts in dark-adapted prothallial cells of the fern *Adiantum* as analysed by microbeam irradiation. *Planta* **198:** 488–493.

Kakimoto, T. 1996. CKI1, a histidine kinase homolog implicated in cytokinin signal transduction. *Science* **274:** 982–985.

Kaneko, T. et al. 1996. Sequence analysis of the genome of the unicellular Cyanobacterium *Synechocystis* sp. strain PCC6803. II. Sequence determination of the entire genome and assignment of potential protein-coding regions. *DNA Res.* **3:** 109–136.

Kehoe, D.M. and Grossman, R. 1996. Similarity of a chromatic adaptation sensor to phytochrome and ethylene receptors. *Science* **273:** 1409–1412.

Kidd, D.G. and Lagarias, J.C. 1990. Phytochrome from the green alga *Mesotaenium caldiorium.* Purification and preliminary characterization. *J. Biol. Chem.* **12:** 7029–7035.

Knight, C.D. 1994. Studying plant development in mosses: the transgenic route. *Plant Cell and Environ.* **17:** 669–674.

Knight, C.D., Futers, T.S. and Cove, D.J. 1991. Genetic analysis of a mutant class of *Physcomitrella patens* in which the polarity of gravitropism is reversed. *Mol. Gen. Genet.* **230:** 12–16.

Krah, M., Marwan, W., Vermeglio, A. and Oesterhelt, D. 1994. Phototaxis of *Halobacterium salinarium* requires a signalling complex of sensory rhodopsin I and its methyl-accepting transducer Htrl. *EMBO J.* **13:** 2150–2155.

Kraml, M. 1994. Light direction and polarisation. In: Photomorphogenesis in plants, Kendrick, R. E. and Kronenberg, G.H.M., (Eds) Kluwer, Dortrecht. pp 417–446.

Kropf, D.L. 1997. Induction of polarity in fucoid zygotes. *Plant Cell* **9:** 1011–1020.

Lagarias, D.M., Wu, S.-H. and Lagarias, J.C. 1995. Atypical phytochrome gene structure in the green alga *Mesotaenium caldoriorum. Plant. Mol. Biol.* **29:** 1127–1142.

Lamparter, T., Lutterbüse, P., Schneider-Poetsch, H.A.W. and Hertel, R. 1992. A study of membrane-associated phytochrome: hydrophobicity test and native size determination. *Photochem. Photobiol.* **56:** 697–707.

Lamparter, T., Podlowski, S., Mittmann, F., Hartmann, E., Schneider-Poetsch, H.A.W. and Hughes, J. 1995. Phytochrome from protonemal tissue of the moss *Ceratodon purpureus. J. Plant Physiol.* **147:** 426–434.

Lamparter, T., Esch, H., Cove, D., Hughes, J. and Hartmann, E. 1996. Aphototropic mutants of the moss *Ceratodon purpureus* with spectrally normal and with spectrally dysfunctional phytochrome. *Plant Cell and Environ.* **19:** 560–568.

Lamparter, T., Mittmann, F., Gärtner, W., Börner, T., Hartmann, E. and Hughes, J. 1997. Characterization of recombinant phytochrome from the cyanobacterium *Synechocystis. Proc. Natl. Acad. Sci. USA.* **94:** 11792–11797.

Lamparter, T., Hughes, J. and Hartmann, E. 1998a. Blue light and genetically-reversed gravitropic response in protonemata of the moss *Ceratodon purpureus. Planta,* **206:** 95–102.

Lamparter, T., Brücker, G., Esch, H., Hughes, J., Meister, A. and Hartmann, E. 1998b. Somatic hybridisation with aphotopropic mutants of the moss *Ceratodon purpureus*: genome size, phytochrome photoreversibility, tip-cell phototropism and chlorophyll regulation. *J. Plant Physiol.* **153:** 394–400.

Manachère, G. 1994. Photomorphogenesis in the fungi. In "Photomorphogenesis in plants" 2nd edition. Kendrick, R.E. and Kronenberg, G.H.M. (Eds) Kluwer Dordrecht. pp 753–782.

Marwan, W., Schäfer, W. and Oesterhelt, D. 1990. Signal transduction in *Halobacterium* depends upon fumarate. *EMBO J.* **9:** 355–362.

Meske, V. and Hartmann, E. 1995. Reorganisation of microfilaments in protonemal tip cells of the moss *Ceratodon purpureus* during the phototropic response. *Protoplasma* **188:** 59–69.

Meske, V., Rupert, V. and Hartmann, E. 1996. Structural basis for the red light induced repolarisation of tip growth in caulonemal cells of *Ceratodon purpureus. Protoplasma* **192:** 189–198.

Mooney, J.L. and Yager, L.N. 1990. Light is required for conidiation in *Aspergillus nidulans. Genes and Dev.* **4:** 1473–1482.

Pratt, L.H. 1994. Distribution and localization of phytochrome within the plant. In "Photomorphogenesis in plants" 2nd. Ed, Kendrick, R.E. and Kronenberg, G.H.M., (Eds) Kluwer, Dordrecht. pp 163–186.

Quail, P.H., Boylan, M.T., Parks, B.M., Short, T.W., Xu, Y. and Wagner, D. 1995. Phytochromes: Photosensory perception and signal transduction. *Science* **268:** 675–680.

Quatrano, R.S. and Shaw, S.L. 1997. Role of the cell wall in the determination of polarity and the plane of division in *Fucus* embryos. *Trends Plant Sci.* **2:** 15–21.

Remberg, A., Lindner, I., Lamparter, T., Hughes, J., Kneip, K., Hildebrandt, P., Braslavsky, S.E., Gärtner, W. and Schaffner, K. 1997. Raman spectroscopic and light-induced-kinetic characterization of a recombinant phytochrome of the cyanobacterium *Synechocystis. Biochemistry* **36:** 13389–13395.

Rudolf, J., Tolliday, N., Schmitt, C., Schuster, S.C. and Oesterhelt, D. 1995. Phosphorylation in halobacterial signal transduction. *EMBO J.* **14:** 4249–4257.

Schaefer, D., Zrÿd, J.-P., Knight, C.D. and Cove, D.J. 1991. Stable transformation of the moss *Physcomitrella patens. Mol. Gen. Genet.* **226:** 418–424.

Schaefer, D. and Zrÿd, J.-P. 1997. Efficient gene targeting in the moss *Physcomitrella patens. Plant J.* **11:** 1195–1206.

Schneider-Poetsch, H.A.W. 1992. Signal transduction by phytochrome: phytochromes have a module related to the transmitter modules of bacterial sensor proteins. *Photochem. Photobiol.* **56:** 839–846.

Schneider-Poetsch, H.A.W., Kolukisaoglu, H.Ü., Clack, T., Hughes, J. and Lamparter, T. 1997. Lower plant phytochromes and the evolution of higher plants. *Physiol. Plantarum* **102:** 612–622.

Schumaker, K.S. and Dietrich, M.A. 1997. Programmed changes in form during moss development. *Plant Cell* **9:** 1099–1107.

Senger, H., Schmidt, W. 1994. Diversity of photoreceptors. In "Photomorphogenesis in plants", 2nd. Ed, Kendrick, R.E. and Kronenberg, G.H.M., (Eds.) Kluwer Dordrecht. pp 301–325.

Siegelman, H.W. and Butler, W.L. 1965. Properties of phytochrome. *Ann. Rev. Plant Physiol.* **16:** 383–392.

Sineshchekov, V., Hughes, J. and Lamparter, T. 1998. Low-temperature fluorescence studies of *Synechocystis* phytochrome. *Photochem. Photobiol.* **67:** 263–267.

Sineshchekov, V., Lamparter, T. and Hartmann, E. 1994. Evidence for the existence of membrane-associated phytochrome in the cell. *Photochem. Photobiol.* **60:** 516–520.

Smith, H. 1994. Sensing the light environment: the functions of the phytochrome family. In "Photomorphogenesis in plants", 2nd. Ed, Kendrick, R.E. and Kronenberg, G.H.M., (Eds.) Kluwer, Dordrecht. pp 377–416.

Smith, H. 1995. Physiological and ecological function within the phytochrome family. *Ann. Rev. Plant Physiol. Plant Mol. Biol.* **46:** 289–315.

Starostzik, C. and Marwan, W. 1995. A photoreceptor with characteristics of phytochrome triggers sporulation in the true slime mould *Physarum polycephalum. FEBS Lett.* **370:** 146–148.

Stock, J.B. and Lukat, G.S. 1991. Bacterial chemotaxis and the molecular logic of intracellular signal transduction networks. *Ann. Rev. Biophys. Biophys. Chem.* **20:** 109–136.

Thümmler, F., Dufner, M., Kreisl, P. and Dittrick, P. 1992. Molecular cloning of a novel phytochrome gene of the moss *Ceratodon purpureus* which encodes a putative light-regulated protein kinase. *Plant Mol. Biol.* **20:** 1003–1017.

von Wettstein, F. 1932. Genetik. In "Manual of bryology", Verdoorn F., (Ed.) Martinus Nijhoff, The Hague. pp 233–272.

Wada, M. and Kadota, A. 1989. Photomorphogenesis in lower green plants. *Ann. Rev. Plant Physiol. Plant Mol. Biol.* **40:** 169–191.

Wada, M., Kanegae, T., Nozue, K. and Fukuda, S. 1997. Cryptogam phytochromes. *Plant Cell and Environ.* **20:** 685–690.

Wada, M. and Sugai, M. 1994. Photobiology of ferns. In "Photomorphogenesis in plants", 2nd. Ed, Kendrick, R.E. and Kronenberg, G.H.M., (Eds.) Kluwer Dordrecht. pp 783–803.

Wagner, T., Cove, D.J. and Sack, F.D. 1997. A positively gravitropic mutant mirrors the wild-type protonemal response in the moss *Ceratodon purpureus. Planta* **202:** 149–154.

Wilde, A., Churin, Y., Schubert, H. and Börner, T. 1997. Disruption of a *Synechocystis* gene with partial similarity to phytochrome genes alters the response to changing light qualities. *FEBS Letters* **406:** 89–92.

Winands, A. and Wagner, G. 1996. Phytochrome of the green alga *Mougeotia:* cDNA sequence, autoregulation and phytogenetic position. *Plant Mol. Biol.* **32:** 589–597.

Yeh, K.-C., Wu, S.-H., Murphy, J.T. and Lagarias, J.C. 1997. A cyanobacterial phytochrome two-component light sensory system. *Science* **277:** 1505–1508.

Zeidler, M., Gatz, C., Hartmann, E. and Hughes, J. 1996. Tetracycline-regulated reporter gene expression in the moss *Physcomitrella patens: Plant Mol. Biol.* **30:** 199–205.

Zeidler, M., Hughes, J.E., Gärtner, W. and Lamparter, T. 1998. Photochemical properties of recombinant phytochrome from the moss *Ceratodon purpureus: Heterologous expression and kinetic analysis of Pr/Pfr conversion. Photochem. Photobiol.* **68:** 857–863.

Concepts in Photobiology: Photosynthesis and Photomorphogenesis
G.S. Singhal, G. Renger, S.K. Sopory, K-D. Irrgang and Govindjee (Eds)

29. The Photoperiodic Control of Plant Reproduction

S.D. Jackson and B. Thomas
Horticulture Research International Wellesbourne, Warwick, CV35 9EF, UK

Summary
Photoperiodism is important in controlling many aspects of plant development and in co-ordinating certain responses, such as flowering, with different times of the year. There are three main photoperiodic response types, short-day, long-day and day-neutral. The responses of SDP and LDP have many features in common such as the importance of the dark period (the critical night length), the involvement of phytochrome and the interaction with a circadian oscillator. There are, however, other features that are peculiar to LDP such as the greater influence of the light period and the possible involvement of a blue-light receptor in the photoperiodic responses of long-day requiring Cruciferae.

In SDP and LDP the mechanism of perception of the photoperiodic stimulus and the production of the transmissible signal(s) in the leaf is the same as has been demonstrated by grafting experiments. There is evidence for both an inducing signal and an inhibitory signal, the natures of which are not yet known. These signals appear to be fairly universal in plants and affect a wide variety of processes. Evidence is accumulating for the involvement of gibberellins in photoperiodic responses and that phytochrome B may be involved in regulating the level of, or sensitivity to, certain types of gibberellins.

Finally, the isolation of genes from flowering-time mutants is going to significantly advance our knowledge about the mechanism of the response to photoperiod, and is going to enable the manipulation of plant photoperiodic responses for agricultural and economic benefit.

1. Introduction

Light is one of the most important environmental factors that affects plant development. All higher plants develop differently in the light than they do in the dark. Development in the light is called photomorphogenesis, and development in the dark is called skotomorphogenesis, or etiolation. Almost all of the key aspects of plant development including germination, plant height, leaf shape, chloroplast development, chlorophyll biosynthesis, and apical dominance can be affected by growing the plant in the light or in darkness. In addition, flowering, tillering, leaf abscission, dormancy, fruit set, senescence and the formation of storage organs are also affected by light, although these are examples of photoperiodic responses which are influenced by the length of time that the plant is in the light.

Photoperiodism, the ability of the plant to detect, and respond to, different lengths of dark and light periods, i.e. to be able to differentiate between short days and long days, is essential for the life-cycle of the plant. The invariable connection between short days and winter, and long days and summer, means that photoperiod provides a robust mechanism by which plants can chart the passing of the year. Together with the plant's response to low temperatures (vernalization), which is used to differentiate between spring and autumn when the days are of a similar length, photoperiodism is used to link the various aspects of plant development with different times of the year in such a way as to enhance the chances of survival of the plant and its progeny. For example, the dropping of leaves, onset of dormancy, development of frost hardiness, and tuberization are promoted by the short days and low temperatures of autumn and thus provide the plants with a means by which they can anticipate, and adopt measures to survive, the freezing temperatures of winter. On the other hand, the breaking of bud dormancy of woody plants, and flowering in many other plants is promoted by the longer days of spring and summer. In some cases separate stages of the same developmental process may have different photoperiodic requirements, e.g. several species of *Bryophyllum* require the longer days of summer for floral induction but then shorter days of autumn for flower development. This is a temperature independent mechanism to ensure that flowers appear in autumn.

Flowering is probably the most extensively studied photoperiodic response because it is so economically important in agriculture and horticulture. Such studies have shown it to be a complex response which is promoted in some species by long days but in other species by short days. A simpler but, as we will see later, a closely related photoperiodic response is tuberization which is promoted by short days and inhibited by long days. For the rest of the chapter we will focus on flowering and tuberization as the principle examples of reproductive development under the control of light.

The discovery of the control over flowering and reproduction that is exerted by photoperiod has had important economic implications. The photoperiodic requirements for all varieties of major crops are now known to enable the best variety to be chosen for a particular latitude, thus crops are now only grown in places where the growing season incorporates days of the necessary length to induce flowering. In addition the season of the crop can be extended by manipulating lighting conditions in the greenhouse. With some species plant breeders can obtain up to three generations of plants per year leading to almost year-round availability, e.g. *Chrysanthemum*, originally an autumn-flowering garden plant, has now become a major horticultural crop because horticulturalists can now induce flowering in all seasons by simply manipulating the length of the light period. It is now possible to time the production of certain plants to coincide with periods of high demand, e.g. the production of poinsettias for Christmas and Easter.

One nice example of the socio-economic impact of understanding photoperiodism dates back to 1947 in Trinidad. On the edge of a swamp paddy field an oil company had built a tall pipe from which waste natural gas from the refinery was burnt. The plants in this paddy field didn't flower in October as they normally did and the farmers were faced with a lost harvest. When scientists discovered that some plants that were shaded by a tree had flowered they concluded that the plant's photoperiodic requirement for short days for flowering had been disturbed by the light from the burning gas pipe at night. The oil company stopped burning the waste gas from that pipe, the plants flowered and the farmers managed a late harvest (Benson and Murray, 1949).

It is important to note, however, that many other factors affect flowering and reproduction. Unusual conditions such as drought, heat stress, or the presence of water in arid environments can induce flowering, which illustrates the fact that infrequently occurring advantageous conditions, or stressful conditions that threaten the survival of the plant, can cause an over-riding of the controls that normally exist over reproduction. There is an obvious evolutionary benefit in having such fail-safe mechanisms which greatly enhance the chances of propagation for a particular plant, although responses vary greatly from species to species. Sucrose can also affect flowering and is partially able to substitute for an inducing light treatment (Friend et al., 1984), high sucrose levels also enhance tuberization (Gregory, 1965). In fact the supply of sucrose and other photoassimilates to the meristem was at one time proposed to be responsible for the control of flowering, long days (even just one) resulting in increased levels of assimilates at the meristems (Abou-Haidar et al., 1985; Britz et al., 1985). However, whilst there is a definite requirement for sucrose and nutrients for the development of reproductive organs, rates of photosynthesis and therefore levels of assimilates can be varied without affecting the induction of flowering and vica-versa, e.g. *Pharbitis nil* seedlings that have been germinated and grown in the dark can be induced to flower by only 5 seconds of red light if they are simultaneously treated with cytokinin (Ogawa and King, 1979), the amount of photoassimilate produced in such a short light treatment would be negligible.

Thus there are many factors involved in several different pathways that all influence flowering, some pathways may overlap and some factors may play a common role in several pathways. It is therefore difficult, and dangerous, to generalize too much and to point to one mechanism as being more important than any others, but in this chapter we will only consider the control of reproduction mediated by light and allude to the principle cases where this control is influenced by other factors such as temperature and the developmental state of the plant.

2. Photoperception

In addition to different lengths of light/dark treatment (photoperiod), plants can also detect different wavelengths of light (light quality) and different amounts of light (light quantity). These all provide the plant with information about its environment and have wide-ranging effects on plant development.

Plants have developed different mechanisms to detect and respond to changes in various aspects of the light environment. The shorter wavelengths of light are mainly detected by the UV-B and blue-light receptors, whereas light at the red and far-red end of the spectrum is detected by the phytochromes with different members of the phytochrome family being involved in different responses. Different intensities of light also lead to different responses such as the Very Low Fluence Response (VLFR), the Low Fluence Response (LFR) and the High Irradiance Response (HIR), some of which have also been attributed to different phytochromes. The LFR occurs in the range of light intensities that normally occur during daylight and it incorporates all photoreversible reactions which includes photoperiodic responses.

Of all the information coming from the light environment, it is photoperiod that has the greatest influence over reproduction, although light intensity and quality have been shown to affect flowering, e.g. high light intensities are known to increase the number of flowers produced, probably by increasing the amount of available photoassimilates. Photoperiodic responses involve detecting normal levels of light at the red and far-red end of the spectrum, and as photoperiodic responses are photoreversible they principally involve phytochrome. There is evidence, however, that in the Cruciferae, which includes *Arabidopsis* and mustard, a blue-light receptor is also involved in the photoperiodic response, this will be discussed in more detail later on.

2.1 The Discovery of Photoperiodism

It was around the beginning of this century when Klebs and Tournois, working independently, proposed that the duration rather than the quantity of light was important in flower induction in certain plants. Garner and Allard in 1918 performed some watershed experiments which established that photoperiodism was a major factor in the control of flowering. They noted that a late maturing strain of soybean, 'Biloxi', flowered at around the same time of year regardless of the time of planting and therefore the age of the plant at flowering. They also noticed that plants grown under the shade of cheesecloth, although they grew more straggly, flowered at the same time as unshaded plants that had received a greater amount of light. In their other work on Maryland Mammoth, which is a variety of tobacco that arose out of a spontaneous single gene mutation and which grew to a much greater size than the rest of the field of Maryland Narrowleaf tobacco, they found that when they grew it indoors in greenhouses over winter it flowered regardless of age; even small young plants produced flowers. They artificially shortened the long days of summer by putting tobacco and soybean plants into a dark chamber before the end of each day and found that the flowering of these plants was greatly accelerated compared to control plants kept outside in the long days (Garner and Allard, 1920). They subsequently extended their observations to other species and to other responses such as branching, tuberization and dormancy.

It is not actually the length of the light period but the length of the dark period that is measured. The length of the light period can be varied within reasonable limits with little or no effect, whereas the dark period has a certain timepoint called the critical night length and the response of the plant depends upon whether the critical night length has been exceeded or not. The necessary period of time spent in the dark to reach the critical night length varies between plants, and so does the response. There are three main photoperiodic response groups;

(i) Short-day plants (SDP). Plants which only flower in, or in which flowering is promoted in, photoperiods where the dark period exceeds the critical night length for that plant.
(ii) Long-day plants (LDP). Plants which only flower in, or in which flowering is promoted in, photoperiods where the dark period is shorter than the critical night length for that plant.
(iii) Day-neutral plants (DNP). Plants in which flowering is not affected by photoperiod.

It is important to note that in some cases LDP, e.g. *Hyoscyamus niger,* have a critical night length (13 h) which is longer than for some SDP, e.g. *Xanthium strumarium* which has a critical night length of 8.5 h. Therefore the LDP *H. niger* will flower in long days of 14 hours that are also short enough for the SDP *X. strumarium* to flower.

Some plants have dual photoperiodic requirements, long-short day plants (LSDP) require long days followed by short days and flower after summer in the autumn, and short-long day plants (SLDP) which flower in the longer days of late spring or summer after the short days of winter. There are also plants that only flower in days of intermediate lengths, e.g. between 12 h and 14 h (nights of 10–12 h), or ambiphotoperiodic plants that flower in both long and short days but not in intermediate daylengths. The effect of daylength on flowering in all these response types is shown in Fig. 1. There can be quite a wide range of photoperiodic requirements within one species which has been developed to adapt the individual plants to the different latitudes in which they are growing. Conversely, some tropical plants experience only very small changes in photoperiod throughout the year yet are still able to perceive, and respond to, those small changes.

2.2 The Involvement of Phytochrome

One line of evidence that implicates phytochrome in the photoperiodic response is that such responses are photoreversible. As phytochrome can be converted from an inactive P_R form to an active P_{FR} form and back again by manipulating the levels of red and far-red light detected by the plant, it is also possible to cancel out the effect of an inducing photoperiod, or to induce a plant in a non-inducing photoperiod. This can easily be demonstrated by use of a night-break. A night-break is a short light treatment which interrupts the dark

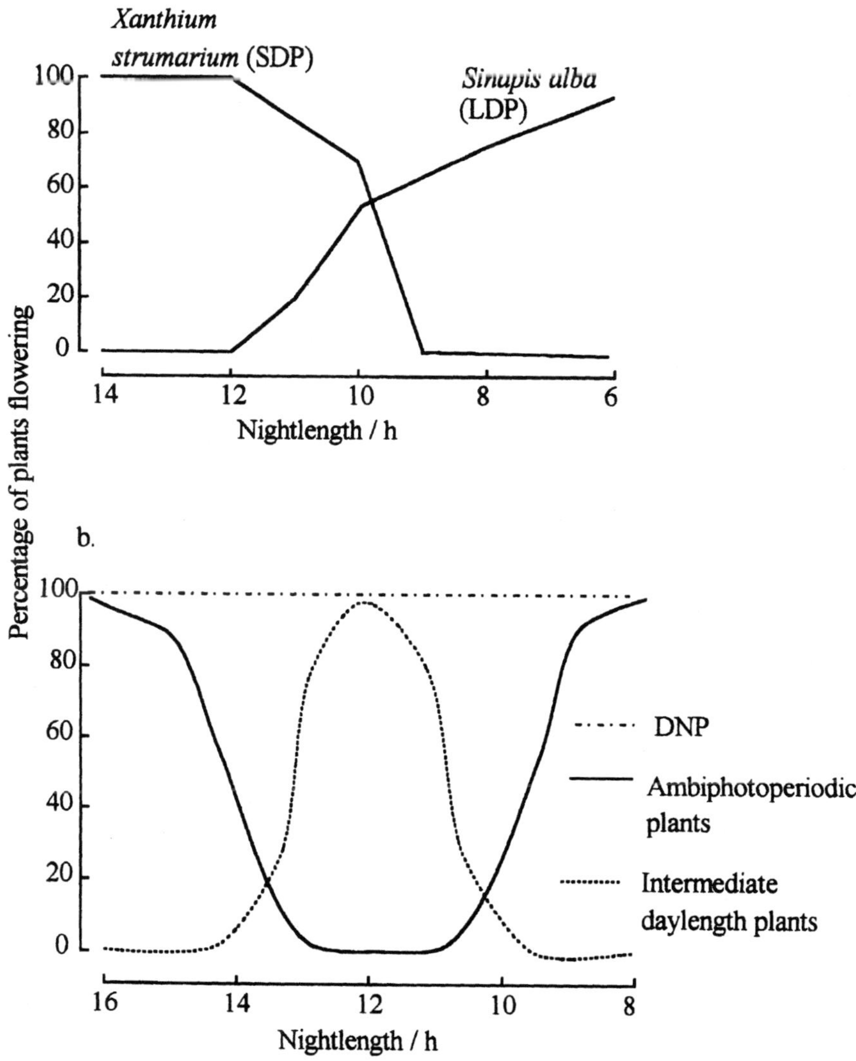

Fig. 1 **(a) The effect of different nightlengths on flowering of *Xanthium strumarium* (a SDP) and *Sinapis alba* (a LDP). (b) The flowering response of day-neutral, ambiphotoperiodic, and intermediate daylength plants in different nightlengths.**

period, thus with SDP a night break would interrupt the dark period before the critical night length was reached and therefore prevent flowering. Interrupting a long light period with a short dark treatment has no effect which is further evidence that the dark period is the most important with respect to photoperiodic responses (Fig. 2a). Some LDP respond to night-breaks, e.g. *Fuchsia*, but not many and usually the night-break treatment has to be much longer than for SDP which may reflect a difference between LDP and SDP. The effect of a night-break varies greatly according to the time at which it is given (Fig. 3). For both SDP and LDP a night-break is most effective when given 8 hours after the start of the dark period regardless of

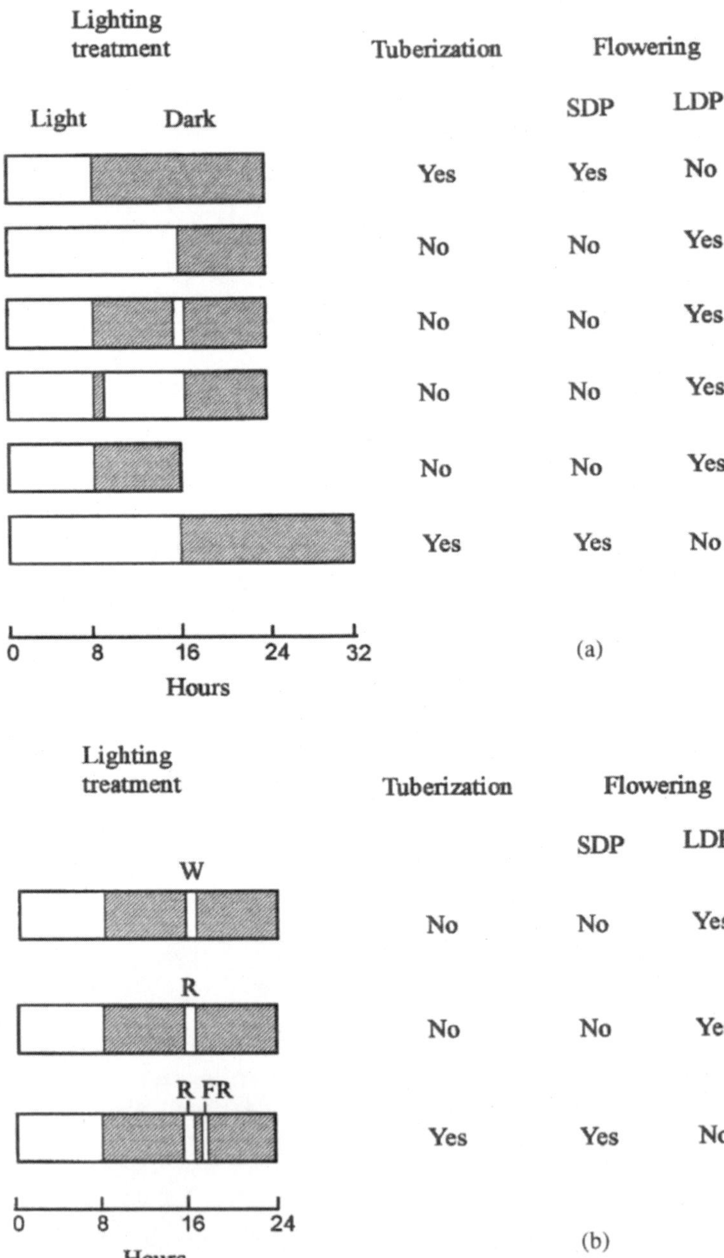

Fig. 2. (a) Flowering and tuberization responses in different photoperiodic conditions. (b) Far-red reversibility of the effect of a red light night break on flowering and tuberization responses.

the total length of the dark period, although for a long night of 16 hours this would be exactly mid-way through (as is the case in Fig. 2). Red light is the most effective wavelength in promoting night-break responses, and the effect of a red light night-break can be completely reversed if far-red light is given immediately afterwards indicating that phytochrome is the only photoreceptor involved in the night-break response. The tuberization of potato leaf cuttings maintained in short days in mist chambers could be inhibited by a night-break of red light, and this inhibition could be alleviated by far-red light given immediately afterwards (Fig. 2b) (Batutis and Ewing, 1982). The inhibition of flowering by a red night-break in SDP such as *Pharbitis* and *Xanthium* could be reversed by far-red light in a similar manner, and so can the induction of flowering by a red night-break in LDPs such as *Hordeum vulgare* and *Hyoscyamus*, although in LDP only partial far-red reversibility has been obtained (Downs, 1956). Phytochrome A mutants of *Arabidopsis* flower at about the same time as wild-type plants in inducing long days and inhibitory short days, but are much less sensitive to an incandescent low-light extension of a short day which in wild-type plants produces the same response as long days (Johnson et al., 1994). Phytochrome A therefore appears to be involved in the perception of daylength in *Arabidopsis*. Phytochrome B *Arabidopsis* mutants, on the other hand, flower earlier than wild-type plants in both short day and long day conditions (Goto et al., 1991).

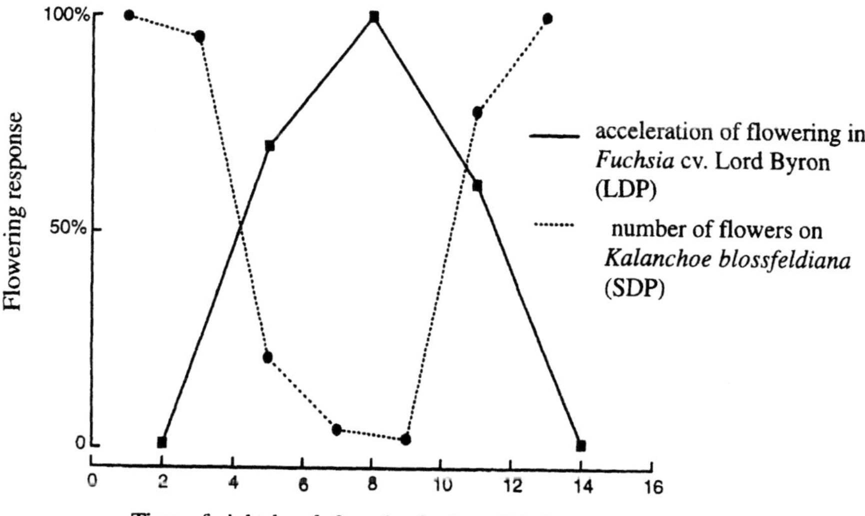

Fig. 3. The varying effect on flowering of *Fuchsia* (a SDP) and *Kalanchoe* (a LDP) of a night break given at different times in the dark period.

Whilst phytochrome is the photoreceptor involved in detecting the night-break and is probably responsible for photoperiodic responses in most plants, there is some evidence that a blue-light receptor also plays a role in the photoperiodic responses of certain long-day requiring Cruciferae such as *Arabidopsis* and mustard. For example blue light night breaks are the most effective for *Sinapis alba* (Hanke et al., 1969), and the addition of blue light to a 16 h low light extension of an 8 h short-day can enhance flowering in *Arabidopsis* to the same extent as control kept in constant light. Plants with the low light extension without the blue light responded as if kept in short days and flowered later (Mozley and Thomas, 1995).

There is only a limited period of time within which far-red reversibility of a red light night-break can take place, usually 0.5–30 minutes, after which subsequent reactions downstream from phytochrome in the signal transduction pathway have taken place and the process has escaped reversibility by far-red light (Downs, 1956). With respect to flowering and tuberization the plant is now said to be induced.

3. Induction

Some plants only need one inducing night period for induction to occur and these are usually the well-described model systems such as *Pharbitis* or *Xanthium*. Most plants however need several inductive cycles, *Chrysanthemum* and *Solanum tuberosum* ssp. *andigena*, for example, both need around two weeks in inducing conditions for flowering and tuberization respectively. A range of responses of SDP is shown in Fig. 4.

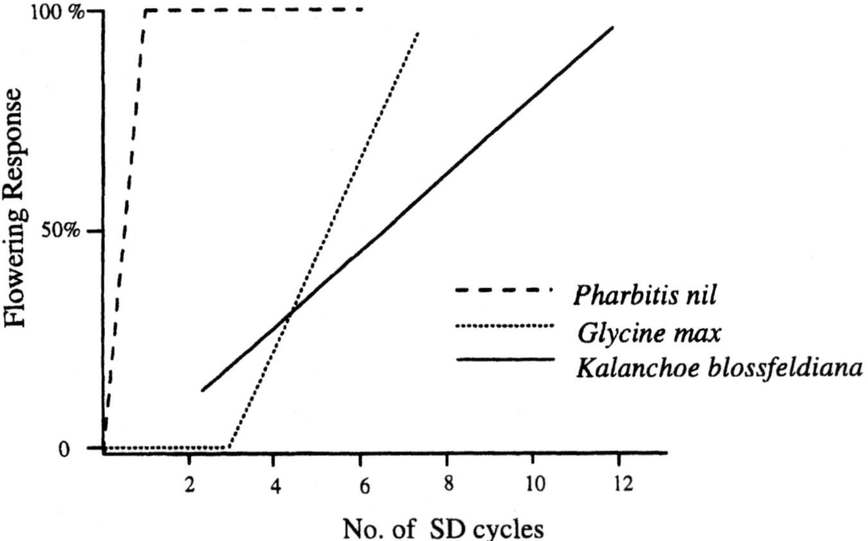

Fig. 4. Different types of responses to inducing conditions, some plants are induced after only one inducing photoperiod (e.g. *Pharbitis nil*), some require several inductive cycles (e.g. *Glycine max*), and others show a response that is proportional to the number of inductive cycles (e.g. *Kalanchoe*).

Despite the main role of the dark period in photoperiodic responses, there is also a requirement for an initial light treatment before the start of the dark period. Without a preceding light treatment a long dark period is not effective in inducing flowering in SDP or in preventing flowering in LDP, e.g. the SDP *Xanthium* needs a light treatment of 3–5 hours before a long dark period will induce flowering (Salisbury, 1965). In addition, the time at which a night-break is most effective (NB_{max}) is not changed by increasing the length of the dark period, e.g. in the SDP *Pharbitis* a night-break given after 8 hours of darkness completely inhibits flowering even if the plants are kept in the dark for a further 40 hours which is much longer than the critical night length (Takimoto and Hamner, 1964). The time of the NB_{max} is therefore always set at a particular time into the dark period starting from the light to dark transition (or dusk signal). The timing mechanism used is the plant's endogenous circadian clock.

3.1 The Involvement of a Circadian Oscillator

The circadian clock is involved in many rhythmic responses in both plants and animals, examples include pupal emergence in *Drosophila*, activity cycles of rodents, sporulation patterns in fungi, and stomatal opening and leaf movement in plants. The rhythmic responses that are coupled to the circadian clock are called circadian rhythms. More than one rhythm can be coupled to one clock, but there is also evidence that more than one clock exists, as leaf movement and photoperiodic rhythms appear to be coupled to different clocks in *Xanthium, Pharbitis* and *Chenopodium* (Salisbury, 1990). Circadian rhythms have been extensively studied and they all have certain features in common, one of which is that they have a free-running period of about 24 hours (about one day, hence the name *circa dia* n) in constant light or constant dark (Fig. 5a). Another is that the period of the rhythms is not affected by changes in temperature. A free-running rhythm in constant conditions becomes dampened, i.e. reduced in amplitude, with time and eventually becomes suspended. The time needed for suspension of the rhythm varies between different rhythms (Lumsden, 1991). An environmental signal, termed a zeitgeber (which is german for time-giver), is needed to restart the rhythm, for example the light to dark transition at dusk (Fig. 5b). The dark to light signal at dawn, and sometimes temperature changes can also act as zeitgebers.

Evidence for the involvement of a circadian clock in photoperiodic responses are the rhythmic changes in the response to light treatments during a long dark period, e.g. in the SDP *Chenopodium rubrum,* the degree of inhibition of flowering by a single night-break varies depending upon when that night-break was given (Fig. 6a). A rhythm is revealed that has a period of about 24 h. Such a rhythmic response has been uncovered in several other photoperiodic species both SDP, e.g. *Glycine max*, and LDP, e.g. *Lolium temulentum,* although in some cases, e.g. *Pharbitis,* the length of the preceding light period needed to be quite short (6 h or less) in order for the rhythm to become apparent (Thomas and Vince-Prue, 1984). Resonance experiments provide further evidence that a circadian clock is involved in the flowering response. These experiments involve giving the plants a short photoperiod and then keeping them in the dark for increasing lengths of time. A rhythm with a period of 24 h in the flowering response of *Chenopodium rubrum* was obtained that was almost identical to the rhythm in night-break sensitivity (Fig. 6b).

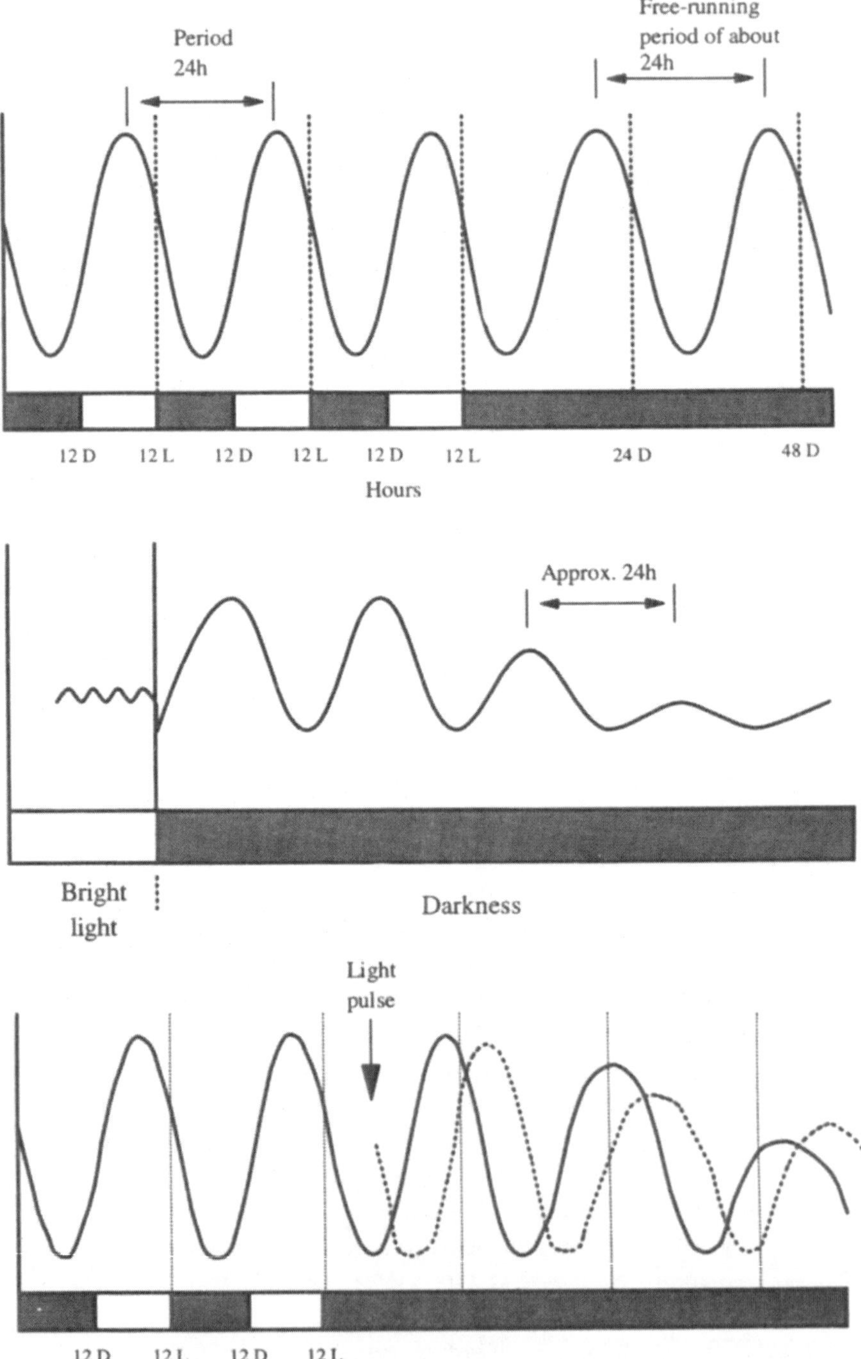

Fig. 5. **(Top) Circadian rhythms have a period of 24 h in diurnal light/dark cycles, and continue to run in constant light or darkness with only a relatively small change in period length. (Middle) In constant conditions rhythms often become dampened but are restarted by a zeitgeber such as transfer to darkness. (Bottom) The phase, but not the period, of circadian rhythms can be changed by a light pulse in the dark period.**

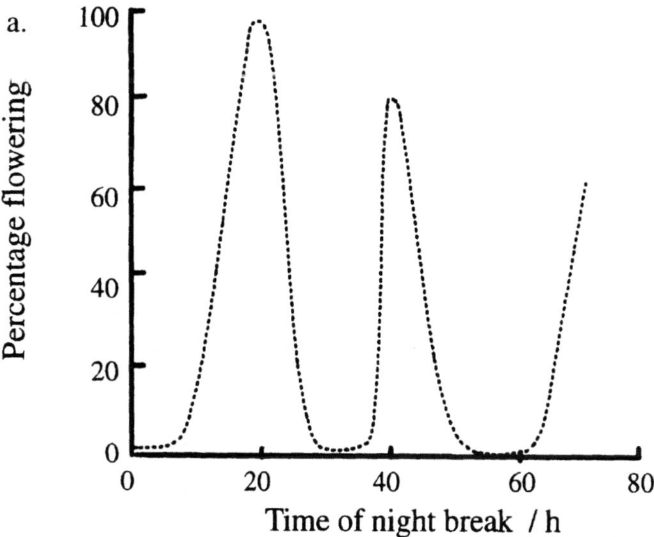

b.

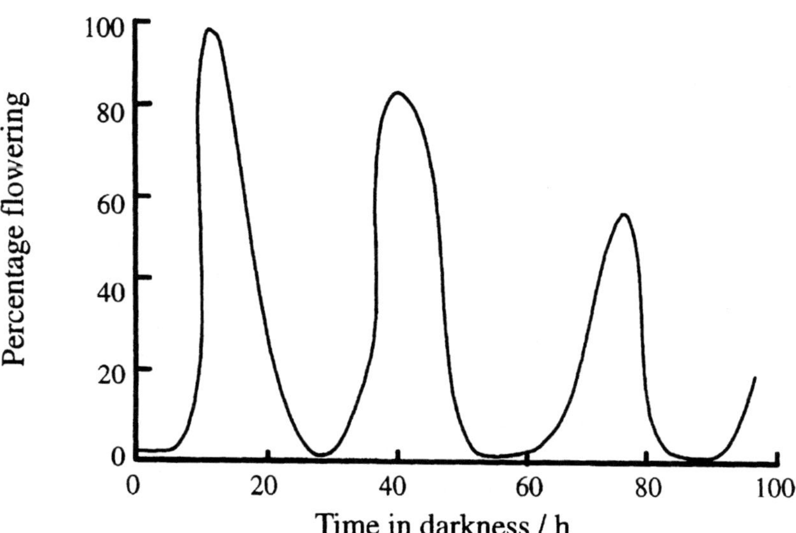

Fig. 6. Circadian rhythms in *Chenopodium rubrum*. (a) The rhythm of sensitivity to a night-break given at different times in the dark period. (b) The rhythm in flowering response with increasing lengths of dark periods.

Both dawn (lights on) and dusk (lights off, or light intensities below the threshold level) signals have been shown to be zeitgebers in the flowering response of the SDP *Pharbitis*. The circadian rhythm is entrained at dawn and runs for up to 6 hr before becoming suspended until dusk whereupon it is released at a particular stage of the cycle (Papenfuss and Salisbury, 1967). As daylengths normally exceed 6 hr this means that plants would almost always use the light-dark transition to start, or release, the rhythm to measure the length of the dark period. Regardless of the changing length of the day in different seasons, because the rhythm is always released at a specific stage of the cycle and because of the constant period of the rhythm, the result is that the critical night length of the plant, and the time of maximum inhibition by a night-break, are always at a specific timepoint after the start of the dark period. This is how the plant can coincide a particular process with a certain time of the day or night.

The phase of these rhythmic responses (but not the period) can be altered, they can be advanced or delayed by a light treatment of white or red light (Figs. 5 (Bottom) and 7), the latter suggesting that phytochrome is able to phase-shift the circadian rhythm in plants. Unfortunately, the inhibitory effect of far-red light on flowering of SDP has complicated the demonstration of far-red reversibility of the phase-shifting of the flowering rhythm although there is a report of this having been achieved (Lumsden, 1991). This far-red reversibility of a red light phase-shift has clearly been shown in other circadian rhythms however, e.g. leaf movement in *Mimosa pudica* (Fondevill et al., 1966). Phytochrome is not the only photoreceptor that is able to shift circadian rhythms, and in some systems, e.g. the unicellular alga *Gonylaux polyedra,*

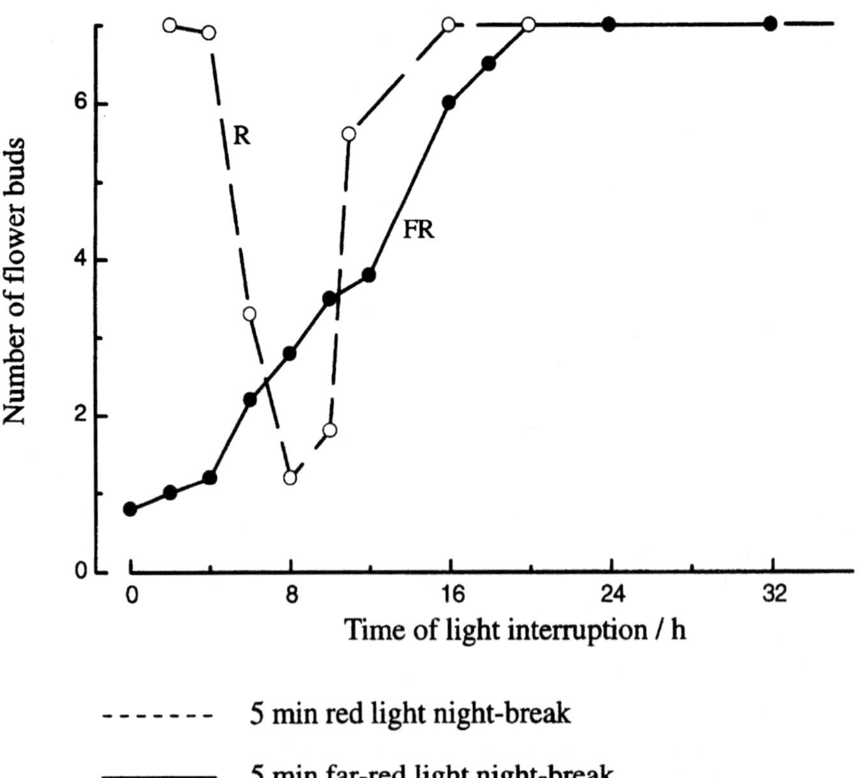

Fig. 7. Characteristics of phase-shifting of a circadian rhythm, e.g. photoperiodic timekeeping in *Chenopodium rubrum,* in plants given a light treatment at various times during the dark period compared to control plants that were not exposed to light.

it is clearly not involved. Blue light plays a role in many organisms, e.g. *Neurospora* and *Drosophila,* and blue light can entrain the leaf movement rhythm in *Samanea* (Satter et al., 1981).

Light and phytochrome are involved in two quite separate processes in the flowering response, firstly in the light to dark transition and setting the phase of the circadian clock which then starts to time the length of the dark period. Once the dark period has exceeded the critical night length for SDP (or before it exceeds the critical night length for LDP), then induction is possible. Because of the rhythmic nature of the clock there is only a certain amount of time (the inducible phase) during which induction is possible before light becomes inhibitory again. The photoperiodic response of a plant depends upon the phase of the rhythm at the time it is exposed to light (dawn or a night-break). The perception of a night break during the dark period illustrates the second process involving phytochrome which is distinct from the first as a night-break given 8 hr into the dark period which greatly inhibits flowering in *Pharbitis* doesn't reset the circadian rhythm (Lumsden and Furuya, 1986). At other times, e.g. 6 hr into the dark period, the rhythm can be shifted with no effect on flowering. These observations show that the inhibition of flowering by a night break is not due to a shift in the rhythm and that these are two distinct actions of light. It has also been demonstrated that these two processes require different amounts of light, the inhibition of flowering 16 hr after a brief photoperiod can be caused by a night-break of 30 $\mu mol \cdot m^{-2}$ whereas to shift the rhythm at the time required 300 $\mu mol \cdot m^{-2}$ (Lee et al., 1987).

In addition, at least two different actions of phytochrome itself can be distinguished in the control of flowering which may reflect the involvement of different types of phytochrome. One is involved in the rhythm of night-break response and is an interaction with the circadian clock as described above. For this phytochrome the conversion of P_R to P_{FR} by a red night-break inhibits flowering in SDP. However, it is also known that the conversion of P_{FR} to P_R by far-red light given at the end of the day, or even in some cases long into the dark period past the point of NB_{max}, also prevents flowering (Cumming et al., 1965). The inhibition of flowering in SDP by a far-red treatment is not a rhythmic response (Fig. 8) and does not affect the timing of the NB_{max} or critical night length, illustrating that these responses are under the control of phytochromes with different properties. Furthermore, there is evidence that the phytochromes involved differ in their stability in the P_{FR} form. Research is currently underway to determine which of the several phytochromes (A-E) identified to date may be involved in the various interactions with the circadian clock, and with the inhibition of flowering by far-red light.

3.2 The Response to Long Days

Most photoperiodic responses occur in one type of photoperiod (e.g. short days) but not in the other (long days), this is exemplified in the tuberization of potato which is enhanced by short days and inhibited by long days. The stringency of the photoperiodic control of tuberization varies greatly so that in some potatoes where the control is very weak the inhibitory effect of long days is negligible and these potatoes can tuberize in almost all daylengths so that they are in effect day-neutral. There are, however, no potatoes that tuberize only in long days and not in short days. Both situations occur with flowering and indicates that there is a greater degree of complexity in the flowering response to photoperiod particularly in the case of long day plants.

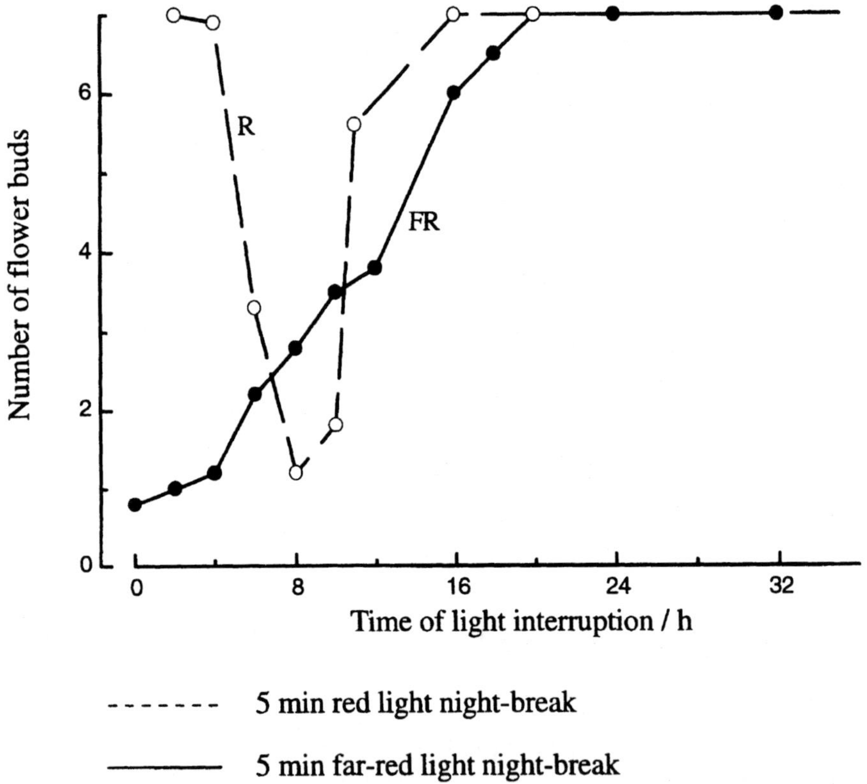

Fig. 8. The differing effects on flowering of *Pharbitis nil* of red light and far-red light treatments given at various times during a dark period.

The fact that the response of most long day plants to different photoperiodic treatments are not just the opposite of the response of short day plants also suggests that there are additional factors involved in the control of flowering of long day plants. The response of most LDP to photoperiod is affected by light intensity much more than the response of SDP. In *Brassica campestris* the flowering response is directly related to the level of light perceived during the day (Friend, 1985), and whilst LDP can be induced to flower by night-breaks during a long dark period the night breaks need to be much longer (at least 1 hr) than is needed for SDP (sometimes less than 5 minutes). Both the night-break response and the flowering response of LDP to increasing dark period lengths show a circadian oscillation as in SDP and there is some evidence that the rhythm in LDP is 12 hr out-of-phase of the rhythm in SDP (Fig. 9) (Vince-Prue, 1994). LDP, however, also respond to far-red light given during the light period which leads to an increased response, e.g. flowering, or bulbing in onion which is also a long day response. The degree of response depends upon when during the light period the far-red light treatment during continual white light was revealed and this rhythm could be phase-shifted (Deitzer et al., 1982). This shows that in LDP a circadian rhythm involved in flowering continues to run in the light period as opposed to SDP in which the rhythm is apparently suspended after a short time in the light.

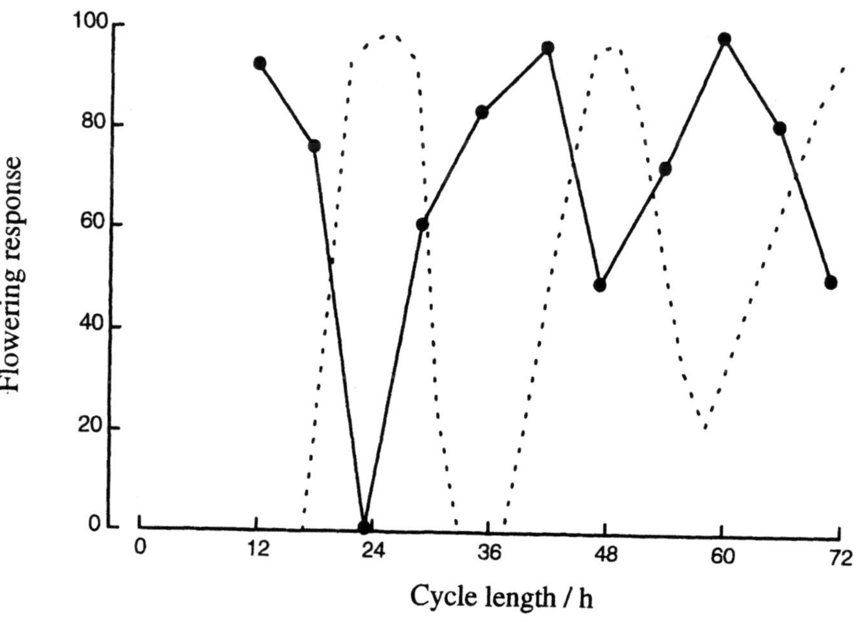

Fig. 9. Circadian rhythms in the flowering response of *Glycine max* (a SDP) and *Hyoscyamus niger* (a LDP) to increasing dark period lengths.

3.3 Interaction with Other Factors

Vernalization is the induction of flowering by a cold treatment given to the plant or to the imbibed seed (dry seeds do not respond to vernalization). A temperature of 1–7°C is the most effective (Lang, 1965) and it appears to be perceived in the meristematic zones of the shoot apex (Thomas and Vince-Prue, 1984) as opposed to the photoperiodic signal which is perceived in the leaves. The response to vernalization varies with age depending upon the plant species, the imbibed seed of winter cereals respond to vernalization whereas other species, such as most biennials which grow as rosettes during the first season after sowing and then flower the following summer, must reach a minimal size/age before they will respond to vernalization (Wellensiek, 1958; Sarkar, 1958).

Photoperiod and vernalization often interact to control flowering, e.g. the dual requirement for long days and cold by plants that flower in the late spring or early summer. In some cases vernalization and photoperiod can substitute, or modify the requirement, for the other, a cold treatment can alter the critical night length requirement of a plant (e.g. *Spinacia,* Vlitos and Meudt, 1955), or even cause the plant to become photoperiodically insensitive. Conversely short days can abolish the vernalization requirement in plants such as *Campanula medium* and *Trifolium repens* (Purvis, 1961; Thomas, 1979). Thus there seems to be some interaction or overlap in the control of flowering by photoperiod and vernalization, it is possible that both mechanisms involve a common step, or component, in the process of flower induction.

The switch from vegetative growth to reproductive growth (e.g. flowering and tuberization) is greatly influenced by plant age. We have described how some biennials need to be of a certain age to respond to vernalization, and the same applies to the photoperiodic induction of flowering. The prevention of flowering in very young, or juvenile, plants is a mechanism by which the plant can delay flowering until it is large enough to support the greatly increased demand on resources that accompanies flower and fruit formation. The juvenile phase of a plant is therefore the early period of growth of the plant during which flowering cannot be induced, the length of the juvenile phase varies between species from a few days in herbaceous plant to several years in woody plants. At a certain developmental stage the plant switches from juvenility to maturity and in some cases this is all that is necessary for the start of flower development. In other cases such as in photoperiodic species where a specific environmental signal is required for the induction of flowering, the mature plant will not flower until this requirement has been met. Plants in such a situation are said to be competent-to-respond, or ripe-to-flower, as they are able to respond (flower/tuberize) and are just awaiting the appropriate inductive conditions to do so. Once a plant becomes mature there is an increasing tendency for it to flower as it gets older, i.e. fewer short days, or long days, are required to induce flowering in older than younger plants (Fig. 10). Often a plant will eventually flower independently of the photoperiod, this is sometimes called autonomous induction. Thus apart from the prevention of flowering in juvenile plants, plant age also affects flower induction in mature plants.

4. Production and Transmission of Stimulus

Although the response can manifest itself at the apex (with flowering) or at the end of underground stolons (with tuberization), the detection of the photoperiodic stimulus occurs in the leaf. This can be demonstrated by only exposing certain parts of the plant to the inducing photoperiod and keeping the rest of the plant in non-inducing conditions. The response is determined by the photoperiodic treatment given to the leaves (in some cases just a single leaf needs to be exposed to an inducing photoperiod for induction to take place, e.g. *Xanthium*), rather than the apex or other parts of the plant.

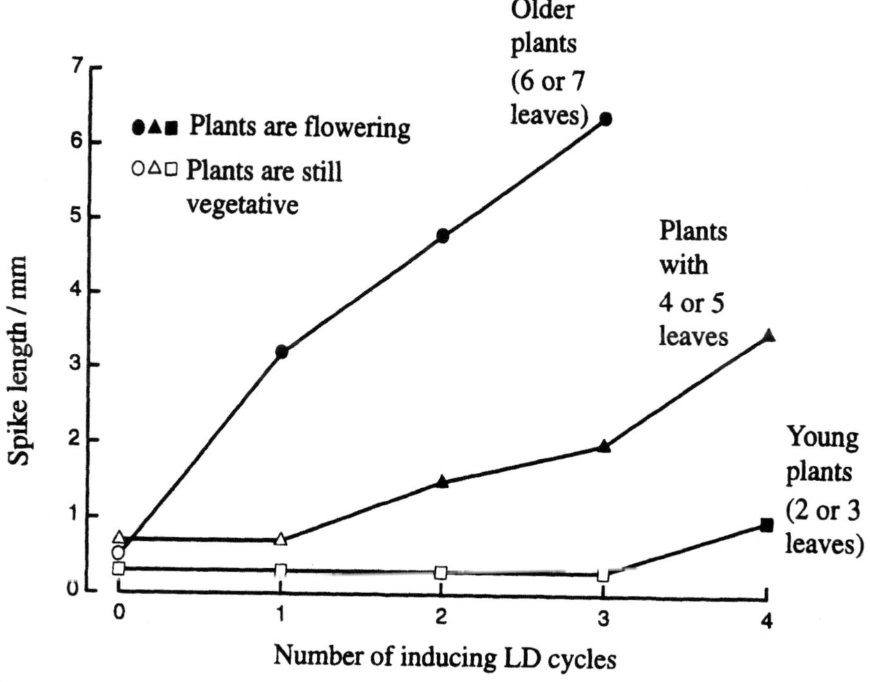

Fig. 10. The increasing sensitivity to photoperiod of flowering in *Lolium* with increasing plant age.

As the photoperiodic signal is perceived in the leaves but results in flower formation at the apex, or tuber formation at the end of stolons, some sort of signal must be produced by the leaves which is transmitted to the other parts of the plant. The existence of such a signal has been demonstrated by grafting experiments where a leaf from an induced plant when grafted onto a non-induced plant can cause this plant to flower even though it is maintained in non-inducing conditions. It is even possible to re-graft such a leaf onto further non-induced plants and it will cause them in turn to flower as well, e.g. *Perilla* (Zeevart, 1958). Detached leaves can be induced photoperiodically and then grafted onto non-induced plants to cause flowering (Zeevart, 1969) illustrating that the photoperiodic induction and production of the signal (called the florigen in the case of flowering) can occur in the leaf alone. In some cases, e.g. *Perilla,* only the leaves exposed to the inducing photoperiod produce the signal, in the others e.g. *Xanthium*, an induced leaf graft can cause induction of the non-induced leaves on the plant to which it has been grafted.

Further evidence for such a signal and for its chemical (e.g. hormonal) rather than physical (e.g. hydrostatic or electrical) nature comes from studying its movement through the plant. If a single leaf is induced by a brief light pulse and then removed at different times, several hours are needed for the production of the signal and for its export from the leaf resulting in flowering of the plant. If it is removed too soon then flowering does not occur (Fig. 11) (Vince-Prue, 1975). The translocation of the signal up the stem has also been monitored by measuring the additional time that an induced leaf must be kept on the plant for the signal to induce flowering in a second meristem situated some distance higher up the stem than the first induced meristem (King et al., 1968). Approximate rates of transport are comparable to the rate of transport of photoassimilates in the phloem, and treatments that restrict phloem transport, e.g. girdling or localised heating, prevent the movement of the signal.

The two-step process, photoperiodic induction followed by production of a chemical signal could explain the dual-phase nature of flowering response curves (Fig. 12) (Salisbury and Ross, 1969). The first part of the curve (A)

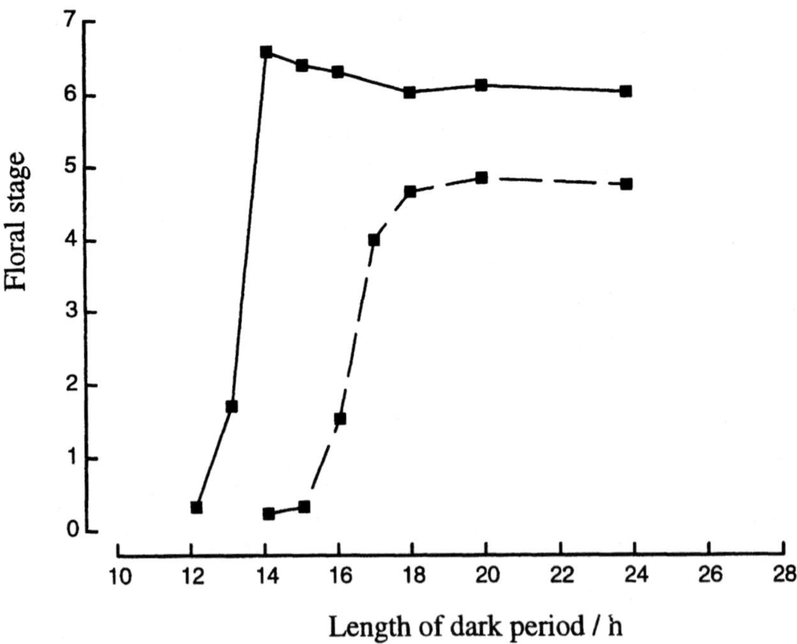

Fig. 11. Flowering response of *Pharbitis*. The effect of increasing night length on intact plants can be contrasted to the response of plants from which the induced cotyledons were removed at the time shown. The cotyledons must be left on the plant for a few hours to allow translocation of the inducing signal before they can be removed and flowering still occur.

is the temperature-insensitive timing component and reflects the increased level of photoperiodic induction once the critical night-length has been exceeded, the second part (B) is temperature sensitive and may reflect the rate of production and transmission of the inducing signal.

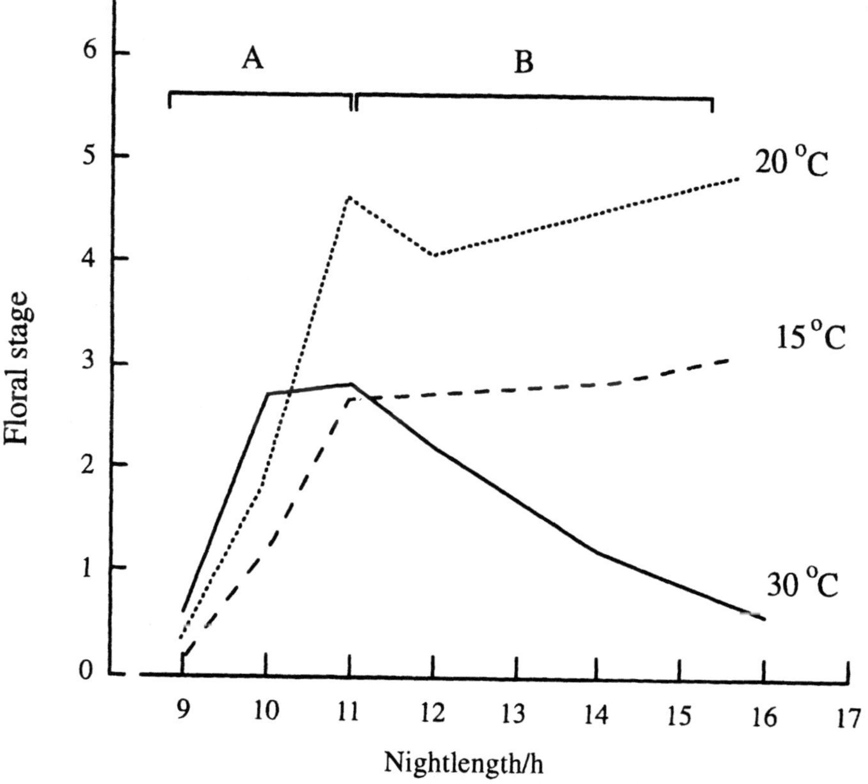

Fig. 12. The effects of temperature and night length of flowering in *Xanthium*. The first part of the curve [A] is relatively temperature insensitive, whereas the second part [B] is affected by temperature.

The signal that is able to induce flowering in SDP is also able to induce flowering is LDP as a leaf from a SDP when grafted onto a LDP grown in short days will cause flowering of the LDP, and vice-versa (see Lang, 1965; Vince-Prue, 1975; Zeevart, 1982). Furthermore, the signals that induce flowering and tuberization are very similar if not the same compound because a leaf from a long-day requiring tobacco (*Nicotiana sylvestris*) that has been induced to flower, will induce tuberization if grafted onto a potato plant kept in non-inducing (long day) conditions (Chailakyan et al., 1981). Similar results have been obtained with grafts involving sunflower and jerusalem artichoke (Nitsch, 1966) so it appears that the same inducing signal is produced in several different species and can induce different responses.

The common nature of the inducing signal and the fact that it is not specific to flowering but affects tuberization and other photoperiodic processes such as sex determination (Takahashi et al., 1982), indicates that the signal is likely to be involved in most, if not all, photoperiodic responses in plants. Apart from an inducing signal there is also evidence for the existence of an inhibitor. Some plants, e.g. the LDP *Hyoscyamus niger* will flower in non-inductive conditions if all their leaves are removed, grafting back a single leaf however will prevent flowering unless it is put into inducing conditions (Lang and Melchers, 1943). In the garden strawberry which requires short days for flowering, if a parent plant is kept in long days but its daughter plants, which are attached to it by runners, are put in short days then flowering of the daughters is greatly reduced (Guttridge, 1959). The identity of these inducing and inhibitory signals are still unknown although many substances including polyamines, stèrols and prostaglandins have been investigated. Most attention, however, has focussed on plant growth factors, but because some plant hormones are probably involved in flower development it is difficult to assess their role in the actual inductive process. Despite some anomalous results it generally appears to be the case that there is no consistent correlation between the levels of auxins, cytokinins, ethylene or ABA and the photoperiodic induction of flowering in SDP and LDP. Evidence is accumulating, however, for a possible role of gibberellins. In many species gibberellin levels increase in stems and leaves under long days, and exogenously applied gibberellin (GA_3) can substitute for the long day requirement (Vince-Prue, 1985). There are exceptions where gibberellins do not promote flowering, often in short-day requiring plants, or that the inductive effect only occurred when gibberellins were applied at certain times.

4.1 The Role of Gibberellins

There are many different types of gibberellin (over a hundred, although only a few are thought to be active), and different gibberellins have different roles. Some are more effective (up to 1, 000 times) in inducing flowering, whilst others are more effective in promoting stem elongation. This difference in flower-inducing activity or stem elongation activity appears to be related to the structure of the gibberellin. Figure 13 shows an outline of the early 13-hydroxylation pathway which is the predominant gibberellin biosynthetic pathway in potato and in most vegetative plant issues (van den Berg et al., 1995a; Sponsel, 1995). One step in this pathway is a 3β-hydroxylation which leads to the formation of 3β-hydroxylated gibberellins such as GA_1. In *Lolium* these 3β-hydroxylated gibberellins are more effective in promoting stem elongation than flowering, whereas gibberellins further back in the pathway before the 3β-hydroxylation step such as GA_{44}, GA_{53}, GA_{19}, and GA_{20} are more active in flowering than stem elongation (Evans et al., 1994). Therefore by manipulating the different steps in the gibberellin biosynthetic pathway to control the relative levels of various gibberellins to each other, the plant can achieve different responses. Photoperiod is known to affect certain steps of the gibberellin biosynthetic pathway. Bolting in Spinach occurs in long days but is prevented in short days because of a reduced activity of a GA_{20}

oxidase that converts GA_{19} to GA_{20} (Gilmour et al., 1986). Senescence in pea is prevented in short days by an increased production of GA_{53} from GA_{12}-aldehyde (Davies et al., 1986).

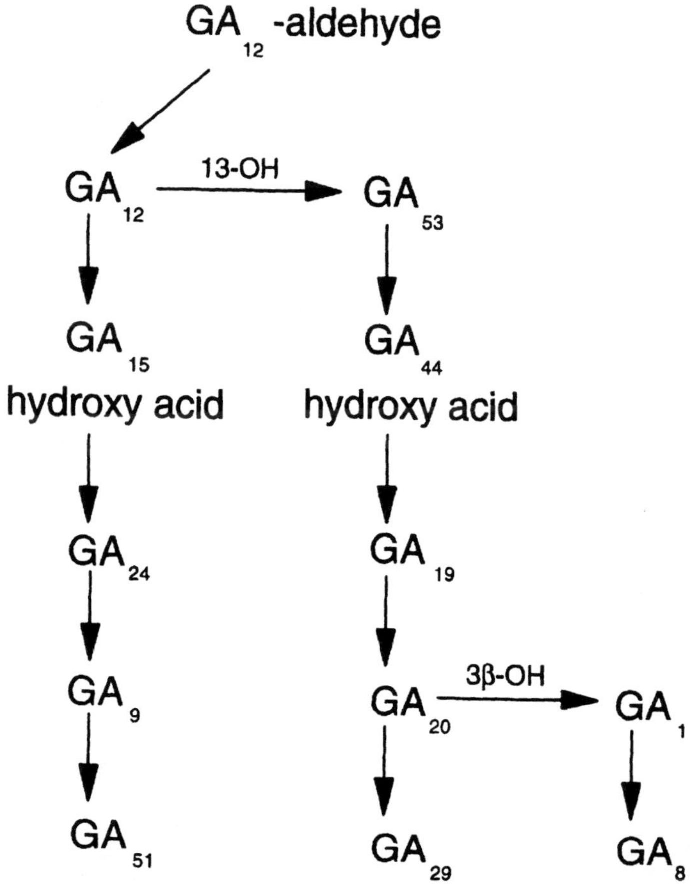

Fig. 13. The early 13-hydroxylation gibberellin biosynthetic pathway leading to the production of the active gibberellins GA_{53}, GA_{44} hydroxy acid, GA_{19}, GA_{20}, and GA_1.

Gibberellins are known to inhibit tuberization and may play a role in the photoperiodic control of tuberization by preventing tuber formation in long days. A dwarf mutant of a quantitatively photoperiodic potato species *Solanum tuberosum* ssp. *andigena* which has a partial block in the step between GA_{12} and GA_{53}, thus causing a reduction in gibberellin levels and resulting in the dwarf phenotype, is able to tuberize in long days as well as short days whereas wild-type plants will only tuberize in short days and not in long days (van den Berg et al., 1995b). Thus the lower levels of gibberellins including, and downstream of, GA_{53} in the pathway, results in the dwarf potato being able to tuberize in long days. This indicates that certain types of gibberellin may play the role of the inhibitor and that the levels of, or response to, these gibberellins may be an important factor in the photoperiodic control of tuberization in potato.

Apart from flowering, bolting, senescence and tuberization, gibberellins are also known to affect sex determination, overcome dormancy and stimulate fruit set, all of which are examples of photoperiodic responses. Furthermore evidence is accumulating that vernalization, which as we have mentioned probably interacts with photoperiodism through some common step or pathway, may also act through gibberellins. It therefore seems increasingly likely that photoperiod and vernalization mediate their control of flowering by modulating the levels of, or response to, specific types of gibberellins.

4.2 The Involvement of Phytochrome B

As outlined previously, phytochrome is involved in the response to photoperiod, and we have just described how photoperiod can affect the levels of, or response to, gibberellins. Phytochrome should therefore have a direct effect on gibberellin levels or sensitivity. This has been shown to be true for phytochrome B. Some phytochrome B mutants were originally identified as gibberellin overproducers as their phenotypes could be mimicked by application of gibberellins to wild-type plants. Increased levels of certain gibberellins have been reported in some phytochrome B mutants, e.g. the ma_3^R Sorghum mutant and the *Brassica ein* mutant (Devlin et al., 1992; Foster et al., 1994), whilst others have increased sensitivity to gibberellins, e.g. the *Arabidopsis phyB* and cucumber *lh* mutants (Weller et al., 1994; Lopez-Juez et al., 1995). Sorghum is a quantitative SDP, the reduction of phytochrome B levels in the ma_3^R mutant has led to a greatly decreased photoperiodic control over flowering. A similar situation has also been observed with respect to the photoperiodic control of tuberization in potato. *S. tuberosum* ssp. *andigena* plants in which the phytochrome B levels had been reduced by antisense inhibition are able to tuberize in long days as well as short days in contrast to the wild-type plants that only tuberized in short days and not in long days (Jackson et al., 1996). Thus the lower levels of phytochrome B in the antisense potato plants and the ma_3^R Sorghum mutant have led to the loss of, or a reduction in, the inhibition of tuberization or flowering by long days. Phytochrome B therefore plays an integral role in photoperiodic responses such as flowering and tuberization, and it is likely to mediate its control over these processes through alterations in gibberellin levels or sensitivity.

5. The Genetics of the Response to Photoperiod

A lot of research has gone into identifying mutants that are altered in flowering and this work has provided much insight into the processes involved in flower formation. We will concentrate on the genes that are involved in the transition from vegetative to reproductive growth as to include all the genes involved in flower formation and organ identity would be beyond the scope of this chapter.

5.1 Arabidopsis

Most of the mutants identified are from the LDP *Arabidopsis* and are delayed in flowering. Of these *fe, fy, fpa, fve* and *fca* are still responsive to photoperiod but show an increased response to vernalization, other mutants, *fd, ft* and *fwa* respond to photoperiod but have a reduced, or no, response to vernalization. Three mutants, *fha, constants (co)* and *gigantea (gi),* flower later in long days but at the same time as wild-type in short days and also have reduced response to vernalization. The *CO* gene encodes a zinc finger protein thought to be a transcription factor which is involved in promoting flowering in long days (Putterill et al., 1995). The product of another gene that causes late flowering when it is mutated, *LUMINIDEPENS (LD),* is also thought to be a transcription factor as it contains two nuclear localization signals and a glutamine-rich region which can act as a transcription activation domain (Lee et al., 1994). The *ld* mutant responds differently in high and low light conditions. In high light intensities it is late flowering but it still responds to differences in photoperiod, whereas it does not respond to a low light day extension that promotes flowering in wild-type plants but not in *ld* which flowers as if in short days.

Mutations causing early flowering have also been identified, *terminal flower (tfl), early flowering (elf1, elf2* and *elf3), tfl, elf1* and *elf2* still respond to photoperiod but *elf3* shows no difference in flowering time when grown in short or long days. There is some evidence that *ELF3* is not involved in a pathway involving phytochrome, but in a blue light signal transduction pathway involved in flowering as the *elf3* mutation is epistatic to the blue light response mutation in h*y4* with respect to flowering (Zagotta et al., 1996). This is consistent with the involvement of a blue-light receptor in photoperiodic responses in Cruciferae. Several *fun* mutants have been isolated which flower earlier than wild-type in short days but at the same time as wild-type in long days (Thomas and Mozley, 1994). There are two groups of *fun* mutants, *fun1* and *fun2* flower at the same time as long day grown wild-type plants regardless of whether they were grown in long days or short days. The second group, *fun3, 4, 5,* and *6,* flower at the same time as wild-type in long days but flower earlier than wild-type in short days. The *embryonic flower (emf)* mutation causes flowering immediately after germination without any vegetative growth, the role of the *EMF* gene being to repress flower formation and promote vegetative growth (Sung et al., 1992). Most early flowering mutants are recessive suggesting that they may affect processes involved in the production or response to an inhibitor of flowering that is produced in LDP such as *Arabidopsis* in short days.

5.2 *Pisum*

The genetic factors influencing flowering in Pea (a LDP) have also been extensively studied (Table 1). The *late flowering (LF)* locus was the first to be identified and several mutations of the dominant *LF* gene have been found which differ in their minimum node number at which flowers are formed. *LF* appears to act at the apex and affects both the response to inductive long days and to vernalization, it is thought that *LF* is involved in establishing the sensitivity at the apex to the transmitted flower-inducing signal (Murfet, 1971). A more extreme phenotype is found in plants with a mutation in the *vegetative (veg)* gene which do not flower at all. The *veg* mutation also acts at the apex where it blocks some step in the flower induction process (Reid and Murfet, 1984).

Table 1 Genes involved in controlling flowering time in *Pisum sativum*

Gene	Description	Function
Sn	Sterile nodes	Interacts with Dne and Ppd to confer daylength sensitivity.
Dne	Day neutral	Interacts with Sn and Ppd to confer daylength sensitivity.
Ppd	Photoperiod	Interacts with Sn and Dne to confer daylength sensitity.
E	Early	Reduces expression of the Sn Dne Ppd system during early development
Hr	High response	Acts later in development to prolong Sn Dne Ppd activity
Gi, Gi^{fsd}	Gigas (flowering short days)	Recessive alleles greatly delay flowering
Lf	Late flowering	Confers sensitivity to flowering stimuli, four naturally occurring alleles

Sn, Dne and *Ppd* are all genes that act in leaves to control the production of an inhibitor of flowering, mutations in these genes result in photoperiodic insensitivity. *Sn* and, *Dne* act in a complementary manner, when both genes are present together they result in delayed flowering in non-inductive short day conditions, but the plants still respond to long days and vernalization (King and Murfet, 1985). The *early (E)* and *high response (Hr)* genes modify the effect of *Sn* and *Dne, E* only operates early on in development to reduce the effects of *Sn* and *Dne* and promote flowering in short days (Murfet, 1971). *Hr* however acts later in development and causes a further delay in flowering in short days.

A gene that may be involved in the production of the flower-inducing signal has also been identified, called *gigas (gi)*. Plants with the recessive alleles *gi* and the more severe gi^{fsd} allele flower much later, or even not at all, in long days (Taylor and Murfet, 1994). Evidence that plants with these recessive alleles have lower levels, or lack, the flower-inducing signal comes from grafting experiments of shoots of these plants onto wild-type stock plants, normal flowering is then restored in the grafts. The effect of the reduced levels of the inducing signal in the *gi* plants can be overcome by a 3–4 week vernalization treatment.

There are many more genes involved in the formation of the flower itself, some of the earliest ones that may be responding directly to the inducing signal are two MADS box genes from *Sinapis alba, saMADS A* and *saMADS B*, and the *KNAT1* gene from *Arabidopsis*. The two MADS box genes are strongly induced in the shoot apical meristem in a rapid response to inducing photoperiods (Menzel et al., 1996). Conversely the *KNAT1* gene is expressed in the vegetative shoot meristem and its expression is down-regulated on the transition to flowering (Lincoln et al., 1994). Furthermore there are also genes that are involved in gibberellin biosynthesis and perception that have an effect on flowering in *Arabidopsis* (Wilson et al., 1992). The gibberellin deficient *gal-3* mutant flowers later than wild-type in long days and is unable to flower in short days unless treated with gibberellin. The gibberellin insensitive (*gai*) mutant flowers as wild-type in long days but is late flowering in short days. This would indicate a requirement for gibberellins for flowering in short days in *Arabidopsis* but not in inducing long days. Unlike *gal-3, gai* shows a response to vernalization by accelerating flowering in short days.

References

Abou-Haidar, S.S., Miginiac, E. and Sachs, R.M. 1985. [^{14}C]-assimilate partitioning in photoperiodically induced seedlings of *Pharbitis nil*. The effect of benzyladenine. *Physiol. Plant.* **64:** 265–270.

Batutis, E.J. and Ewing, E.E. 1982. Far-red reversal of red light effect during long night induction of potato (*Solanum tuberosum* L.) tuberization. *Plant Physiol.* **69:** 672-674.

Benson, E.G. and Murray, D.E. 1949. Light effects. Caribbean Commission Monthly Information Bulletin. December issue.

Britz, S.J., Hungerford, W.E. and Lee, D.R. 1985. Photoperiodic regulation of photosynthetic partitioning in leaves of *Digitaria decumbens* Stent. *Plant Physiol.* **78**: 710–714.

Chailakyan, M. Kh., Yanina, L.I., Devedzhyan, A.G. and Lotova, G.N. 1981. Photoperiodism and tuber formation in grafting of tobacco onto potato. *Dokl. Acad. Nauk SSSR* **257:** 1276–1280.

Cumming, B.G., Hendricks, S.B. and Borthwick, H.A. 1965. Rhythmic flowering responses and phytochrome changes in a selection of *Chenopodium rubrum. Can. J. Bot.* **43:** 825–853.

Davies, P.J., Birnbergg, P.R., Maki, S.L. and Brenner, M.L. 1986. Photoperiod modification of (1^{4}C)gibberellin A_{12} aldehyde metabolism in shoots of pea, line G2. *Plant Physiol.* **81:** 991–996.

Deitzer, G.F., Hayes, R.G. and Jabben, M. 1982. Phase-shift in circadian rhythm of floral promotion by far-red energy in *Hordeum vulgare L. Plant Physiol.* **69:** 597–601.

Devlin, P.F., Rood, S.B., Somers, D.E., Quail, P.H. and Whitelam, G.C. 1992. Photophysiology of the elongated internode (*ein*) mutant of *Brassica rapa. Plant Physiol.* **100:** 1442–1447.

Downs, R.J. 1956. Photoreversibility of flower initiation. *Plant Physiol.* **31:** 279–284.

Evans, L.T., King, R.W., Mander, L.N. and Pharis, R.P. 1994. The relative significance for stem elongation and flowering in Lolium temulentum of 3β-hydroxylation of gibberellins. *Planta* **192:** 130–136.

Fondeville, J.C., Borthwick, H.A. and Hendricks, S.B. 1966. Leaflet movement of *Mimosa pudica* L. Identification of phytochrome action. *Planta* **69:** 357–364.

Foster, K.R., Miller, F.R., Childs, K.L. and Morgan, P.W. 1994. Genetic regulation of development in *Sorghum bicolor* IX. the ma_3^R allele disrupts diurnal control of gibberellin biosynthesis. *Plant Physiol.* **105:** 941–948.

Friend, D.J.C. 1985. Brassica. In Handbook of Flowering vol. II, ed. Halevy, CRC press. Boca Raton.

Friend, D.J.C., Bodson, M. and Bernier, G. 1984. Promotion of flowering in *Brassica campestris* L. cv Ceres by sucrose. *Plant Physiol.* **75:** 1085–1089.

Garner, W.W. and Allard, H.A. (1920). Effect of the relative length of day and night and other factors of the environment on growth and reproduction in plants. *J. Agric. Res.* **18:** 553–606.

Gilmour, S.J., Zeevart, J.A.D., Schwenen, L. and Greabe, J.E. 1986. Gibberellin metabolism in cell-free extracts from spinach leaves in relation to photoperiod. *Plant Physiol.* **82:** 190–195.

Goto, N., Kumagai, T. and Koornneef, M. 1991. Flowering responses to light-breaks in photomorphogenesis mutants of *Arabidopsis thaliana,* a long-day plant. *Physiol. Plant.* **83:** 209–215.

Gregory, L.E. 1965. Physiology of tuberization in plants. [Tubers and tuberous roots]. *Handbuch Pflanzenphysiol.* 1328–1354.

Guttridge, C.G. 1959. Further evidence for a growth-promoting and flower-inhibiting hormone in strawberry. *Ann. Bot. N.S.* **23:** 612–621.

Hanke, J., Hartmann, K.M. and Mohr, H. 1969. Die Wirkung von 'Storlicht' auf die Blüten-bildung von *Sinapis alba* L. *Planta* **86:** 235–249.

Jackson, S.D., Heyer, A., Dietze, J. and Prat, S. 1996. Phytochrome B mediates the photoperiodic control of tuber formation in potato. *Plant J.* **9:** 159–166.

Johnson, E., Bradley, M., Harberd, N.P. and Whitelam, G.C. (1994). Photoresponses of light-grown *phyA* mutants of *Arabidopsis. Plant Physiol.* **105:** 141–149.

King, R.W., Evans, L.T. and Wardlaw, I.F. 1968. Translocation of the floral stimulus in *Pharbitis nil* in relation to that of other assimilates. *Z. Pflanzenphysiol.* **59:** 377–385.

King, W.M. and Murfet, I.C. 1985. Flowering in *Pisum.* A sixth locus, Dne. *Ann. Bot.* **56:** 835–846.

Lang, A. 1965. Physiology of flower initiation. In Encyclopedia of Plant Physiology, vol. XV/1, ed. Ruhland, Springer-Verlag, Berlin.

Lang, A. and Melchers, G. 1943. Die photoperiodische Reaktion von *Hyoscyamus niger. Planta* **33:** 653–702.

Lee, H.S.J., Vince-Prue, D. and Kendrick R.E. 1987. Phase shifting effects in the photoperiodic rhythm of flowering in dark-grown seedlings of *Pharbitis nil* Choisy. *Plant Cell Physiol.* **28:** 93–100.

Lee, I., Aukerman, M.J., Gore, S.L., Lohman, K.N., Michaels, S.D., Weaver, L.M., John, M.C., Feldmann, K.A. and Amasino, R.M. 1994. Isolation of *LUMINIDEPENS*—a gene involved in the control of flowering time in *Arabidopsis. Plant Cell* **6:** 75–83.

Lincoln, C., Long, J., Yamaguchi, J., Serikawa, K. and Hake, S. 1994. A *knotted1*-like homeobox gene in *Arabidopsis* is expressed in the vegetative meristem and dramatically alters leaf morphology when overexpressed in transgenic plants. *Plant Cell* **6:** 1859–1876.

Lopez-Juez, E., Kobayashi, M., Sakurai, A., Kamiya, Y. and Kendrick, R.E. 1995. Phytochrome, gibberellins, and hypocotyl growth. *Plant Physiol.* **107:** 131–140.

Lumsden, P.J. 1991. Circadian rythms and phytochrome. *Ann. Rev. Plant Physiol. Plant Mol. Biol.* **42:** 351–371.

Lumsden, P.J. and Furuya, M. 1986. Evidence for two actions of light in the photoperiodic induction of flowering in *Pharbitis nil. Plant Cell Physiol.* **27:** 1541–1551.

Menzel, G., Apel, K. and Melzer, S. 1996. Identification of two MADS box genes that are expressed in the apical meristem of the long-day plant *Sinapis alba* in transition to flowering. *Plant J.* **9:** 399–408.

Mozley, D. and Thomas, B. 1995. Developmental and photobiological factors affecting photoperiodic induction in *Arabidopsis thaliana* Heynh. Landsberg erecta. *J. Exp. Bot.* **46:** 173–179.

Murfet I.C. 1971. Flowering in *Pisum:* reciprocal grafts between known genotypes. *Aust. J. Biol. Sci.* **24:** 1089–1101.

Nitsch, J.P. (1966). Photoperiodisme et tuberisation. *Bul. Soc. Franc. Physiol. Véget.* **12:** 233–246.

Ogawa, Y. and King, R.W. 1979. Establishment of photoperiodic sensitivity by benzyladenine and a brief red irradiation on dark grown seedlings of *Pharbitis nil* Chois. *Plant Cell Physiol.* **20:** 119–122.

Papenfuss, H.D. and Salisbury, F.B. 1967. Properties of clock re-setting in flowering of *Xanthium. Plant Physiol.* **42:** 1562–1568.

Purvis, O.N. 1961. The physiological analysis of vernalisation. In Encyclopedia of Plant Physiology **16**, Ruthland, W (ed.) Springer-Verlag, Berlin.

Putterill, J., Robson, F., Lee, K., Simon, R. and Coupland, G. 1995. The *CONSTANS* gene of *arabidopsis* promotes flowering and encodes a protein showing similarities to zinc finger transcription factors. *Cell* **80:** 847–857.

Reid, J.B. and Murfet, I.C. 1984. Flowering in *Pisum* a fifth locus, *Veg. Ann. Bot.* **53:** 369–382.

Salisbury, F.B. 1965. Time measurement and the light period in flowering. *Planta* **66:** 1–26.

Salisbury, F.B. and Ross, C. 1969. Plant Physiology, Wadsworth Publishing Co. Inc., Belmont, CA, USA.

Salisbury, F.B. 1990. *Xanthium strumarium.* In Handbook of Flowering vol. IV, (ed. Halevy), CRC Press, Boca Raton.

Sarkar, S. 1958. Versuche zur Physiologie der Vernalization. *Biol. Zbl.* **77:** 1–49.

Satter, R.L., Guggino, S.E., Lonergan, T.A. and Galston, A.W. 1981. The effect of blue and far-red light on rhythmic leaf movements in *Samanea* and *Albizia. Plant Physiol.* **67:** 965–968.

Sponsel, V.M. 1995. Gibberellin biosynthesis and metabolism. In Plant Hormones: Physiology, Biochemistry and Molecular Biology, Davies, R. (ed.) Kluwer Academic Publishers, Dordrecht, The Netherlands.

Sung, Z.R., Belachew, A., Shunong, B. and Bertrand-Garcia, R. 1992. EMF, an *Arabidopsis* gene required for vegetative shoot development. *Science* **258:** 1645–1647.

Takahashi, H., Saito, T. and Suge, H. 1982. Intergeneric translocation of floral stimulus across a graft in monoecious Cucurbitaceae with special reference to the sex expression of flowers. *Plant and Cell Physiol.* **23:** 1–9.

Takimoto, A. and Hamner, K.C. 1964. Effect of temperature and pre-conditioning on photoperiodic response of *Pharbitis nil. Plant Physiol.* **39:** 1024–1030.

Taylor, S.A. and Murfet, I.C. 1994. A short-day mutant in pea is deficient in the floral stimulus. *Flowering Newsletter* **18:** 39–43.

Thomas, R.G. 1979. Inflorescence initiation in *Trifolium repens* L.: influence of natural photoperiods and temperatures. *New Zealand J. Bot.* **17:** 287–299.

Thomas, B. and Mozley, D. 1994. Isolation and properties of mutants of *Arabidopsis thaliana* with reduced sensitity to short days. In Molecular and Cellular Aspects of Plant Reproduction, eds. Scott and Stead, SEB seminar series. Cambridge University Press, Cambridge.

Thomas, B. and Vince-Prue, D. 1984. Juvenility, photoperiodism and vernalisation. In Advanced Plant Physiology, Wilkins, M.B. (ed.) Pitman Publishing Ltd., London.

van den Berg, J.H., Davies, P.J., Ewing, E.E. and Halinska, A. 1995a. Metabolism of gibberellin A_{12} and A_{12}-aldehyde and the identification of endogenous gibberellins in potato (*Solanum tuberosum* ssp. *Andigena*) shoots. *J. Plant Physiol.* **146:** 459–466.

van den Berg, J.H., Simko, I., Davies, P.J., Ewing, E.E. and Halinska, A. 1995b. Morphology and (^{14}C)gibberellin A_{12} metabolism in wild-type and dwarf *Solanum tuberosum* ssp. *andigena* grown under long and short photoperiods. *J. Plant Physiol.* **146:** 467–473.

Vince-Prue, D. 1975. Photoperiodism in Plants, McGraw-Hill & Co., London.

Vince-Prue, D. 1994. The duration of light and photoperiodic responses. In Photomorphogenesis in plants, 2nd edition, Kendrick, R.E. and Kronenberg, G.H.M. (eds.) Kluwer Academic Publishers, The Netherlands.

Vlitos, A.J. and Meudt, W. 1955. Interactions between vernalisation and photoperiod in spinach. *Contrib. Boyce Thompson Inst. Plant Res.* **18:** 159–166.

Welensick, S.J. 1958. Vernalisation and age in *Lunaria biennis*. *Proc. K. ned. Akad. Wet.* **C61:** 552–560.

Weller, J.L., Ross, J.J. and Reid, J.B. 1994. Gibberellins and phytochrome regulation of stem elongation in pea. *Planta* **192:** 489–496.

Wilson, R.N., Heckman, J.W. and Somerville, C.R. 1992. Gibberellin is required for flowering in *Arabidopsis thaliana* under short days. *Plant Physiol.* **100:** 403–408.

Zagotta, M.T., Hicks, K.A., Jacobs, C.I., Young, J.C., Hangarter, R.P. and Ry Meeks-Wagner, D. 1996. The *Arabidopsis ELF3* gene regulates vegetative photomorphogenesis and the photoperiodic induction of flowering. *Plant J.* **10:** 691–702.

Zeevart, J.A.D. 1958. Flower formation as studied by grafting Meded. LandbHoogesch. Wageningen **58:** 1–88.

Zeevart, J.A.D. 1969. *Perilla*. In the induction of flowering, ed. Evans, MacMillan, Melbourne.

Zeevart, J.A.D. (1982). Transmission of the floral stimulus from a long-short-day plant, *Bryophyllum daigremontianum*, to the short-long-day plant *Echeveria harmsii*. *Ann. Bot.* **49:** 549–552.

Concepts in Photobiology: Photosynthesis and Photomorphogenesis
G.S. Singhal, G. Renger, S.K. Sopory, K-D. Irrgang and Govindjee (Eds)

30. Light Signal Transduction and Gene Expression

S. K. Sopory[1], Neeti Sanan[2] and R. Oelmüller[3]

[1] International Centre for Genetic Engineering and Biotechnology, Aruna Asaf Ali Marg, New Delhi -110067, India

[2] School of Life Sciences, Jawaharlal Nehru University, New Delhi -110067, India

[3] Institut für Allgemeine Botanik, Abt. Pflanzenphysiologie, Dornburgerstr. 159, 07743 Jena, Germany

Summary

Light is one of the most important regulators of plant growth and development. Genetic, biochemical and physiological studies in plants have shown the existence of multiple photoreceptors for red, far-red, blue and UV light. The fact that a large number of photoresponses are regulated by various photoreceptors indicates the presence of different, intimately interconnected, cellular signal transduction pathways regulating the expression of different genes. To decipher the signal components involved in light mediated gene expression, a number of photomorphogenetic mutants have been analysed and this work has been covered in Chapter 31. In this chapter we elucidate the cellular and molecular approach which has provided evidence for the participation of GTP-binding proteins, inositol lipids and calcium/calmodulin. In addition, evidence for a role played by the kinases/phosphatases in light-regulated gene expression has been elucidated. Phosphorylation has been shown to modulate the activity of various proteins, from the photoreceptor to the transcription regulators. Several *cis* elements have been identified in light responsive genes which are regulated by transcription factors whose function may be mediated by phosphorylation events. Although lot of work has been initiated to understand the link between photoreceptors and gene expression, it is only now that the sequential events of light signal transduction are beginning to be unfolded for various physiological/developmental photoresponses.

1. Introduction

Photoreception represents the first step in a series of events that occur between stimulus (light) and the activation of light-regulated responses. The stimulus-response reaction involves recognition of the external environment by a plant cell and transduction of the message to the gene expression machinery, through generation of second messengers. This finally leads to the manifestation of photomorphogenetic responses. The basic mechanism by which light-mediated information is sorted and conveyed to the interior of the plant cell is unknown, although recent studies have led to a partial understanding of the process.

By studying the wavelength dependence of different responses, three major classes of photoreceptors have been implicated in the photomorphogenetic responses viz. RL/FRL photoreceptor, phytochrome (see Chapter 25), the blue/UV photoreceptors (see Chapters 26, 27), which include the flavin based proteins and the UV-B photoreceptor. In addition, existence of multigene family for phytochrome has been shown and it is likely that multiple alleles might exist for genes encoding the other photoreceptors.

The multiplicity of photoreceptors suggests that the light signalling process in plants comprises a complicated network. The existence of the same or different signalling pathways for each individual photoreceptor or photoreceptor family is dependent on the photomorphogenic responses and developmental stage of the plants. The true nature and number of light signalling mechanisms existing in different plant systems remain unclear.

2. Beyond Photoperception

In an attempt to understand the biochemical pathways by which the light signal is transduced to the interior of the plant cell, animal models have been used to identify the second messengers involved. The overall scheme operative in animal systems is that an exogenous stimulus activates membrane associated G-proteins, the α subunit of which acts as a part of the transducing complexes involved in the signal transduction from photoreceptor to the various effector systems such as adenylate cyclase and phospholipase C [PLC] (Gilman, 1987). The cleavage of phosphatidylinositol 4,5-bisphosphate (PIP_2) by phosphodiesterase (PLC), releases the second messengers inositol tri-phosphate (IP_3) and diacylglycerol (DAG) (Berridge and Irvine, 1989; Downes and Michell, 1985; Berridge, 1987). When IP_3 is released into the cytosol, it stimulates the mobilisation of calcium from endogenous stores like endoplasmatic reticulum. The increased levels of calcium results in the activation of enzymes-kinases, phosphatases—either directly or through calcium—binding proteins like calmodulin, which initiate and modulate physiological responses. The other messenger, DAG, either activates a set of kinases in the presence of calcium, called protein kinases C (PKC), or it gets metabolized into other biochemical pathways. The activation of kinases and phosphatases is an important step for amplification of the signal and these eventually bring about the activation of regulatory proteins in the nucleus, thereby influencing the genetic machinery.

In plants, physiological, biochemical and genetic studies have provided evidence for the participation of G-proteins, cAMP, cGMP, calcium/calmodulin (CaM), protein kinases and phosphatases and inositol phospholipids, in the photocontrol of gene expression. However, a direct demonstration of all the components has not been achieved in any one system. The identification of various signalling intermediates has been either through identification based on biochemical properties or antibody cross-reactivity or indirect evidence based on activator and inhibitor studies. The study of signalling pathway has also been aided by isolating mutants which are defective in the photoreceptor *per se* or in the components constituting the signal transduction chain (see Chapter 31). Similarly, the transgenic plants either overexpressing a given gene or carrying a reporter gene fused to the upstream (promoter) sequences of a photoregulated gene, have been useful to study the expression pattern of the gene and understanding its mode of regulation.

2.1 GTP-binding Proteins

Proteins binding to GTP have been identified as early intermediates of the signal transduction pathway in animal cells. There are two classes of regulatory G-proteins: the monomeric, small molecular weight proteins, and the heterotrimeric proteins. The heterotrimeric G-proteins consist of three subunits: α, β and γ. Once GTP is bound, the Gα subunit is separated from G$\beta\gamma$ subunits and gets activated. The activated Gα interacts with important metabolic enzymes like PLC or adenylcyclase and activates them. Once the stimulus is removed the intrinsic GTPase activity hydrolyses GTP to GDP and returns the G-proteins to their inactive state.

Hasunuma and Funadera (1987) using [^{35}S] GTPγS, a non-hydrolysable analogue of GTP that maintains G-proteins in their activated from, showed that several G-proteins were present in fractions isolated from *Lemna* and that these could be ADP-ribosylated by the pertussis toxin. Since then various reports on the presence and involvement of G-proteins and their cloning and characterisation have appeared from different plant tissues (see, Ma, 1994; Palme, 1996). The participation of G-proteins has been demonstrated by testing the effects of agonists and/or antagonists on the light mediated responses. There is no direct evidence yet on how G-protein is coupled to the photoreceptor. It is also not clear if all photoresponses do indeed require G-protein activation.

Romero et al. (1991b) studied the relationship between phytochrome and G-proteins using three different techniques. They demonstrated the presence of RL/FRL modulated specific GTP binding proteins in crude extracts from etiolated oat seedlings. This was also shown by western analysis, using antibodies that recognize GTP-binding sites of several α-subunits (GA/1) which detected a 24 kDa GTP-binding protein in oat seedlings. By using labelled GTPγS it was further shown that 24 kDa protein was present among the various polypeptides that were efficiently ADP-ribosylated by cholera toxin in the presence of GTP.

There is also an indication to show that the blue light (BL) modulation of photoresponses occurs through G-proteins. Warpeha et al. (1991) identified a heterotrimeric G-protein in plasma membrane fractions of pea apical buds, which was activated by low fluence BL. This protein exhibited GTPase activity on irradiation with BL, but not with RL. Though these reports suggest the involvement of GTP-binding proteins in both RL and BL induced signalling pathways, it is not clear whether the same or different proteins are involved in these phototransduction processes.

In addition to trimeric G-proteins, the monomeric, small molecular weight

GTP-binding proteins, homologous to the *myb*, *ras* and *src* oncogenes have been detected in plants and a role for these in photo-signalling pathway has been shown. Clark et al. (1993) have reported a RL and FRL activation of monomeric G-protein activity in nuclear envelope of pea cells. These proteins have been proposed to play a role in the GTP-dependent fusion of membrane vesicles to the nuclear envelope during its re-assembly after mitosis and in the regulation of transport across nuclear pores. Phytochrome mediated down regulation of expression of genes encoding small GTP-binding proteins belonging to the *ras* superfamily has been seen in *Pisum sativum* (Yoshida et al., 1993). These studies do suggest that monomeric as well as trimeric G-proteins, as discussed above, could play key roles in light-induced cellular activities. More data which gives further indication of their involvement in signal-response coupling is given in later section in this chapter.

2.2 Calcium

The role of calcium in mediating signal transduction processes in plants has been reviewed (Poovaiah and Reddy, 1993). Over the last decade a number of papers have also been published on calcium-mediated phototransduction pathways (see, Tretyn and Kendrick, 1991; Roux, 1984). The evidence for the involvement of calcium in photosignalling pathways has come from studies involving measurement of ^{45}Ca uptake or release, and using calcium chelators, calcium channel blockers like verapamil, nifedipine, lanthanum, ruthenium red, and calcium ionophores (which increase calcium uptake into cells). A number of other methods have also been used for measuring calcium fluxes in cells like colorimetric assays with murexide and atomic absorption spectroscopy measurements of total cellular calcium. Also free calcium concentration have been measured by fluorometric assays with calcium-quin-2 complexes and cytoplasmic assays have been employed for intracellular calcium localisation. The development of transgenic systems for expression of a calcium binding protein, aequorin has greatly facilitated the monitoring of calcium transients in plant cells following specific stimuli. The change in calcium flux has been inferred from studies on several phytochrome-mediated responses using various methods as summarized in Table 1.

Haupt and Weisenseel (1976) for the first time proposed that phytochrome responses may be mediated by increasing intracellular calcium concentration. Dryer and Weisenseel (1979) demonstrated that RL stimulates an instantaneous influx of ^{45}Ca into *Mougeotia* cells and this phenomenon is FRL reversible. It was proposed that the influx of calcium was due to phytochrome mediated alteration of membrane permeability. The link was established when Serlin and Roux (1984) showed that chloroplast movement in *Mougeotia*, which is under phytochrome control, could be induced in dark by calcium ionophore, A23187. Hale and Roux (1980) measured phytochrome induced calcium fluxes in protoplasts of oat coleoptiles and reported that RL induced an efflux of calcium into the medium, that can be reverted by subsequent FRL

Table 1. Role of calcium ions in phytochrome regulated developmental processes in higher plants

Spore germination		
	Onoclea sensibilis	Wayne and Hepler (1984)
	Adiantum capillus-veneris	Iino et al. (1989)
	Dryopteris paleacea	Scheuerlein et al. (1991)
Turion germination		
	Spirodela polyrhiza	Appenroth and Augsten (1990)
Leaf unrolling		
	Wheat	Tretyn and Kendrick (1990)
	Barley	Viner et al. (1988)
	Sorghum bicolor	Sanan and Sopory (1998)
Leaf movement		
	Albizzia julibrissin, *A. lophantha*	Moysset and Simon (1989)
	Cassia fasciculata	Robin et al. (1989)
	Samanea saman	Roblin et al. (1990)
Flower induction		
	Lemna perpusilla	Halaban and Hillman (1970)
	Pharbitis nil	Friedman et al. (1989)

irradiation. Similar studies were done in *Vallisneria* (Takagi et al., 1990). However, studies using ^{45}Ca in etiolated seedling protoplasts of maize (Das and Sopory, 1985) and wheat (Bossen et al., 1988) showed a phytochrome-dependent calcium influx. Chae et al. (1990) were the first to directly demonstrate that RL but not FRL can induce an increase in cytoplasmic free calcium concentration in oat cells. Mehta et al. (1993) monitored the calcium fluxes in wheat leaf protoplasts using murexide and radiolabelled calcium and demonstrated that RL promoted while subsequent FRL reversed an influx of calcium into protoplasts. This RL mediated influx could be inhibited by verapamil, whereas FRL mediated efflux could be inhibited by vanadate (an ATPase inhibitor). They also demonstrated a role of PI metabolism in the phytochrome-mediated calcium influx into the protoplasts. These studies demonstrated the involvement of ion channels and pumps in phytochrome controlled calcium fluxes. Earlier phytochrome was also found to effect Ca^{2+}-ATPase of mitochondria (Serlin et al., 1984) which may have a role in calcium efflux from internal stores (Roux et al., 1981).

The transduction of calcium signal also involves the calcium-binding protein, calmodulin (CaM). There are reports of cloning and characterization of calmodulin from both soluble and membrane fractions (Jena et al., 1989; Ling et al., 1991). Experiments with CaM inhibitors have shown to block light responses. These aspects are further discussed in a section on signal response coupling.

In summary there are evidences for changes in intracellular calcium concentration in response to light and presence of calcium binding proteins like CaM and calcium- and CaM- dependent kinases and the presence of calcium channels in both higher and lower plants has been shown (Roux et al., 1986; Bush, 1993). Thus, all elements of a calcium-based phototransduction are found in plant cells. It is becoming clear that calcium and CaM occupy an important position in light-mediated signal transduction pathway.

2.3 Inositol Lipids

The phosphoinositides play an important role as an early step in signal transduction in animals. They are present in the membranes and may comprise upto 10% of the total acyl lipids. Upon stimulation of cells hydrolysis of phosphatidylinositol 4, 5-biphosphate (PIP_2) occurs. The products of this phospholipase C cleavage, IP_3 (inositol 1, 4, 5-triphosphate) as well as DAG (diacylglycerol), function as second messengers within the cell and eventually lead to various responses.

All components of the polyphosphoinositide cycle (PPI) are reported to be present in plant cells. However, their exact role as second messengers in signal transduction pathways in plant cells is not clear (Drϕbak, 1992, 1993; Gilroy and Trewavas, 1994). Several reports indicate that plant inositol phospholipids are potential sources of second messengers and they are involved in light-mediated signal transduction (Sopory and Chandok, 1996).

That the phosphoinositide cycle may be involved in light-mediated responses was indicated when 5-hydroxytryptamine (5HT, serotonin), an activator of the PPI cycle in animal cells, was shown to replace the light requirement for phytochrome regulation of calcium fluxes in maize protoplasts (Das and Sopory, 1985). Morse et al. (1987) were however, the first to study the effect of light on the turnover of phosphoinositides in *Samanea saman* pulvini. The leaf movements in this plant are under phytochrome control. Irradiation of pulvini with white light (WL) for 5-30 sec resulted in the incorporation of radiolabel into IP_2 and IP_3. Reddy et al. (1987), who studied the gravitropic response, also reported a light-mediated increase in IP_3 in maize roots while Basu et al. (1990) showed similar changes in *Brassica* cultures and hypocotyl. Memon and Boss (1990) showed that PIP_2 levels decreased if sunflower hypocotyls were irradiated with WL. These reports indicated that light affects inositol turnover in plants.

The evidence that phytochrome affects inositol turnover in plants was given by Guron et al. (1992) in maize leaves. Using myo-[^{3}H]inositol and [^{32}P] labelling they showed a red light (RL)-induced increase in PIP_2 levels could be reversed by far-red light (FRL); however the levels of PIP and PI did not change. This means that an increase in PIP_2 is probably due to an increase in the PIP kinase activity and a decrease in the PLC activity. Alternatively, light may cause a decrease in activity of PIP_2 phosphatase resulting in an accumulation of PIP_2. This study also suggests that one of the initial steps involved in Pfr-mediated signal transduction is to affect the inositol turnover.

Whether phytochrome affects any component of PI cycle either directly or through some intermediate like G-proteins, is not understood. Also, the pathway followed by inositol phosphates and the identity of second messengers involved in the transduction of the photosignal are not clear. There are however, different reports showing the presence of PLC activity and its regulation by calcium in various tissues like dark grown shoots from wheat seedlings (Melin et al., 1987), plasma membrane of protonemal tip cells of *Ceratodon purpureus* (Hartmann and Pfaffmann, 1990), but there are no reports on direct effects of light on PLC to regulate phosphoinositide turnover.

As for the role of the two messengers released by PLC activity, there are reports on light induced changes in levels of IP_3 and DAG. However, the precise mechanism by which IP_3 releases calcium from intracellular stores and the unequivocal demonstration of DAG stimulated kinases (PKC) was not shown. Recently a receptor for IP_3 has been isolated from microsomal fractions of *Vigna* (Biswas et al., 1995) and the presence of PKC type kinases in plant cells has also been indicated (Chandok and Sopory, 1992). We have purified from maize a PKC homologue, which showed PMA-binding and activation by okadaic acid (Chandok and Sopory, 1996; Chandok and Sopory, 1998). This is also recent reports on the existence of DAG kinases (Katagiri et al., 1996) and PLC (Shi et al., 1995) which suggest that the lipid signalling pathway via PI cycle may be operative in higher plants and therefore could be involved in light-mediated signalling.

2.4 Protein Phosphorylation

Protein phosphorylation plays an important role in signal transduction. In photosignalling it was shown that the photoreceptor itself could get phosphorylated or it could lead to generation of various signals, as mentioned above, which in turn would transduce and amplify the signals through a kinase cascade resulting in activation/inhibition of various proteins including transcription factors for regulating gene expression.

As far the phosphorylation of the photoreceptor is concerned there is evidence for and against an intrinsic or associated kinase activity of phytochrome (see Miller et al., 1994; Roux, 1994). Wong et al. (1986) showed that Pr form was phosphorylated more than the Pfr form. They also showed that purified oat phytochrome itself has a kinase activity. Wong et al. (1989) further characterized the protein kinase activity in *Avena* phytochrome preparations and phosphorylation of N-terminal serine residues was shown *in vitro* (McMichael and Lagarias, 1990). However studies by Kim et al. (1989) and Grimm et al. (1989) showed that kinase activity does not reside in the phytochrome molecule *per se* and it was separated out on further purification. Later, sequence analysis by Schneider-Poetsch et al. (1991) revealed the existence of protein-kinase like domains as an integral part of the phytochrome gene. Using phytochrome antibodies Biermann et al. (1994) demonstrated that protein kinase activity coprecipitates with phytochrome. They proposed that the kinase activity is either intrinsic to the phytochrome molecule or associated with it by high affinity interactions. Moreover this protein kinase was quantitatively affected by light-induced conformational changes that occurred *in vivo*. Their studies also showed that the kinase was Pr-specific and it could distinguish between light-induced conformational states in a manner similar to second-messenger regulated protein kinases.

Whether this kinase acivity is in the photoreceptor molecule itself or in another peptide associated with it remains controversial. Nevertheless, the various studies show that kinase stimulation may represent an important early step in light signal transduction chain. Thümmler et al. (1992) showed that a moss, *Ceratodon,* phytochrome (145 kDa) has a C-terminal domain similar to light sensor protein kinases and this is catalytically active *in vitro*. Moreover, this protein seems to have an autophosphorylation activity that is modulated by light (Algarra et al., 1993). Recently, Thümmler et al. (1995) have shown that the C-terminus of phytochrome from higher plants has some sequence similarity to eukaryotic serine/threonine protein kinases and the bacterial histidine kinases. The N-terminal domain, which undergoes light induced conformational changes, has amino acid sequence similarity also to a portion of the *Aspergillus nidulans* BRLA protein, which is involved in transcriptional activation of asexual sporulation (Griffith et al., 1994). The details of phytochrome phosphorylation are discussed in Chapter 28.

Just like the data about the phosphorylation of phytochrome, it itself possessing a kinase activity are not resolved, the results with BL receptor are also not very clear. As mentioned earlier, the putative BL receptors identified are flavoproteins known as cryptochromes. However, recently genetic analysis of photomorphogenesis has resulted in identification of an *Arabidopsis* mutant, *hy4* (long hypocotyl mutant), which showed no inhibition towards RL/FRL treatment and did not respond to BL irradiation. Recently, the *hy4* gene has been cloned and identified as the gene encoding a cryptochrome, *cry*1 (Ahmad and Cashmore, 1993). This photoreceptor is apparently a soluble protein of 75 kDa with its N-terminus showing homology to microbial DNA photolyases, but it lacks the DNA repair activity.

Earlier, Briggs group (Short et al., 1993) had identified a 120 kDa plasma membrane protein, from the growing regions of pea stems, which undergoes a BL-dependent phosphorylation. Since the Triton- or CHAPS-solubilised membrane preparations retained the BL-induced phosphorylation, it was suggested that there may be a tight relationship between the photoreceptor, the kinase and the substrate. Further, the 120 kDa protein was found to have nucleotide binding site, indicating that the polypeptide is autophosphorylated. The spectral response characteristics and spatial location of this protein correspond with its role as a photoreceptor or an early step in the BL-signalling pathway, for phototropism. For example, this protein was deficient in an *Arabidopsis* mutant, JK224, which does not respond to BL. Liscum and Briggs (1995) showed that this protein was also absent in the NPH1 mutants, which have a null phenotype in all light qualities tested and moreover the *nph*1 gene was shown to be genetically distinct from *cry*1 (Liscum et al., 1992). Just like in pea and *Arabidopsis,* even in maize a 100 kDa plasma membrane protein has also been identified in the coleoptile region which was phosphorylated rapidly upon BL irradiation. It is not clear if this is a putative photoreceptor having an autophosphorylating kinase domain (Hager, 1996).

Whether photoreceptors are phosphorylated or not, it is, however, clear that they do alter the phosphorylation status of other proteins in response to light, though the identity of proteins that get phosphorylated has not been established in most of the cases. Datta et al. (1985) demonstrated that phytochrome regulates phosphorylation of several nuclear proteins in a calcium-dependent manner and these could also be blocked by CaM inhibitor, chlorpromazine. The light irradiations and phosphorylation reactions were performed *in vitro* using highly purified nuclei from pea. Otto and Schäfer (1988) found that a brief RL irradiation of *Avena* coleoptiles, followed by *in vivo* phosphorylation resulted in increased incorporation in a 30 kDa polypeptide and decreased incorporation in two polypeptides (32 and 29 kDa). This response was reversed by FRL. Park and Chae (1989) showed increased phosphorylation of 15 different soluble fraction proteins in RL irradiated oat protoplasts following *in vivo* phosphorylation. They demonstrated the involvement of both calcium dependent kinases and PKC, as H-7 and Li^+ were found to reduce the light stimulated phosphorylation of the proteins.

Doshi et al. (1992) have also demonstrated *in vitro* phosphorylation of a 55 kDa protein and also of other proteins in *Sorghum* shoot tips in response to RL and the effect of RL was found to be reversible by FRL. Fallon et al. (1993) have shown that, in wheat leaf protoplasts the phosphorylation pattern of proteins change very rapidly in response to RL and also following an increase in the calcium concentration, which was achieved either by the release from caged calcium or from intracellular stores by IP_3 action. Harter et al. (1994) have shown that short-time irradiation with different fluences of RL or WL results in rapid changes in the phosphorylation pattern of several cytosolic proteins in evacuolated parsley protoplasts. Romero et al. (1991a) reported a RL/FRL modulated phosphorylation of 60 and 75 kDa nuclear proteins isolated from etiolated oat seedlings. These light-dependent phosphorylations were significantly blocked by cholera toxin and could be brought about even in the absence of light by the activation of G-proteins. This clearly indicates a role for kinases downstream to the G-protein in phototransduction pathways.

In pea seedlings, Hamada et al. (1996) observed that phosphorylation of a 15 kDa protein in the membrane fraction was stimulated by RL irradiation that was reversed by FRL. This polypeptide was found to contain nucleotide diphosphate kinase activity. Sharma et al. (1997b) have demonstrated that WL irradiation of 1 hour or more effected the phosphorylation status of soluble fraction proteins of 4-day-old etiolated wheat seedlings. The short term RL, FRL or BL treatments had no effect. The phosphorylation of many of these proteins were inhibited by EGTA and CaM antagonists, which indicated the involvement of Ca^{2+}/CaM-dependent kinases in this light-regulated phenomenon.

Besides RL, even in response to BL irradiation phosphorylation of several proteins, especially in membranes, with an apparent molecular mass between 114-130 kDa from several plant species like tomato, zucchini, maize, barley, wheat, oat and *Arabidopsis* has been shown (Short and Briggs, 1994). Sharma et al. (1997a) demonstrated the presence of BL-sensitive phosphopolypeptides in shoot portions of etiolated wheat seedlings. These polypeptides were associated with the plasma membrane and their phosphorylation represented one of the early steps in BL-elicited signal transduction chain. Moreover, these proteins were found to be more abundant in the leaf tissue than in the coleoptile sheath.

In summary, reversible protein phosphorylation controls several intracellular events and the involvement of kinases and phosphatases in the signal transduction chain have been well documented. Several kinases have been identified and cloned in higher plants (Redhead and Palme, 1996) and these have been classified by sequence similarity and presence of functionally related groups into families viz. calcium-dependent protein kinases, calcium/CaM-dependent protein kinases, calcium/phospholipid-dependent protein kinases, protein kinase C (PKC), phosphatidyl inositol and phosphatidyl inositol phosphate kinases, mitogen-activated protein kinases, receptor-like kinases, cAMP protein kinases (PKA). The presence of a variety of kinases indicates that phosphorylation/dephosphorylation of proteins provide an important regulatory switch in governing the growth responses, including photomorphogenic responses. These kinases may be regulated either directly or through the signalling intermediates. Infact the light-dependent regulation of the expression of kinases has been reported which provides additional evidence for their involvement in the signalling pathway. Studies have demonstrated differential accumulation of transcripts encoding protein kinase homologs in greening pea seedlings (Lin et al., 1991).

Although, as mentioned earlier the substrates that are phosphorylated in response to light treatments have not been identified, yet in a number of studies phosphorylation of *trans* acting factors has been shown, and further found to regulate their binding to the promoters of light-regulated genes. A light regulated protein, LRF-1, was shown to bind to an upstream region of a *RbcS* gene of *Lemna* in gel shift assays which were performed using light and dark tobacco extracts (Buzby et al., 1990). Similarly, Sarokin and Chua (1992) reported that expression of light regulated gene, *RbcS*-3A, requires the presence of phosphoproteins, 3AF5 and 3AF3. Of these, binding of 3AF3 to its target site is dependent upon light induced phosphorylation state, although 3AF5 maintains its ability to bind even when dephosphorylated. Datta and Cashmore (1989) have shown that an N II type-casein kinase may regulate the binding of a pea nuclear protein, AT-1, to the promoters of certain photoregulated genes. The DNA-binding activity was inhibited by phosphorylation. Later, Klimczak et al. (1992) demonstrated that phosphorylation by casein kinase II stimulated the DNA-binding activity of the *Arabidopsis* G-box binding factor, GBF1. They partially purified the 128 kDa kinase from broccoli and showed that it had a high affinity for GBF1. The properties of this kinase were found to be similar to those of a chromatin associated casein kinase II purified from pea nuclei (Li and Roux, 1992).

The steady state level of phosphorylation status is governed by the activity of kinases and also phosphatases. There are evidences that protein dephosphorylation also plays a very important role in signal transduction pathway. The involvement of phosphatases has been demonstrated in light induction of chlorophyll accumulation and photosynthetic gene expression. In etiolated maize leaves okadaic acid, a potent inhibitor of protein phosphatase 1 and 2A, prevented gene expression and greening process while no inhibition was observed in presence of protein kinase inhibitors (Sheen, 1993). It is evident now, from the above mentioned reports, that protein phosphorylation and dephosphorylation is an important event in light-mediated processes, including gene expression.

3. Gene Expression

It was in the late sixties that Hans Mohr suggested that phytochrome may be regulating the expression of genes, either positively or negatively. Subsequently, by the use of metabolic inhibitors and density labelling techniques it was shown that the Pfr-regulated enzymes are synthesized *de novo* and that their level could be blocked by inhibitors of transcription and translation. Still later using cDNA probes and developing northern blots for detecting transcript levels of *RbcS* and *LhcB* genes, it was demonstrated in a number of plants that Pfr does increase the level of mRNA. Using a probe for protochlorophyllide oxidoreductase it was shown that Pfr can also down regulate mRNA level. Infact Pfr also negatively regulates its own mRNA. Thus, it was conclusively shown that Pfr can exert its influence at the transcription level.

In the mid-eighties upstream regions of a number of light regulated genes were fused with bacterial reporter genes and transferred *via Agrobacterium* into plants. In such plants it was shown that the property of a gene to be regulated by light resides often in specific *cis*-elements that are located 5' upstream of a gene and these were called light-regulatory elements (LRE). Many such elements were discovered later as also the corresponding transcription factors which bind to these DNA sequences. However, screening the promoter elements of many more genes during the last few years has indicated that sequences downstream to transcription start site, which include sequences within the coding region of the gene and the untranslated leader regions, are also involved in photoregulation of specific genes. In the following section we describe their nature and present concept of LRE.

3.1 *cis*-elements (LRE) Located Upstream of the Transcription Start Site

A light-regulatory element (LRE) is defined as a small DNA segment that is present upstream of the transcription initiation site and is sufficient to confer light-regulated expression onto the homologous or a heterologous TATA box. It was found initially that if upstream regions were attached to some reporter genes like chloramphenicol transferase (CAT) and β-glucuronidase (GUS) their expression could be regulated by light irradiation. By deletion analysis, specific *cis* elements could be defined.

Although an enormous amount of experimental evidence has confirmed that light regulation is mediated by upstream regulatory elements, this paradigm is mainly based on the analysis of members of a few gene families, such as those encoding the small subunit of ribulose-1, 5-bisphosphate carboxylase (gene: *RbcS*), the major photosystem II light-harvesting chlorophyll a/b-binding protein (gene: *LhcB*1 or *cab*), the chalcone synthase or the light-labile phytochrome (Tobin and Silverthorne, 1985; Manzara and Gruissem, 1988; Gilmartin et al., 1990). In all of these genes, the elements responsible for the light-regulated expression are located upstream of the transcription start site. The pea *RbcS*-3A promoter has been used as a paradigm and many of the *cis*-elements are also found in promoters of other light-regulated genes.

The best characterised *cis*-elements in the *RbcS*-3A promoter are the GT-1 boxes. Site-directed mutagenesis of the GT-1 binding site in the context of a functional promoter results not only of the loss of the GT-1 binding *in vitro*, but also to the loss of transcription *in vivo*. cDNAs encoding GT-1 have been isolated and characterised from various organisms. A second well characterised binding site in the promoters of various light-regulated genes is represented by the G-box. It interacts with the G-box-binding factors, which belong to the class of basic leucine zipper proteins. However, site-directed mutagenesis of various ACGT motifs essential for the binding of G-box-binding proteins suggested that these components are not directly involved in transducing the light signal to the responsive *cis*-elements. The major *cis*-elements identified are listed in Table 2 and for more detailed information about light-regulated expression of *RbcS* and *LhcB* genes, one can refer to a recent review by Terzaghi and Cashmore (1995).

3.2 *cis*-Regulatory Elements (iLRE) Located within the Transcribed Part of a Gene

Recently, regulatory elements of various light-regulated genes, mainly for plastid polypeptides, have been investigated in more detail. These studies uncovered novel regulatory mechanisms in which *cis*-elements for light regulation are located within the parts of the genes (Table 3).

Internal control elements were first discovered for the pea ferredoxin gene, *Fed*-1. Ferredoxin is reversible attached at the stromal part of photosystem I

Table 2. LRE identification of upstream elements and *trans*-acting factors (TAF) in light-regulated gene expression

Plant	Gene	Region identified	TAF	Reference
Pea	*cab* (AB80)	247 bp (-347 to -89)	–	Simpson et al. (1986)
	rbcs-E9	235 bp (-317 to -82)	–	Fluhr et al. (1986)
	rbcs-3A	box II (-151 to -138)	GT-1	Gilmartin et al. (1990) Green et al. (1987)
		box III (-125 to 114)	3AF5, 3AF3	Sarokin & Chua (1992)
		3AF1 site (-51 to -31)	3AF1	Lam et al. (1990)
Wheat	*cab*-1	268 bp (-357 to -89)	–	Nagy et al. (1987)
Tobacco	*cab*-E	PRE-1 (-1554 to -1182)	–	Castresana et al. (1988)
		PRE-2 (-747 to -516)	–	
Lemna	*cab*-AB19	70 bp (-174 to -104)	–	Kehoe (1992)
	rbcs-3A	GATA box (-165 to -140)		Sarokin & Chua (1992)
Tomato	*rbcs*-3A	GATA box (-211 to -201)	GBF	Giuliano et al., (1988)
		AT-1 site (-288 to -27)	AT-1	Ueda et al. (1989), Datta & Cashmore (1989)
Arabidopsis	*cab*-165	78 bp (-111 to -33)	–	Sun et al. (1993)
	cab-140	39 bp (-138 to -99)	CA-1	Sun et al. (1993)
	rbcs-3A	G-box (-247 to -237)	GBF	Giuliano et al. (1988)
	rbcs SS3.6	900 bp (-973 to -90)		Timko et al. (1985)
	rbcs-1A	GATA box (-236 to -228)	GAF1	Gilmartin et al. (1990)

Table 3. Genes where internal light regulatory elements have been shown or where regulatory elements are located immediately upstream of the transcription start site

Gene	Plant	Region identified	Reference
Fed genes			
Fed-1	Pea	5′ untranslated leader 5′ portion of gene	Dickey et al. (1992, 1994)
Fed-A	*Arabidopsis*	5′ untranslated leader, first part of coding region	Bovy et al. (1995)
Psa genes			
PsaD	spinach	intron sequences	Bolle et al. (1997)
*Psa*Db	tobacco	5′ untranslated leader	Yamamoto et al. (1995)
Pet genes			
*Pet*E	Pea	5′ untranslated region	Helliwell & Gray (1995)
*Pet*E	spinach	5′ untranslated region	Bolle et al. (1994a)
Atp genes			
*Atp*C *Atp*D		first 100 nt upstream of transcription start site	Bolle et al. (1996b)

reaction center and transfers electron from the reaction center to ferredoxin-dependent soluble acceptors in the stroma. Initially, photophysiological studies indicated that the *Fed*-1 mRNA accumulates with exceptional rapidity, and that the accumulation was readily reversible when the plants were returned to darkness (Kaufman et al., 1986). Subsequent work with transgenic tobacco revealed that although the intact gene had a strong light response, neither 5' flanking nor 3' flanking sequences, alone, could confer light responsiveness in reporter gene fusions. On the other hand, only constructs containing the transcribed region of *Fed*-1 showed strong light responses even when driven by the constitutive viral 35S RNA CaMV promoter, indicating that response determinants were located within the coding region and could act independently.

The above observation established the existence of a novel gene regulatory mechanism, and suggested that post-transcriptional events might play an important role in regulating the abundance of *Fed*-1mRNA (Elliott et al., 1989a, b; Thompson et al., 1991). Infact, Dickey et al. (1992, 1994) could show that the iLRE requires sequences located both in the untranslated leader and in the first part of the coding region. The response mediated by the iLRE is blocked when translation of the chimeric mRNA is prevented by inserting a stop codon at any of several positions (Dickey et al., 1994). These results indicate that light alters the rate of mRNA decay and that this process may involve translation-dependent changes in mRNA stability (Thompson, personal communication).

Such unusual features as given above were also observed for the ferredoxin gene from *Arabidopsis* (*Fed*-A). In light-grown *Arabidopsis* seedlings, the abundance of the *Fed*-A message was approximately 20-fold higher than in etiolated seedlings. However, run-on transcription assays with isolated nuclei revealed only a two-fold increase in the transcriptional activity in light (Vorst et al., 1993). Further analysis revealed that similar to pea *Fed*-1, full light-regulated expression of *Arabidopsis Fed*-A requires sequences upstream and downstream of the transcription initiation site (Bovy et al., 1995).

Not only genes for peripheral components associated with the photosystem I reaction center, but also a component associated with the reaction center itself, such as *Psa*D, exhibits unusual light-regulatory features. *Psa*D encodes the subunit II, a polypeptide at the stromal site of the reaction center, which is involved in the electron transfer from the reaction center to ferredoxin. It was found that when the *Psa*D promoter and the complete 5'-untranslated leader region was fused to the GUS reporter gene and expressed in transgenic tobacco seedlings, the expression was not properly light regulated. Infact full light-regulated expression of *Psa*D was observed only if the promoter, the entire transcription unit including an intron was transformed into *N. tabacum* (Flieger et al., 1994). A significant light response, measured on the *Psa*D transcript level, was also observed if the expression was driven by the constitutive 35S CaMV promoter, again suggesting that sequences downstream of the transcription start site in combination with elements in the promoter/

leader are essential (Bolle et al., 1994, 1997). Since removal of the intron sequences results in a constitutive expression (Bolle et al., 1997), this sequence appears to confer light-regulated expression onto the *Psa*D mRNA. Surprisingly, sequences within the transcribed region are also essential for the expression of one of the two *Psa*D genes from *N. sylvestris* (*Psa*Db; Yamamoto et al., 1995). However, in contrast to the spinach gene, the tobacco gene lacks an intron suggesting that *cis*-determinants for the light response should be located in the coding region or in either their 5'- or 3'-flanking sequences.

iLREs were also observed for the pea plastocyanin gene (*Pet*E). Plastocyanin is a lumenal protein and transfers electrons from the cytochrome b_6f-complex to the photosystem I reaction center. The complete *Pet*E gene with 1 kb of 5' upstream sequence and 2 kb of 3' downstream sequence shows normal light regulation in transgenic tobacco (Last and Gray, 1990; Pwee and Gray, 1993), but chimeric genes consisting of *Pet*E promoter regions fused to the GUS gene do not (Pwee and Gray, 1993). However, the transcribed region of the *Pet*E gene when fused to the constitutive 35S RNA CaMV promoter gave full light-regulated expression in transgenic tobacco seedlings (Helliwell and Gray, 1995). Gene constructs producing transcripts containing 14 bp of the 35S RNA CaMV leader at the 5' end of the *Pet*E mRNA did not show full light regulation (Helliwell and Gray, 1995), suggesting that the correct 5' end of the *Pet*E mRNA may be necessary for light-regulated expression. These few examples highlight importance of *cis*-elements within the transcribed regions of light-regulated photosynthesis genes.

3.3 *cis*-element within the 5' Untranslated Region and in the Vicinity of the Transcription Start Site

Several reports demonstrated the importance of *cis*-determinants within the 5'-untranslated leader region (UTRs), however the function of these elements remains enigmatic. UTRs of several light-regulated genes harbour translational enhancer, which are not involved in the light signalling (Caspar and Quail, 1993; Yamamoto et al., 1995). Bolle et al. (1994 and 1996a) investigated UTRs of several photosynthesis genes and proposed that they operate at the transcriptional level, again in a light-independent manner. In contrast, nucleotides within the 5'-untranslated leader of the pea *Fed*-1 gene appear to be directly involved in the light response (Dickey et al., 1994).

In contrast to similarities between *Fed* and *Psa*D genes from different organisms, expression and light-regulation of *Pet*E differ among different species. Expression of the spinach and *Arabidopsis Pet*E genes is mediated by upstream regulatory elements and does not require sequences downstream of the translation start site, as has been reported for the pea gene. This is surprising, because *Pet*E is a single-copy gene in all three organisms (Last and Gray, 1989; Lübberstedt et al., 1994) and should derive from a single ancestor. This indicates that regulated expression of the *Pet*E genes might be the result of independent evolutionary processes.

Unusual features for the light-regulated expression have also been shown for other photosynthesis related genes. Lübberstedt et al. (1994) reported the presence of two adjacent elements in the promoter of the spinach ferredoxin-$NADP^+$-oxidoreductase (*Pet*H) gene, each of which is sufficient to confer light-regulated expression onto the heterologous 90 bp 35S RNA CaMV promoter. Since these two segments exhibit no sequence similarities the expression of *Pet*H appears to be coupled to the light signal pathway *via* at least two different *trans*-factors. The spinach *Pet*H promoter does not exhibit similarities to promoter sequences of the equivalent gene from *Arabidopsis*.

More recently, two other light-regulated genes, *Atp*C and *Atp*D, have been investigated and both exhibit new expression characteristics. *Atp*C and *Atp*D encode two of the three nuclear-encoded components of the chloroplast ATP synthase (the subunits γ and δ). Both genes harbour their essential *cis*-regulatory elements within the first 100 nucleotides upstream of the transcription start site. However, *Atp*D differs from *Atp*C in that its enhancer-like elements are present upstream of position-100, whereas this region appears to be of minor importance in the *Atp*C promoter. The crucial *cis*-element in the *Atp*C promoter was located immediately upstream of the CAAT box. If this region is mutagenized by site-directed mutagenesis, the transgene is no longer light regulated. Surprisingly, however the absence of a significant light response is caused by an increase in the reporter gene activity in darkness rather than a decline in light suggesting that the mutation has destroyed a *cis*-regulatory element that represses high expression levels in darkness rather than stimulates gene expression in light. In addition, this example demonstrates that *cis*-regulatory elements for the light response can be located at positions which are usually occupied by components constituting the basal transcription apparatus. The obvious question as to how binding of light-regulatory factors can activate the assembly of the basal transcription apparatus needs to be addressed. Thus *Atp*C established again a novel mechanism for light regulation and demonstrates that the generally accepted concepts, in which conserved and universal regulatory elements mediate light regulated expression of responsive genes is by far too narrow.

3.4 Light as one of Several Regulations of Nuclear Gene Expression

Light-regulated genes are usually also responsive to many other stimuli (e.g. circadian clock, phytohormones, developmental stage of the plastids in a particular tissue, tissue- and organ-specific factors). This raises the question whether they operate *via* separate *cis*-element and *trans*-acting factors, or whether these signals interact prior to gene regulation. One way of identifying the precise sequence for a given response utilizes specific nucleotide exchanges in the context of a functional promoter segment, which are introduced by site-directed mutagenesis.

The *Atp*C gene from spinach mentioned above is an excellent example to demonstrate the complexity of the signalling systems and their interaction with short and defined DNA sequences in promoters of light-responsive genes. It was found that mutation of the short sequence in the *Atp*C promoter immediately upstream of the CAAT box not only destroyed light responsiveness, it also uncoupled the expression of the transgene from the developmental stage of the plastids, from the circadian rhythm and from signals which prevent *Atp*C expression in roots. Since the crucial nucleotides within the *Atp*C *cis*-element are no longer than 5 nucleotides, it is unlikely that this sequence is the binding site for more than one regulatory factor.

Three further examples may illustrate that interactions of defined *cis*-elements for light-regulation are probably not exclusively involved in this response. Anderson and Kay (1995) analysed the *Arabidopsis cab*2 promoter and found that a 78 base pair segment is sufficient for both phytochrome- and circadian clock-regulated transcription. This segment contains two separate binding sites for DNA-binding proteins, designated CUF-1 and CGF-1. CUF-1 is a basic leucine zipper protein and belongs to the class of G-box binding proteins, whereas CGF-1 binds a repeated GATA motif. Inactivation of the CUF-1 binding site by site-directed mutagenesis demonstrates that this protein is not required to generate the phytochrome and circadian clock response and functions as a positive factor which stimulates the expression. In contrast, inactivation of the CGF-1 site in the *Arabidopsis cab2* promoter causes an altered response pattern towards phytochrome action and circadian clock response. This example again confirms that different regulators can operate *via* the same *cis*-element and that the light response can be only one of the regulators of expression of a nuclear gene and that signals from different pathways may converge prior to gene regulation.

Lübberstedt et al. (1994), Kusnetsov et al. (1996) and Bolle et al. (1997) have analysed the G-box sequence in the spinach *Pet*H and *RbcS*-1 promoter. In both promoters, site-directed mutagenesis within the G-box sequence causes a significant reduction in the GUS level without a notable change in the light response. The responsiveness of the *Pet*H and *RbcS*-1 promoters to other regulatory pathways such as circadian rhythm, developmental stage of the plastids and phytohormones was also unaffected by these mutations. This confirms that the G-box binding factors in plants operate primarily quantitatively and they are not directly involved in transducing the light signal to responsive *cis*-elements. Since only short DNA fragments are usually used for these studies, it is difficult to assume that they interact with a large number of DNA-binding proteins which become activated by a variety of signalling systems.

Finally, a comparative analyses of *cis*-regulatory elements involved in the plastid-dependent expression revealed that it was not possible to separate them from the LRE (Kusnetsov et al., 1996; Bolle et al., 1997). This suggests that signal pathways originating from quite different activators converge prior to gene regulation, so that various stimuli become integrated and operate *via* the same (DNA- or RNA-binding) regulatory proteins.

4. Signal Response Coupling

In the previous sections we showed that light stimulus, be it red light or blue light, can induce the expression of second messengers and also lead to changes in the phosphorylation status of proteins, some of which could be *trans* acting factors. We also learnt that there are specific *cis* element be these be present upstream, in the gene, or in the UTR, which functions as light regulatory elements. The basic question is whether the fast responses of generating signals are coupled to gene expression and other physiological or developmental responses and if so, what is the sequence of signal transduction process. Lately, several mutants have been identified that are thought to encode gene products representing key elements of the signal transduction network and they have been used to identify a number of sequential steps in both light-regulated gene expression and photomorphogenesis. The mutants can be assigned to two broad categories. The first includes the phenotypes that do not show normal light-regulated responses in the presence of light. These are usually deficient in the photoreceptors and hence show an arrested phenotype. The second class includes those which show a constitutive expression of photomorphogenic characters even in the absence of light signal, i.e. they are de-etiolated in dark. More details of the study with photomorphogenic mutants is given in Chapter 31.

4.1 Regulation of Gene Expression

Even though the photoregulated genes may be induced by integration of many signals, there must exist at least some unique signalling steps if not completely independent pathways, for specific photoreceptors, to control expression of different light regulated genes. Also, signals from different photoreceptors can be integrated to control expression of a particular gene or developmental process or that multiple signalling steps may converge at some point. Infact *in vivo* gene expression is subjected to regulation by multiple factors such as plant growth regulators, stress and so on. It is expected that molecular and genetic studies will aid in the identification of elements that mediate multiple controls. In the meanwhile several biochemical studies have provided evidence towards understanding the steps involved in photocontrol of gene expression.

Romero et al. (1991b) demonstrated that it was possible to stimulate a RL response in absence of light by activating G-proteins. They could induce expression of *cab* genes and inhibit *phy*A by treating etiolated oat seedlings with cholera toxin, that stabilizes G-proteins in their GTP-activated from. Similarly, Romero and Lam (1993) showed involvement of G-proteins in early steps of light regulated *Cab* gene expression in the light responsive soybean cell cultures. The G-proteins are associated with the membranes and on activation they result in the production of new secondary signalling molecules. As pointed out earlier, calcium/CaM are known to act downstream in the photosignalling pathway and are likely candidates as the messengers, generated following the activation of G-proteins.

The studies in photomixotrophic soybean cell lines also demonstrate a role of calcium and CaM as a mediator of RL induced *Cab* gene expression (Lam et al., 1989). Treating the cells with CaM inhibitors, TFP and W-7, inhibited the gene expression even in the presence of light, while treating dark-adapted cells with calcium ionophore, ionomycin induced gene expression. Further, W-7 could inhibit the cholera toxin stimulated gene expression in the dark-adapted cells. These experiments shed light on the order of signalling intermediates in the pathway, indicating that calcium/CaM act downstream to the G-proteins.

Neuhaus et al. (1993) while working on phytochrome deficient hypocotyl cells of *aurea* mutant of tomato demonstrated that the effect of phytochrome in activating *Cab* gene expression, anthocyanin synthesis and chloroplast development could be mimicked by GTPγS or cholera toxin. The cells microinjected with both PHY A and GDPβS did not show the light response. The multiple phytochrome dependent responses could also be observed by varying the cellular concentrations of calcium or activated CaM.

Further experiments using the single cell assay system of *aurea* mutants showed that the photoresponses initiated by cholera toxin could be inhibited by treating the cells with calcium/CaM inhibitors. During microinjection experiments, the participation of cGMP in phytochrome-mediated signal transduction pathways has also been reported. cGMP alone was able to trigger the production of anthocyanins but could not induce the expression of *Cab*::GUS reporter genes. These results were exactly opposite to that obtained on injections of Ca^{2+}, thereby indicating that in addition to calcium, cGMP plays a role in the phototransduction pathway and that the two components represent two functionally independent signalling mechanisms. Recently, Wu et al. (1996) have used *aurea* single cell assays to identify light-responsive elements that are regulated by calcium and cGMP. Using GUS reporter gene constructs they have shown that Box II of pea *RbcS-3A* gene and Unit I of parsely chalcone synthase gene are targets of calcium and cGMP pathways, respectively. The GUS gene expression in both cases can also be obtained on treatment with GTPγS, an activator of G-proteins, which can activate both the signalling chains.

The experiments on the soybean cell lines as well as single cell microinjection studies have also provided evidence that both cGMP and Ca^{2+}/CaM are necessary for the expression of GUS gene linked to promoter of FNR, which encodes the ferredoxin-NADPH-oxidoreductase associated with Photosystem I (PS I). Similarly cGMP alone can bring about only partial chloroplast maturation but the co-presence of Ca^{2+}/CaM and cGMP resulted in formation of fully developed chloroplasts. Using the same system,

Neuhaus et al. (1997) showed that down regulation of aspargine synthetase gene by light is also modulated via Ca^{2+} and cGMP signal transduction pathway.

Based on the available evidence a model has been suggested for the mechanism of light regulation of gene expression (Figure 1). The model suggests that the activation of the G-proteins by the photoreceptor is the first step in signal generation. Upon G-protein activation there is bifurcation of the signalling pathway; one involving Ca^{2+}/CaM to bring about the expression of photoregulated genes like those participating in the synthesis of photosynthetic complexes. The other pathway does not require calcium and is mediated through cGMP to bring about responses such as anthocyanin biosynthesis. Coaction of both is required for inducing a third category of genes like members of PS I and the cytochrome b_6f-complex. In all these cases Ca^{2+}/CaM and/or cGMP act downstream to G-proteins and these bring about activation of downstream regulatory molecules like kinases to activate the transcription machinery. However, it is not known whether all the photoresponses are mediated through G-proteins and if so, the number of G-proteins involved.

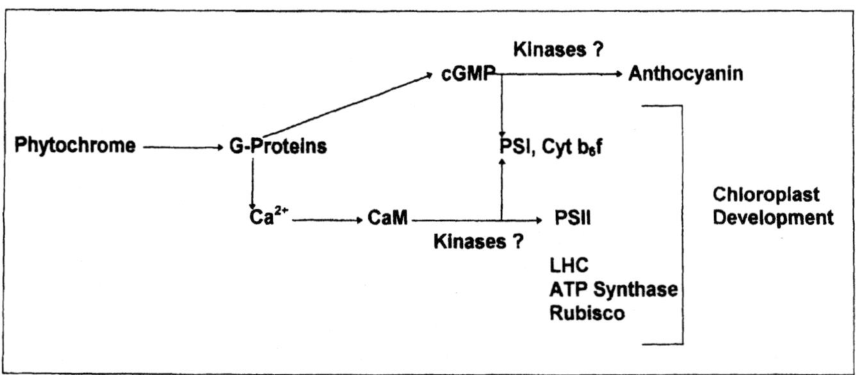

Fig. 1. Biochemical pathways for phytochrome-A signal transduction. On activation of G-proteins by phytochrome there is bifurcation of the signalling pathway. Expression of genes participating in chloroplast development is mediated through Ca^{2+}/CaM, downstream to G-proteins, while anthocyanin synthesis does not require Ca^{2+} and is mediated through cGMP. Coaction of cGMP and Ca^{2+} is required for inducing members of PS I and Cytochrome b_6f gene families (after Neuhaus et al., 1993).

In the model given above we learnt that how one photoreceptor could regulate expression of multiple genes. In another study of chalcone synthase (Chs) gene expression it was found how multiple photoreceptors regulate expression of a single gene. The expression of *chs* gene is regulated by RL/FRL, UVL and BL. The phytochrome regulation is through cGMP pathway as discussed above. Recently, studies by Christie and Jenkins (1996) in *Arabidopsis* cell suspension culture showed that UV-B and UV-A/BL signalling pathways are different from the phytochrome signal transduction pathway in regulating *chs* expression. Both UV-B and UV-A/BL phototransduction processes involve reversible protein phosphorylation and calcium, although the elevation of cytosolic calcium is insufficient to stimulate *chs* expression on it's own. Moreover, UV-A/BL induction of *chs* expression does not appear to involve CaM, whereas UV-B response does. This indicates that distinct signal transduction pathways control expression of the same gene under different light conditions (Figure 2). These studies show that there is crosstalk between signalling pathways thus indicating the nature of complexity of signalling network.

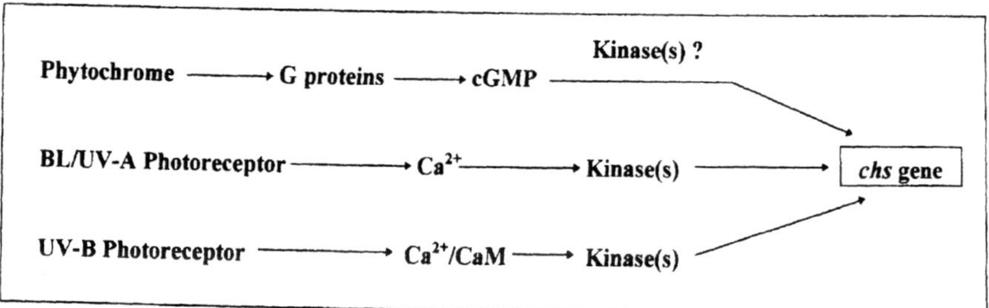

Fig. 2. Regulation of gene expression by multiple photoreceptors. *chs* gene expression is regulated by red, blue and UV light through different signalling pathways. The phytochrome transduction involves cGMP whereas blue and UV phototransduction involves calcium. Moreover, only the UVB signalling requires CaM. Reversible protein phosphorylation is involved in all cases, however it is not known whether the signal is transduced through same or different kinases.

Using light regulation of nitrate reductase (NR) gene expression as a model system evidence has been provided for the involvement of PI-cycle in photo-signalling mechanism. NR, a key enzyme in nitrogen metabolism is strongly up regulated by light, in the presence of nitrate. There are several reports demonstrating that nitrate triggers the synthesis of this enzyme while light regulates its expression either at transcriptional or post-transcriptional levels. Sharma and Sopory (1984) and Sharma et al. (1994) showed that Pfr induces a transmitter which can interact with nitrate to induce NR. The nature of the transmitter and it's site of action is not known, but the light signalling mechanism has been shown to involve PI cycle and PKC (Chandok and Sopory, 1994; Raghuram and Sopory, 1995; Chandok and Sopory, 1996). It was demonstrated that PMA, an activator of PKC, and 5HT, an activator of PI cycle, could mimic RL effect in enhancing the transcript levels of NR gene and inhibiting *phy*I gene, in etiolated maize leaves. This indicates that PI cycle transduces the signal through a PKC type kinase (Raghuram and Sopory, 1995). Based on these studies an additional pathway for light signal transduction was proposed which involves activation of G-proteins, PI cycle and PKC leading to phosphorylation of *trans*-acting factors that regulates gene expression - in this case, stimulation of NR gene and inhibition of *phy*I gene (Figure 3). This also suggests that G-proteins may not be the only common point in phototransduction pathways and that divergence could also be brought about by phosphorylation-dependent activation/inactivation of specific *trans*-acting factors.

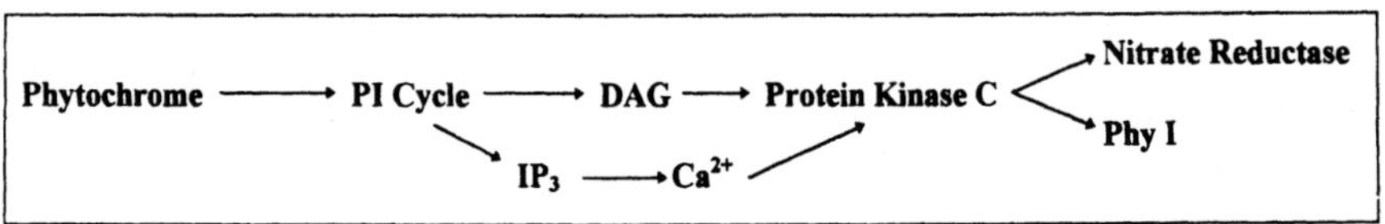

Fig. 3. Role of PI cycle in phytochrome signal transduction. The light signal transduction pathways involves PI cycle and Protein Kinase C (PKC) in regulation of *NR* and *PhyI* gene expression. The photosignal brings about PLC-mediated cleavage of PIP_2 to release second messengers IP_3 and DAG. While IP_3 stimulates the mobilization of Ca^{2+} from endogenous stores, DAG activates PKC in the presence of calcium leading to phosphorylation of *trans*-acting factors that regulate gene expression.

It is well known that phytochrome belongs to a multigene family and all members are conserved in terms of their basic structure and molecular properties (see Chapter 25). It is possible that these different photoreceptors utilise totally different signal transduction pathways. Recent studies have demonstrated that *phy*B, which requires a higher light fluence, may be acting directly in the nucleus and apparently does not require the usual signal transduction components. Using *phy*B-GUS fusion proteins produced in transgenic *Arabidopsis* plants it was demonstrated that *phy*B has multiple nuclear localization signals within its C-terminal region. Moreover, immunoblot analysis of nuclei as well as protoplasts has revealed that *phy*B is localised to the nucleus in a light-dependent manner (Sakamoto et al., 1996). The total level of cellular *phy*B remained constant while the level of nuclear *phy*B was greatly reduced by dark adaption of plants, indicating that the nucleo-cytoplasmic partitioning of the phytochrome changes in response to light. The targets of *phy*B in the nucleus are not known but it is hypothesized that it may regulate gene expression through interactions with nuclear transcription regulators.

4.2 Regulation of Physiological and Developmental Responses

Experiments have also been done to understand the signalling mechanisms involved in bringing about the physiological and developmental responses in plants, like leaf unrolling and/or leaf movements which are known to be regulated by light. In order to understand the mechanisms behind these processes experiments have been performed on phytochrome-controlled swelling of protoplasts obtained from cereal leaves.

Bossen et al. (1990) demonstrated that swelling of etiolated wheat leaf protoplasts had an absolute requirement for calcium in the surrounding medium and that protoplast swelling was inhibited in medium containing EGTA or on addition of calcium channel blockers like verapamil and lanthanum. However, it could be stimulated in absence of light by addition of calcium ionophore. The role of calcium fluxes was also demonstrated by Shacklock et al. (1992) using fluorescent dye imaging techniques. They loaded wheat leaf protoplasts with calcium indicator dye, Fluo-3, and demonstrated a transient increase in cytosolic calcium as well as protoplast swelling, on RL treatment, which returned to the resting levels within one minute. Similar results were observed on releasing calcium from caged compounds. Bossen et al. (1990) demonstrated a role for CaM in this process as its inhibitor, W7 induced protoplast swelling in darkness or after FRL irradiation but did not influence the RL response. These observations suggest participation of a CaM-activated plasma membrane ATPase for increasing the cytoplasmic calcium concentration.

Besides calcium the involvement of G-proteins has also been demonstrated in light induced protoplast swelling. Protoplasts that were electroporated with GTPγS, swelled in absence of light or after FRL irradiation while protoplasts electroporated with GDPβS, an inhibitor of G-proteins, did not show any swelling in response to RL. Experiments by Shacklock et al. (1992) suggested that internal calcium released via IP_3 may be involved in phytochrome responses. They showed that release of caged IP_3 inside wheat protoplasts could substitute for RL in inducing calcium-dependent swelling response in these cells. Since IP_3 is produced by action of PLC, neomycin, a PLC inhibitor, as well as lithium chloride, an inhibitor of phosphoinositide cycle were tested. Both these prevented RL and GTPγS induced swelling of wheat leaf protoplasts. Myo-inositol, the precursor of phosphoinositide pathway, nullified the inhibitory effect of lithium. These results suggested that there is a role of G-proteins, PI cycle and calcium in light regulated protoplast swelling. In fact the effect of cAMP and the participation of PKC in regulation of phytochrome-controlled protoplast swelling have also been demonstrated.

Instead of single cell systems, experiments are now being performed in intact plant systems to understand the phototransduction pathways. Borochov et al. (1993) have studied the involvement of signal transduction pathway components in linking photoperiodic stimulus with the flowering response in *Pharbitis nil*. They have shown that changes in cytosolic pH and Ca^{2+} levels in cotyledons can modulate the early photoperiodic flower response. Also treatment of cotyledons with PMA prior to the inductive dark period enhanced the flowering response. Similar results were obtained with GTPγS while GDPβS had not effect. However both PMA and GTPγS were not effective when applied under non-inductive conditions. Their experiments also demonstrate a role for kinases but, phosphorylation of lipids rather than proteins is postulated.

Our own studies on the formation of primary leaves in *Sorghum bicolor* have demonstrated the involvement of several components of signal transduction in this phytochrome-mediated response (Sanan and Sopory, 1998). Primary leaves are not produced in *Sorghum* seedlings even after 10 days of germination in absence of light. However, irradiation with a short but saturating pulse of RL or WL given to 5 day old etiolated seedlings results in development of etiolated leaves. This effect was not seen with FRL (Rajasekhar et al., 1981). Interestingly, exogenous application of Ca^{2+} and PMA (an activator of protein kinase) could induce leaf formation in absence of light and was found to enhance the RL response (Figure 4). However, the leaf development response could be arrested at the stage of leaf emergence or leaf expansion by the addition of inhibitors of G-proteins, by calcium channel blockers and protein kinase inhibitor H-7, in a concentration-dependent

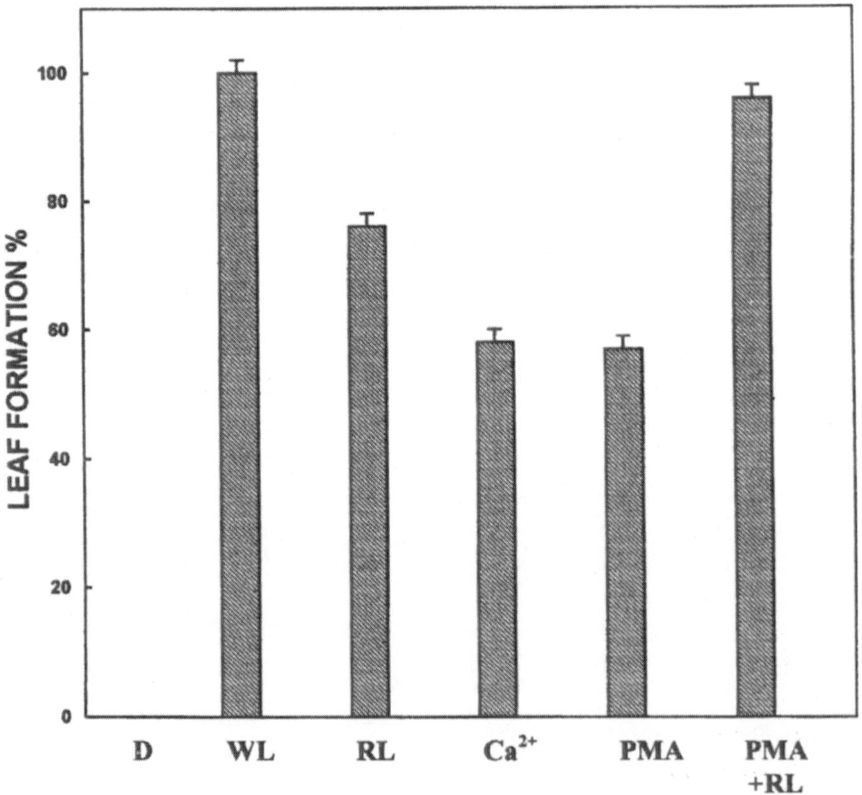

Fig. 4. Effect of light, calcium and PMA on leaf formation in *Sorghum bicolor*. There is no leaf formation in dark grown (D) *Sorghum* seedlings. Leaf emergence and expansion is observed on irradiating 5-day old etiolated seedlings with 5 min white light (WL) or red light (RL) or by supplying exogenous calcium (optimum concentration: 4 mM). Treating the seedlings with 1 μg PMA enhances the RL-induced response. Leaf emergence could also be obtained in absence of light on treating the seedlings with PMA, however these leaves do not undergo complete expansion and remain partially open. The percentage response obtained in each case is plotted.

manner. This suggests a role for G-proteins, calcium and PKC-type enzymes in phototransduction during primary leaf formation.

5. Conclusions

The current upsurge in interest in identifying the biochemical nature of the components of signal transduction and cloning of their genes (Redhead and Palme, 1996) reflect the great importance of this area of research. Various approaches are being adapted to analyze light signalling in different plants. Mutational and cellular approaches have yielded important information in recent years (Quail, 1994; Chamovitz and Deng, 1996; Staub and Deng, 1996; see Chapter 31). Evidence for the role of known signalling intermediates such as G-proteins, components of PI cycle, calcium/CaM and cGMP has been provided in light signal transduction. In the last few years analysis of promoters of a large number of genes has indicated that at the far-end of the signalling chain specific *cis*-elements regulate the transcription of light regulated genes by getting modulated by *trans*-acting factors. It is now shown, however, that some of the light regulatory *cis* elements also reside in the transcribed region and some of these may respond to other signals also. We are now beginning to understand the nature of crosstalk between various signals and also about the exact nature of *cis* elements in a single gene which respond differently to various signalling intermediates. It is now being revealed how a single photoreceptor can modulate the expression of a large number of genes by employing different signal molecules and intermediary pathways. On the other hand similar second messengers are utilized by different photoreceptors which end up in regulating the same gene. It is understood, it will take more time to get a clear picture with respect to the exact *in vivo* situation where multiple photoreceptors operate simultaneously and regulate multiple genes.

More work is awaited to functionally correlate various signalling intermediates with the steps defined by photomorphogenic loci. Also ions and molecules like calcium, cGMP, phospholipids cannot provide absolute specificity to a stimulus, therefore it is expected that other cellular molecules, like protein kinases and phosphatases and their specific substrates will play an important role in signal transduction.

At present the knowledge about light signalling pathway is in a developing phase and a lot more intensive work is expected in the years to follow. One point is clear that there appears to be a tremendous redundancy in the photoregulatory machinery and a closer understanding will help in appreciating its adaptive value to plants in the natural conditions.

References

Ahmad, M. and Cashmore, A.R. 1993. *HY4* gene of *A. thaliana* encodes a protein with characteristics of a blue-light photoreceptor. *Nature* **366:** 162-166.

Algarra, P., Linder, S. and Thümmler, F. 1993. Biochemical evidence that phytochrome of the moss *Ceratodon purpureus* is a light regulated protein kinase. *FEBS Lett.* **315:** 69-73.

Anderson, S.L. and Kay, S.A. 1995. Functional dissection of circadian clock and phytochrome regulated transcription of *Arabidopsis cab*2 gene. *Proc. Natl. Acad. Sci. USA* **92:** 1500-1504.

Appenroth, K.J. and Augsten, H. 1990. Photophysiology of turion germination in *Spirodella polyrhiza* L. Schleiden. V. Demonstration of calcium requiring phase during phytochrome mediated germination. *Planta* **174:** 94-100.

Basu, A., Sethi, U. and Guha-Mukherjee, S. 1990. Phytochrome in control of differentiation and phosphatidylinositol turnover in *Brassica oleracea* cultures. *Phytochem.* **29:** 1539-1541.

Berridge, M.J. 1987. Inositol triphosphate and diacyglycerol. Two interacting second messengers. *Annu. Rev. Biochem.* **56:** 159-193.

Berridge, M.J. and Irvine, R. F. 1989. Inositol phosphates and cell signalling. *Nature* **341:** 197-205.

Biermann, B.J., Pao, L.I. and Feldman, L.J. 1994. Pr-specific phytochrome phosphorylation *in vitro* by a protein kinase present in anti-phytochrome maize immunoprecipitates. *Plant Physiol.* **105:** 243-251.

Biswas, S., Dalal, B., Sen, M. and Biswas, B. B. 1995. Receptor for myo-inositol triphosphate from microsomal fraction of *Vigna radiata. Biochem. J.* **306:** 631-636.

Bolle, C., Herrmann, R.G. and Oelmüller, R. 1997. Intron sequences are involved in the plastid- and light- dependent expression of the spinach *Psa*D gene. *Plant J.* **10:** 919-924.

Bolle, C., Herrmann, R.G. and Oelmüller, R. 1996a. Differential sequences for 5′ untranslated leaders of nuclear genes for plastid proteins affect the expression of the *β*-glucuronidase gene. *Plant Mol. Biol.* **32:** 861-868.

Bolle, C., Kusnetsov, V.V., Herrmann, R.G. and Oelmuller, R. 1996b. The spinach *AtpC* and *AtpD* genes contain elements for light-regulated, plastid-dependent and organ-specific expression in the vicinity of the transcription start sites. *Plant J.* **9:** 21-30.

Bolle, C., Sopory, S.K., Lübberstedt, T. H., Herrmann, R.G. and Oelmüller, R. 1994a. Segments encoding 5′-untranslated leaders of genes for thylakoid proteins contain *cis*-elements essential for transcription. *Plant J.* **6:** 513-523.

Bolle, C., Sopory, S.K., Lübberstedt, T.H., Klösgen, R.B., Herrmann, R.G. and Oelmüller, R. 1994b. The role of plastids in the expression of nuclear genes for thylakoid proteins studied with chimeric glucuronidase gene fusions. *Plant Physiol.* **105:** 1355-1364.

Borochov, A., Spiegelstein. H. and Halevy, A.H. 1995. Involvement of signal transduction pathway components in photoperiodic flower induction in *Pharbitis nil. Physiol. Plant.* **95:** 393-398.

Bossen, M.E., Dassen, H.H.A., Kendrick, R.E. and Verdenberg, W.J. 1988. The role of calcium ions in phytochrome-controlled swelling of etiolated wheat (*Triticum aestivum L.)* protoplasts. *Planta* **174:** 94-100.

Bossen, M.E., Kendrick, R.E. and Verdenberg, W.J. 1990. The involvement of a G-protein in phytochrome-regulated, Ca^{2+}-dependent swelling of etiolated wheat protoplasts. *Physiol. Plant.* **80:** 55-62.

Bovy, A., Van Den Berg, C., De Vrieze, G., Thompson, W.F., Weisbeek, P. and Smeekens, S. 1995. Light-regulated expression of the *Arabidopsis thaliana* ferredoxin gene requires sequences upstream and downstream of the transcription initiation site. *Plant Mol. Biol.* **27:** 27-39.

Bowler, C. and Chua, N.-H. 1994. Emerging themes of plant signal transduction. *The Plant Cell* **6:** 1529-1541.

Bush, D.S. 1993. Regulation of cytosolic calcium in plants. *Plant Physiol.* **103:** 7-13.

Buzby, J.S., Yamada, T. and Tobin, E.M. 1990. A light regulated DNA binding activity interacts with a conserved region of *Lemna gibba RbcS* promoter. *The Plant Cell* **2:** 805-814.

Chae, Q., Park, H.J. and Hong, S.D. 1990. Loading of Quin 2 into the oat protoplast and measurement of cytosolic calcium ion concentration changes by phytochrome action. *Biochem. Biophys. Acta* **1051:** 115-122.

Caspar, T. and Quail, P.H. 1993. Promoter and leader regions involved in the expression of the *Arabidopsis* ferredoxin A gene. *Plant J.* **3:** 161-174.

Castresana, C., Garcia-Luque, I., Alonso E., Malik, V.S. and Cashmore, A.R. 1988. Both positive and negative regulatory elements mediate expression of a photoregulated *cab* gene from *Nicotiana plumbaginifolia. EMBO J* **7:** 1929-1936.

Chamovitz, D.A. and Deng, X.-W. 1996. Light signalling in plants. *Crit. Rev. Pl. Sci.* **15:** 455-470.

Chandok, M.R. and Sopory, S.K. 1992. Phorbol myristate acetate replaces phytochrome-mediated stimulation of nitrate reductase in maize. *Phytochem.* **31:** 2255-2258.

Chandok, M.R. and Sopory, S.K. 1994. 5-Hydroxytryptamine affects turnover of polyphosphoinositides in maize and stimulates nitrate reductase in the absence of light. *FEBS Lett.* **36:** 39-42.

Chandok, M.R. and Sopory, S.K. 1996. Phosphorylation/dephosphorylation steps are key events in the phytochrome-mediated enhancement of nitrate reductase mRNA levels and enzyme activity in maize. *Mol. Gen. Genet.* **251:** 599-608.

Chandok, M.R. and Sopory, S.K. 1998. ZmcPkc 70, a protein kinase C enzyme from maize. Biochemical characterization, regulation by phorbol 12-myristate 13-acetate and its possible involvement in nitrate reductase gene expression. *J. Biol. Chem.* **273:** 19235–19242.

Christie, J.M. and Jenkins, G.I. 1996. Distinct UV B and UV A/blue light signal transduction pathways induce chalcone synthase gene expression in *Arabidopsis* cells. *The Plant Cell* **8:** 1555-1567.

Clark, G.B., Memon, A.R., Tong, C.G., Thompson Jr., G.A. and Roux, S.J. 1993. Phytochrome regulates GTP-binding protein activity in the envelope of pea nuclei. *Plant J.* **4:** 399-402.

Das, R. and Sopory, S.K. 1985. Evidence of regulation of calcium uptake by phytochrome in maize protoplasts. *Biochem. Biophys. Res. Commun.* **128:** 1455-1460.

Datta, N. and Cashmore, A.R. 1989. Binding of a pea nuclear protein to promoters of certain photo-regulated genes is modulated by phosphorylation. *The Plant Cell* **1:** 1069-1077.

Datta, N., Roux, S.J. and Chen, Y.R. 1985. Phytochrome and calcium stimulation of protein phosphorylation in isolated pea nuclei. *Biochem. Biophys. Res. Commun.* **128:** 1403-1408.

Dickey, L.F., Gallo-Meagher, M. and Thompson, W.F. 1992. Light regulatory sequences are located within the 5' portion of the *Fed*-1 message sequence. *EMBO J.* **11:** 2311-2317.

Dickey, L.F., Nguyen, T.-T., Allen, G.C. and Thompson, W.F. 1994. Light modulation of ferredoxin mRNA abundance requires an open reading frame. *The Plant Cell* **6:** 1171-1176.

Doshi, A., Aneeta and Sopory, S.K. 1992. Regulation of protein phosphorylation by phytochrome in *Sorghum bicolor. Photochem. Photobiol.* **55:** 465-468.

Downes, C.P. and Michell, R.H. 1985. Inositol phospholipid breakdown as a receptor-controlled generator of second messengers. In: P. Cohen and M.D. Houslay (eds.) Molecular Mechanisms of Transmembrane Signalling, pp. 3-56. Elsevier, Amsterdam.

Drøbak, B.K. 1992. The plant phosphoinositide system. *Biochem. J.* **288:** 697-712.

Drøbak, B.K. 1993. Plant phosphoinositides and intracellular signalling. *Plant. Physiol.* **102:** 705-709.
Dryer, E.M. and Weisenseel, M.H. 1979. Phytochrome-mediated uptake of calcium in *Mougeotia* cells. *Planta* **146:** 31-39.
Elliott, R.C., Dickey, L.F., White, M.J. and Thompson, W.F. 1989a. *cis*-acting elements for light regulation of pea ferredoxin I gene expression are located within transcribed sequences. *The Plant Cell* **1:** 691-698.
Elliott, R.C., Pedersen, T..J., Fristensky, B., White, M.J., Dickey, L.F. and Thompson, W.F. 1989b. Characterisation of a single-copy gene encoding ferredoxin I from pea. *The Plant Cell* **1:** 681-690.
Fallon, K.M., Shacklock, P.S. and Trewavas, A.J. 1993. Detection *in vivo* of very rapid red light-induced calcium-sensitive protein phosphorylation in etiolated wheat (*Triticum aestivum*) leaf protoplasts. *Plant Physiol.* **101:** 1039-1045.
Flieger, K., Wicke, A., Herrmann, R.G. and Oelmüller, R. 1994. Promoter and leader sequences of the spinach *Psa*D and *Psa*F genes direct an opposite light response in tobacco cotyledons: *Psa*D sequences downstream of the ATG codon are required for a positive light response. *Plant J.* **6:** 359-368.
Fluhr, R., Kuhlemeier, C., Nagy, F. and Chua, N.-H. 1986. Organ-specific and light-induced expression of plant genes. *Science* **232:** 1106-1112.
Friedman, H., Goldschmidt, E.E. and Haley, A.H. 1989. Involvement of calcium in the photoperiod flower induction process in *Pharbitis nil. Plant. Physiol.* **89:** 530-534.
Gilman, A.G. 1987. G-proteins: Transducers of receptor-generated signals. *Annu. Rev. Biochem.* **56:** 615-649.
Gilmartin, P. M., Sarokin, L., Memelink, J. and Chua, N.-H. 1990. Molecular light switches for plant genes. *The Plant Cell* **2:** 369-378.
Gilroy, S. and Trevawavas, T. 1994. A decade of plant signals. *Bioessays* **16:** 677-682.
Giuliano, G., Pichersky, E., Malik, V.S., Timko, M.P., Scolnik, P.A. and Cashmore, A.R. 1988. An evolutionarily conserved protein binding sequence upstream of a light-regulated gene. *Proc. Natl. Acad. Sci. USA* **85:** 7098-7093.
Griffith, G.W., Jenkins, G.I., Milner-White, E.J. and Clutterbuck, A.J. 1994. Homology at the amino acid level between plant phytochromes and a regulator of asexual sporulation in *Emericella (=Aspergillus) nidulans. Photochem. Photobiol.* **59:** 252-256.
Grimm, R., Gast, D. and Rüdiger, W. 1989. Characterisation of a protein kinase activity associated with phytochrome from etiolated oat (*Avena sativa*) seedlings. *Planta* **178:** 199-206.
Green, P.J., Kay, S.A. and Chua, N.-H. 1987. Sequence-specific interactions of a pea nuclear factor with light-responsive elements upstream of the *rbcs* -3A gene. *EMBO J.* **6:** 2543-2549.
Guron, K., Chandok, M.R. and Sopory, S.K. 1992. Phytochrome-mediated rapid changes in the level of phosphoinositides in etiolated levels of *Zea mays. Photochem. Photobiol.* **56:** 691-695.
Hager, A. 1996. Properties of a blue-light absorbing photoreceptor kinase localised in the plasma membrane of the coleoptile tip region. *Planta* **198:** 294-299.
Halaban, R. and Hillman, W.S. 1970. Response of *Lemna perpusilla* to periodic transfer to distilled water. *Plant. Physiol.* **46:** 641-644.
Hale, C.C.H. and Roux, S.J. 1980. Photoreversible calcium fluxes induced by phytochrome in oat coleoptiles cells. *Plant Physiol.* **65:** 658-662.
Hamada, T., Tanaka, N., Noguchi, T., Kimura, N. and Hasunuma, K. 1996. Phytochrome regulates phosphorylation of a protein with characteristics of a nucleotide diphosphate kinase in the crude membrane fraction from stem sections of etiolated pea seedlings. *J. Photochem. Photobiol.* **33:** 143-151.

Harter, K., Frohnmeyer, H., Kircher, S., Kunkel, T., Muhlbauer, S. and Schäfer, E. 1994. Light induces rapid changes of the phosphorylation pattern in the cytosol of evacuolated parsely protoplasts. *Proc. Natl. Acad. Sci. USA* **91:** 5038-5042.

Hartmann, E. and Pfaffman, H. 1990. Phosphatidylinositol and phytochrome-mediated phototropism of moss protonemal tip cells. In: D.J. Morre, W.F. Boss and F.A. Loewus (eds.) Inositol Metabolism in plants, pp. 259-276. Wiley-Liss, New York.

Hasunuma, K. and Funadera, K. 1987. GTP-binding protein(s) in green plant, *Lemna paucicostata. Biochem. Biophys. Res. Commun.* **143:** 908-912.

Haupt, W. and Weisenseel, M.H. 1976. Physiological evidence and some thoughts on localised responses, intracellular localisation and action of phytochrome. In: H. Smith (ed.) Light and Plant development, pp. 63-74. Butterworths; London.

Helliwell, C.A. and Gray, J.C. 1995. The sequence surrounding the translation initiation codon of the pea plastocyanin gene increases translational efficiency of a reporter gene. *Plant. Mol. Biol.* **29:** 621-626.

Iino, M., Endo, M. and Wada, M. 1989. The occurrence of a Ca^{2+}- dependent period in the red-light-induced late G1 phase of germinating *Adiantum* spores. *Plant. Physiol.* **91:** 610-616.

Jena, P.K., Reddy, A.S.N. and Poovaiah, B.W. 1989. Molecular cloning and sequencing of a cDNA for plant calmodulin: Signal induced changes in the expression of calmodulin. *Proc. Natl. Acad. Sci. USA* **86:** 3644-3648.

Katagiri, T., Mizoguehi, T. and Shinozaki, K. 1996. Molecular cloning of a cDNA encoding diacylglycerol kinase (DGK) in *Arabidopsis thaliana. Plant. Mol. Biol.* **30:** 647-653.

Kaufman, L.S., Roberts, L.R., Briggs, W.R. and Thompson, W.F. 1986. Phytochrome control of specific mRNA levels in developing pea buds. Kinetics of accumulation, reciprocity and escape kinetics of the low fluence response. *Plant Physiol.* **81:** 1033-1038.

Kehoe, D.M. 1992. Phytochrome regulation of *Lemna gibba cab AB*19 transcripts. Ph.D. thesis, University of California, Los Angeles, California.

Kim, I.S., Bai, U. and Song, P.S. 1989. A purified 124 kda oat phytochrome does not possess a protein kinase activity. *Photochem. Photobiol.* **49:** 319-323.

Klimczak, L.J., Schindler, U. and Cashmore, A.R. 1992. DNA binding activity of *Arabidopsis* G-box binding factor GBF1 is stimulated by phosphorylation by casein kinase II from broccoli. *The Plant Cell* **4:** 87-98.

Kusnetsov, V.V., Bolle, C., Lübberstedt, T., Sopory, S.K., Herrmann, R.G. and Oelmüller, R. 1996. Evidence that the plastid signal and light operate via the same *cis*-elements in the promoters of nuclear genes for plastid proteins. *Mol. Gen. Genet.* **252:** 631-639.

Lam, E., Benedyk, M and Chua, N.-H. 1989. Characterisation of phytochrome regulated gene expression in photoautotrophic cell-suspension: possible role for calmodulin. *Mol. and Cell. Biol.* **9:** 4819-4823.

Lam, E., Kano-Murakami, Y., Gilmartin, P., Niner, B. and Chua, N.-H. 1990. A metal-dependent DNA binding protein interacts with a constitutive element of a light-responsive promoter. *Plant Cell* **2:** 857-866.

Last, D.I. and Gray, J.C. 1989. Plastocyanin is encoded by a single-copy gene in the pea haploid genome. *Plant Mol. Biol.* **12:** 655-666.

Last, D.I. and Gray, J.C. 1990. Synthesis and accumulation of pea plastocyanin in transgenic tobacco plants. *Plant Mol. Biol.* **14:** 229-238.

Li, H. and Roux, S.J. 1992. Purification and characterisation of a casein kinase 2-type protein kinase from pea nuclei. *Plant Physiol.* **99:** 686-692.

Lin, X., Feng, X.-H. and Watson, J.C. 1991. Differential accumulation of transcripts encoding protein kinase homologs in greening pea seedlings. *Proc. Natl. Acad. Sci. USA* **88:** 6951-6955.

Ling, V., Prerea, I. and Zielinski, R.E. 1991. Primary structure of *Arabidopsis* calmodulin isoforms deduced from the sequences of cDNA clones. *Plant Physiol.* **96:** 1196-1202.

Liscum, E., and Briggs, W.R. 1995. Mutations in the *NPH1* locus of *Arabidopsis* disrupt the perception of phototropic stimuli. *The Plant Cell* **7:** 473-485.

Liscum, E., Young, J.C., Poff, K.L. and Hangarter, R.P. 1992. Genetic separation of phototropism and blue light inhibition of stem elongation. *Plant Physiol.* **100:** 267-271.

Lübberstedt, T., Oelmüller, R., Wanner, G. and Herrmann, R.G. 1994. Interacting *cis* elements in the plastocyanin promoter from spinach ensure regulated high-level expression. *Mol. Gen. Genet.* **242:** 602-613.

Ma, H. 1994. GTP-binding proteins in plants: new members of an old family. *Plant Mol. Biol.* **26:** 1611-1636.

Manzara, T. and Gruissem, W. 1988. Organisation and expression of the genes encoding ribulose-1, 5-bisphosphate carboxylase in higher plants. *Photosynth. Res.* **16:** 117-139.

McMichael Jr, R.W. and Lagarias, J.C. 1990. Phosphopeptide mapping of *Avena* phytochrome phosphorylated by protein kinases *in vitro. Biochem.* **29:** 3872-3878.

Mehta, M., Malik, M.K., Khurana, J.P. and Maheshwari, S.C. 1993. Phytochrome modulation of calcium fluxes in wheat (*Triticum aestivum* L.) protoplasts. *Plant Growth Reg.* **12:** 293-302.

Melin, P. M., Sommarin, M., Sandelius, A.S. and Jergil, B. 1987. Identification of Ca^{2+} stimulated polyphosphoinositide phospholipase C in isolated plant plasma membrane. *FEBS Lett.* **223:** 87-91.

Memon, A.R. and Boss, W. F. 1990. Rapid light induced changes in phosphoinositide kinases and H^+ATPase in plasma membrane of sunflower hypocotyl. *J. Biol. Chem.* **265:** 14817-14821.

Miller, A.J., McGarth, R.B. and Chua, N.-H. 1994. Phytochrome phototransduction pathways. *Annu. Rev. Genet.* **28:** 325-349.

Morse, M.J., Crain, R.C. and Satter, R.L. 1987. Light-stimulated inositol phospholipid turnover in *Samanea saman* leaf pulvini. *Proc. Natl. Acad. Sci. USA* **84:** 7075-7078.

Moysset, L. and Simon, E. 1989. Role of calcium in phytochrome-controlled nyctinastic movements of *Albizzia lophantha* leaflets. *Plant Physiol.* **90:** 1108-1114.

Nagy, F., Boutry, M., Hsu, M. Y., Wong, M. and Chua, N.-H. 1987. The 5′ proximal region of the wheat *cab-1* gene contains a 268-bp enhancer-like sequence for phytochrome response. *EMBO J.* **6:** 2537-2542.

Neuhaus, G., Bowler, C., Hiratsuka, K., Yawagata, Y. and Chua, N.-H. 1997. Phytochrome-regulated repression of gene expression requires calcium and cGMP. *EMBO J.* **16:** 2554-2564.

Neuhaus, G., Bowler, C., Kern, R. and Chua, N.-H. 1993. Calcium/calmodulin-dependent and independent phytochrome signal transduction pathways. *Cell* **73:** 937-952.

Otto, V. and Schäfer, E. 1988. Rapid phytochrome controlled protein phosphorylation and dephosphorylation in *Avena sativa* L. *Plant Cell Physiol.* **29:** 1115-1121.

Palme, K. 1996. A role for the heterotrimeric G-protein switch in higher plants. In: M. Smallwood, J.P. Knox, D.J. Bowles (eds.) Membranes: Specialised Functions in Plants, pp. 151-163. BIOS Scientific Publ., UK.

Park, M.H. and Chae, Q. 1989. Intracellular protein phosphorylation in oat (*Avena sativa* L.) protoplasts by phytochrome action. Measurement of action spectra for protein phosphorylation. *Biochem. Biophys. Res. Commun.* **162:** 9-14.

Poovaiah, B.W. and Reddy, A.S.N. 1993. Calcium and signal transduction in plant. *Crit. Rev. Plant Sci.* **12:** 85-211.

Pwee, K-H. and Gray, J.C. 1993. The pea plastocyanin promoter directs cell-specific but not full light-regulated expression in transgenic tobacco plants. *Plant J.* **3:** 437-449.

Quail, P.H. 1994. Photosensory perception and signal transduction in plants. *Curr. Opin. Gen. Dev.* **4:** 652-661.

Raghuram, N. and Sopory, S.K. 1995. Evidence for some common signal transduction events for opposite regulation of nitrate reductase and phytochrome-I gene expression by light. *Plant Mol. Biol.* **29:** 25-35.

Rajasekhar, V.K., Rao, L.V.M., Guha-Mukherjee, S. and Sopory, S.K. 1981. Phytochrome control of carotenoid accumulation in *Sorghum bicolor. Plant Cell Physiol.* **22:** 773-780.

Reddy, A.S.N., McFadden, J.J., Friedmann, M. and Poovaiah, B.W. 1987. Signal transduction in plants. Evidence for the involvement of calcium and turnover of inositol phospholipids. *Biochem. Biophys. Res. Commun.* **149:** 334-339.

Redhead, C.R. and Palme, K. 1996. The genes of plant signal transduction. *Crit. Rev. Pl. Sci.* **15:** 425-440.

Roblin, G., Fleurat-Lessard, P. and Bonmort, J. 1989. Effects of compounds affecting calcium channels on phytochrome- and blue-pigment mediated pulvinar movements of *Cassia fasciculata. Plant Physiol.* **90:** 697-701.

Roblin, G., Fleurat-Lessard, P., Everat-Bourbonlon, A., Bonmort, J. and Moyen, C. 1990. Phytochrome and blue pigment-mediated leaf movements: experimental approach in the signal transduction. *Photochem. Photobiol.* **52:** 197-202.

Romero, L.C., Biswal, B. and Song, P.S. 1991a. Protein phosphorylation in isolated nuclei from etiolated *Avena seedlings.* Effect of red/far-red light and cholera toxin. *FEBS Lett.* **282:** 347-350.

Romero, L.C. and Lam, E. 1993. Guanine nucleotide binding protein involvement in early steps of phytochrome-regulated gene expression. *Proc. Natl. Acad. Sci. USA* **90:** 1465-1466.

Romero, L.C., Sommer, D., Gotor, C. and Song, P.S. 1991b. G-proteins in etiolated *Avena* seedlings. Possible phytochrome regulation. *FEBS Lett.* **282:** 341-346.

Roux, S.J. 1984. Calcium and phytochrome action in plants. *Bioscience* **34:** 25-29.

Roux, S.J. 1994. Signal transduction in phytochrome responses. In: R.E. Kendrick and G.M.H. Kronenberg (eds.) Photomorphogenesis in plants, pp.187-209 Kluwer Academic Press, The Netherlands.

Roux, S.J., McEntire, K., Slovin, R. D., Cedel, T.E. and Hale, C.C. 1981. Phytochrome induces photoreversible calcium fluxes in purified mitochondria fractions from oats. *Proc. Natl. Acad. Sci. USA* **78:** 283-287.

Roux, S.J., Wayne, R.C. and Datta, N. 1986. Role of calcium ions in phytochrome responses: an update. *Physiol. Plant.* **66:** 344-348.

Sakamoto, K. and Nagatani, A. 1996. Nuclear localisation activity of phytochrome B. *Plant J.* **10:** 859-868.

Sanan, N. and Sopory, S.K. 1998. A role of G proteins and calcium in light regulated primary leaf formation in *Sorghum bicolor. J. Expt. Bot.* (in press).

Sarokin, L.P. and Chua, N.-H. 1992. Binding sites for two novel phospho-proteins, 3AF5A and AF3, are required for *rbcs*-3A expression. *The Plant Cell* **4:** 473-483.

Scheuerlein, R., Schmidt, K., Poenie, M. and Roux, S.J. 1991. Determination of cytoplasmic calcium concentration in *Dryopteris* spores. A developmentally non-disruptive technique for loading of the calcium indicator fura-2. *Planta* **184:** 166-174.

Schneider-Poetsch, H.A.W., Braun, B., Marx, S. and Schaumburg, A. 1991. Phytochromes and bacterial sensor proteins are related by structural and functional homologies. *FEBS Lett.* **281:** 245-249.

Serlin, B.S. and Roux, S.J. 1984. Modulation of chloroplast movement in the green alga *Mougeotia* by the Ca^{2+} ionophore A23187 and by calmodulin antagonists. *Proc. Natl. Acad. Sci. USA* **81:** 6368-6372.

Serlin, B.S., Sopory, S.K. and Roux, S.J. 1984. Modulation of oat mitochondrial ATPase activity by Ca^{2+} and phytochrome. *Plant Physiol.* **74:** 827-833.

Shacklock, P.S., Read, N.D. and Trewavas, A.J. 1992. Cytosolic free calcium mediates red light induced photomorphogenesis. *Nature* **358:** 753-755.

Sharma, A.K., Raghuram, N., Chandok, M.R., Das, R. and Sopory, S.K. 1994. Investigations on the nature of the phytochrome-induced transmitter for the regulation of nitrate reductase in etiolated leaves of maize. *J. Expt. Bot.* **45:** 485-490.

Sharma, A.K. and Sopory, S.K. 1984. Independent effect of phytochrome and nitrate on nitrate reductase and nitrite reductase activities in maize. *Photochem. Photobiol.* **39:** 491-493.

Sharma, V.K., Jain, P.K., Meheshwari, S.C. and Khurana, J.P. 1997a. Rapid blue-light-induced phosphorylation of plasma-membrane-associated proteins in wheat. *Phytochem.* **44:** 775–780.

Sharma, V.K., Jain, P.K. Malik, M.K. Maheshwari, S.C. and Khurana, J.P. 1997b. Light- and calcium-modulated phosphorylation of proteins from wheat seedlings. *Phytochem.* **44:** 781–786.

Sheen, J. 1993. Protein phosphatase activity is required for light-inducible gene expression in maize. *EMBO J.* **12:** 3497-3505.

Shi, J., Gonzales, R.A. and Bhattacharya, M.K. 1995. Characterisation of a plasma membrane associated phosphoinositide-specific phospholipase C from soybean. *Plant J.* **8:** 381-390.

Short, T.W. and Briggs, W.R. 1994. The transduction of blue light signals in higher plants. *Annu. Rev. Plant Physiol. Plant Mol. Biol.* **45:** 143-171.

Simpson, I., Schell, J., Van Montagu, M. and Herrera-Estrella, L. 1986. The light-inducible and tissue specific expression of pea LHCP gene involves an upstream element combining enhancer and silencer-like properties. *Nature* **323:** 551-553.

Short, T.W., Reymond, P. and Briggs, W.R. 1993. A pea plasma membrane protein exhibiting blue light-induced phosphorylation retains photosensitivity following Triton solubilization. *Plant Physiol.* **101:** 647-655.

Sopory, S.K. and Chandok, M.R. 1996. Light-induced signal transduction pathway involving inositol phosphates. In: Myoinositol phosphates, phosphoinositides and signal transduction, B.B. Biswas and S. Biswas (eds.) Subcellular Biochemistry, Vol 26, pp. 345-370. Plenum Press, New York.

Staub, J.M. and Deng, X.-W. 1996. Light-signal transduction in plants. *Photochem. Photobiol.* **64:** 897-905.

Sun, L., Doxsee, R.A., Harel, E. and Tobin, E.M. 1993. CA-1, a novel phosphoprotein, interacts with the promoter of the *cab 140 gene* in *Arabidopsis* and is undetectable in det1 mutant seedlings. *Plant Cell* **5:** 109-121.

Takagi, S., Yamamoto, K.T., Furuya, M. and Nagai, R. 1990. Co-operative regulation of cytoplasmic streaming and Ca^{++} fluxes by Pfr and photosynthesis in *Vallisneria* mesophyll cells. *Plant Physiol.* **94:** 1702-1708.

Terzaghi, W.B. and Cashmore, A.R. 1995. Light-regulated transcription. *Annu. Rev. Plant. Physiol. Plant. Mol. Biol.* **46:** 445-474.

Thompson, W.F., Elliott, R., Dickey, L., Gallo, M. and Pedersen, T. 1991. In: B. Thomas (ed.) Phytochrome Properties and Biological Acion, Vol. (in press). Springer-Verlag, Berlin.

Thümmler, F., Algarra, P. and Fobo, G.M. 1995. Sequence similarities of phytochrome to protein kinases: implication for the structure, function and evolution of the phytochrome gene family. *FEBS Lett.* **357:** 149-155.

Thümmler, F., Dufner, M., Kreisl, P. and Dittrich, P. 1992. Molecular cloning of a novel phytochrome gene of moss *Ceratodon purpureus* which encodes a putative light regulated protein kinase. *Plant Mol. Biol.* **20:** 1003-1017.

Timko, M.P., Kausch, A.P., Castresana, C., Fassler, J., Herrera-Estrella, H. 1985. Light regulation of plant gene expression by an upstream enhancer-like element. *Nature* **318:** 579-582.

Tobin, E.M. and Silverthorne, J. 1985. Light regulation of gene expression in higher plants. *Annu. Rev. Plant. Physiol.* **36:** 569-593.

Tretyn, A. and Kendrick, R.E. 1990. Induction of leaf unrolling by phytochrome and acetylcholine in etiolated wheat seedlings. *Photochem. Photobiol.* **52:** 123-129.

Tretyn, A. and Kendrick, R.E. 1991. The role(s) of calcium ions in phytochrome action. *Photochem. Photobiol.* **54:** 1135-1155.

Ueda, T., Pichersky, E., Malik, V.S. and Cashmore, A.R. 1989. Level of expression of the tomato *rbcs*-3A gene is modulated by a far-upstream promoter element in a developmentally regulated manner. *Plant Cell* **1:** 217-227.

Viner, N., Whitelam, G. and Smith, H. 1988. Calcium and phytochrome control of leaf unrolling in dark-grown barley seedlings. *Planta* **175:** 209-213.

Vorst, O., Van Dam, F., Weisbeek, P. and Smeekens, S. 1993. Light-regulated expression of the *Arabidopsis thaliana* ferredoxin A gene involves both transcriptional and post-transcriptional processes. *Plant J.* **3:** 793-803.

Warpeha, K.M.F., Hamm, H.E., Rasenick, M.M. and Kaufman, L.S. 1991. A blue light activated GTP binding protein in the plasma membrane of etiolated pea. *Proc. Natl. Acad. Sci. USA* **88:** 8925-8929.

Wayne, R. and Hepler, P.K. 1984. The role of calcium ions in phytochrome-mediated germination of spores of *Onoclea sensibilis* L. *Planta* **160:** 12-20.

Weisenseel, M.H. and Ruppert, H.K. 1987. Phytochrome and calcium ions are involved in light induced membrane depolarisation in *Nitella. Planta* **137:** 225-229.

Wong, Y.S., Cheng, H.C., Walsh, D.A. and Lagarias, L.C. 1986. Phosphorylation of *Avena* phytochrome *in vitro* as a probe of light induced conformational changes. *J. Biol. Chem.* **61:** 12089-12097.

Wong, Y.S., McMichael Jr. R.W. and Lagarias, J.C. 1989. Properties of polycation-stimulated protein kinase associated with purified *Avena* phytochrome. *Plant Physiol.* **91:** 709-718.

Wu, Y., Hiratsuka, K., Neuhaus, G. and Chua, N.-H. 1996. Calcium and cGMP target phytochrome-responsive elements. *Plant J.* **10:** 1149-1154.

Yamamoto, Y.Y., Tsuji, H. and Obokata, J. 1995. 5'-leader of a photosystem I gene in *Nicotiana sylvestris, psaDb,* contains a translational enhancer. *J. Biol. Chem.* **270:** 12466-12470.

Yoshida, K., Nagano, Y., Murai, N. and Sasaki, Y. 1993. Phytochrome regulated expression of the genes encoding the small GTP-binding proteins in peas. *Proc. Natl. Acad. Sci. USA* **90:** 6636-6640.

Concepts in Photobiology: Photosynthesis and Photomorphogenesis
G.S. Singhal, G. Renger, S.K. Sopory, K-D. Irrgang and Govindjee (Eds)

31. Use of Mutants and Transgenics in Understanding Photomorphogenesis

Rameshwar Sharma[1] and Richard E. Kendrick[2]

[1]School of Life Sciences, University of Hyderabad, Hyderabad-500046, India

[2]Department of Plant Physiology, Wageningen Agricultural University, Arboretumlaan 4, NL-6703 BD Wageningen, The Netherlands

Summary

The nature of the photoreceptor and their mechanism of action in regulating various developmental responses is not fully understood. In this chapter we describe the behaviour of various photomorphogenetic mutants that have been isolated and are deficient in various photochrome species, both in apoprotein and in chromophore, and in blue light receptor. In addition, various mutants showing defects in signal transduction events are also described. It is emphasized that in addition to the biochemical and molecular approach the use of mutants and transgenic plants either expressing the photopigments or blocked due to antisense constructs has a great potential in revealing the mysteries of light regulation of plant gene expression and development.

1. Introduction

Plants being sessile organisms have evolved an intricate array of mechanisms by which they sense the ambient environment. These sensing mechanisms, that regulate both physiology and metabolism of plants, endow them with a built-in safeguard to overcome environmental exigencies and ensure their survival. Among the several sensing mechanisms used by plants, the ability to detect and interpret their light environment is the foremost. Plants have evolved intricate mechanisms to detect several facets of the light environment, such as its direction, duration, solar angle and spectral quality. To be able to detect these complex and varied parameters, plants have armed themselves with a battery of several photoreceptors, which detect the light environment independently, and in cooperation throughout the life cycle of the plant.

Studies on the influence of varying the spectral quality of light on several plant growth responses have revealed that plants principally detect their light environment using photoreceptors detecting specific wavebands such as UV-B (280–320 nm), UV-A/blue (320–500 nm), and red (R)/far-red light (FR) (600–750 nm) (Kendrick and Kronenberg, 1994) (Table 1). It has been further revealed that photoreceptors absorbing a particular waveband may elicit a specific response; for example, the phototropic orientation of plants is controlled by a photoreceptor absorbing in the UV-A/blue light (BL) region; whereas, the detection of the photoperiod is by a R-FR absorbing photoreceptor (Table 1). Since the specific wavebands of light elicit specific photoresponses, a considerable effort has been made to identify the molecular nature of the putative photoreceptors. These studies have unequivocally identified phytochrome, a biliprotein, as photoreceptor for perceiving R/FR wavebands. The molecular-genetic analyses have revealed that phytochrome is encoded by a small multi-gene family in higher plants. In *Arabidopsis,* at least five phytochrome genes have been identified, namely, *PHYA-E* encoding distinct apoproteins (Quail, 1991; Furuya, 1993; Pratt, 1995; Clack et al., 1994). Among these, the *PHYA* gene encodes a phytochrome species which accumulates predominantly in etiolated seedlings, and is rapidly down-regulated on exposure to light (Table 2). The remaining phytochrome species are relatively more stable in the light, therefore are predominantly present in light-grown mature plants. Recently a gene has been cloned which encodes a putative photoreceptor for some BL-mediated responses (Ahmad and Cashmore, 1993; Lin et al., 1995). The molecular identity of UV-B photoreceptors is not yet known.

Table 1. Photoresponses regulated by different photoreceptors

Photoresponse	Major Photoreceptor
Phototropism	UV-A/blue photoreceptor
Photoperiodism	Phytochrome
Stomatal opening	UV-A/blue photoreceptor
Perception of neighbors proximity	Phytochrome
Seed germination	Phytochrome
Hypocotyl elongation	Phytochrome, UV-A/blue photoreceptor
Cotyledon expansion	Phytochrome, UV-A/blue photoreceptor
Shade avoidance	Phytochrome
Solar tracking	UV-A/blue photoreceptor

Table 2. Differences in molecular properties and physiological responses of phyA and phyB species of phytochrome

	phyA	phyB
Molecular properties		
Transcription	Repressed by light	Constitutive
Pfr-decay	Rapid, by ubiquitination	Slow
Subunits	Homodimer	Not Known
Physiological Responses		
Effective wavelength for the high irradiance response	Far-red light	Red light
Effect of end of the day far-red light on growth	Inactive	Active
Red/far-red reversible	No	Yes
Very low fluence response	Active	Inactive

The molecular nature of the photoreceptor(s) associated with a particular photoresponse has been mainly inferred from the studies involving action spectroscopy. The similarity in action spectra of known photoresponses with the absorption spectra of known biomacromolecules has been used as preliminary evidence regarding the molecular nature of the putative chromophores likely to be covalently associated with the apoprotein. However, evidence obtained from such studies has been inconclusive because, often more than one photochromic molecule has a similar spectrum. For example, the photoreceptor participating in the BL-induced response has been inferred as a flavin, a carotenoid, or a pterin; all of which exist in plant cells. Moreover, these putative photoreceptor molecules are also found associated with the likely intracellular site of action such as the plasma membrane. Furthermore, the range of molecules absorbing in the spectral region of the UV-B is so large that not even a cursory suggestion has been made for a likely candidate for the putative photoreceptor.

Conventional biochemical approaches have so far failed to definitely identify the photoreceptors responsible for perceiving UV-A/blue and UV-B. To date, only the R-FR detecting photoreceptor—phytochrome has been purified and extensively characterized biochemically. Similarly, the steps leading from the perception of the light and execution of response, such as expression of genes, are still not fully resolved using conventional biochemical techniques. This impasse in the identification of photoreceptors, and the components associated with the signal transduction mechanism, has recently been slowly overcome, by an alternate strategy involving a combination of the techniques from genetics, molecular biology, physiology and biochemistry. Often the starting point of such an effort is the isolation of a mutant which is either defective in, or because of mutation, has constitutively gained a particular photoresponse. This review highlights the information obtained using mutants to understand the mechanisms involved in photoperception and signal transmission.

2. Usefulness of Mutants

Traditionally, mutants have been used to identify the genes responsible for a definite phenotype. Mutants have also been useful in understanding gene function and regulation, which enables the dissection of biochemical and developmental pathways. Several benefits of using mutants to understand the photophysiology of plants are:

1. Phenotypic characterization of mutants in presence and absence of light can provide information on the role played by a particular wave-band of light in the photophysiology of higher plants.
2. Mutants can be used to study the nature of the photoreceptors and signal transmission pathway activated after perception of light.
3. Mutants can be used as test plants to make an *in vivo* analysis of the

activity of modified and chimeric photoreceptor species and components of signal transduction chains.

4. Most importantly, mutants can be used to isolate and characterize the genes encoding photoreceptors and components of their signal transduction chains.

Similar to mutants, transgenic plants also offer advantages in understanding the function of photoreceptors or associated signal chain components. Transgenics can also be essentially considered as mutants which possess an excess dosage of genes or bear an altered gene with an aberrant expression pattern. Use of the above methods in recent years has provided much new information about the photophysiology of plants.

3. Photomorphogenic Mutants

The generation of photomorphogenic mutants was pioneered by the classic work of M. Koornneef in Wageningen, the Netherlands, who isolated *Arabidopsis* photomorphogenic mutants. Seedlings derived from mutagenized seeds were screened under white light, and six mutants which had a hypocotyl substantially longer than that of the wild type were isolated (Koornneef et al., 1980). This report was followed by isolation and characterization of several other mutants which displayed aberrant phenotypes or photoresponses on exposure to light. A reverse of the Koornneef screening protocol was used by J. Chory and X.W. Deng, who screened for mutant seedlings in darkness, which had a phenotype resembling light-grown seedlings. Photomorphogenic mutants have been reported in a number of species, including *Arabidopsis, Sorghum, Brassica,* tobacco, tomato and pea. In essence, the photomorphogenic mutants can be classified either as deficient or defective in the photoreceptor, or altered in some aspect of the signal transmission chain.

4. Photoreceptor-Deficient Mutants

The first mutant confirmed as deficient in phytochrome was the *aurea* (*au*) mutant of tomato (Koornneef et al., 1985). The seedlings of the *au* mutant lack the R-mediated inhibition of hypocotyl elongation, and are deficient in chlorophyll, and anthocyanin (Adamse et al., 1988). Most importantly, no spectrally active phytochrome can be detected in the dark-grown mutant seedlings, whereas high levels are detectable in the wild type seedlings (Parks et al., 1987). Searches for other photoreceptor-deficient mutants and investigations using molecular biological, biochemical and immunological techniques have enabled mutants to be broadly classified, depending on their loss of capacity to detect spectral quality, such as phytochrome, BL and UV mutants. In the case of phytochrome the mutation could lie in the genes regulating chromophore biosynthesis or in that determining the apoprotein. Table 3 shows some photoreceptor-deficient mutants.

4.1 Phytochrome Chromophore Mutants

The biological activity of phytochrome is governed by its characteristic property of existing in two spectrally reversible forms; by virtue of a covalently linked open tetrapyrrole chromophore. Biosynthesis of the phytochrome chromophore, phytochromobilin, shares the biosynthetic pathway with chlorophyll up to formation of protoporphyrin IXα, and thereafter, the pathway leading to phytochromobilin diverges (Terry et al., 1993). Plants severely depleted in chromophore level after the treatment with an inhibitor of the pathway such as gabaculine and 4-amino-5-hexynoic acid accumulated spectrally inactive phytochrome apoprotein (Gardner and Gorton, 1985; Elich and Lagarias, 1988). The above inactive apoprotein, however, could be rendered spectrally active by feeding the treated plants with a precursor such as biliverdin, or the phytochrome chromophore analogue, phycocyanobilin (Elich et al., 1989).

Immunochemical analysis of *hy1* and *hy2* mutants of *Arabidopsis* showed the presence of phytochrome A apoprotein (PHYA) at a level similar to that of the wild type, whereas, the spectral analysis of these two mutants showed a virtual absence of phytochrome, making a strong case in favor of these being chromophore mutants (Parks et al., 1989). This view was confirmed by experiments, in which the spectral activity of the phytochrome was restored *in vivo* by feeding the plants with a chromophore precursor, biliverdin. The spectral rescue of phytochrome was also accompanied by a rescue of the phenotype which was similar to the wild-type control. The rescued phytochrome species obtained by feeding biliverdin had absorption spectra similar to the wild type. These studies also indicated that the mutation in the biosynthetic pathway lies before formation of biliverdin (Parks and Quail, 1991). Similar to *hy1* and *hy2*, pea *pcd1* mutant has a block in its chromophore biosynthesis, and is unable to convert heme to biliverdin IXα (Weller et al., 1996). The mutant seedlings possess spectrally inactive phytochrome A apoprotein at a level equivalent to wild type and it could be reconstituted *in vitro* by incubation with phycocyanobilin.

The tomato *au* mutant on further investigation was also found to be a chromophore-deficient mutant similar to the *hy1* and *hy2* mutants, but different in several respects (Sharma et al., 1993). The *au* mutant possesses spectrophotometrically inactive PHYA apoprotein, but only 20% of the level in the wild type. Compelling evidence in favor of *au* being a chromophore mutant was obtained by crossing it with transgenic tomatoes over-expressing oat phytochrome *PHYA3*, and selecting the *au*/oat *PHYA3* double mutants. Although these synthesized oat PHYA, they bore no spectral activity (Kendrick et al., 1994; Van Tuinen et al., 1996). The endogenous phytochrome in the *au* is recalcitrant to spectral rescue, and unlike *hy1* and *hy2*, feeding of seedlings with biliverdin and phycocyanobilin failed to rescue it (R.P. Sharma, unpublished results). Microinjections of phycocyanobilin in *au* hypocotyl cells also failed to restore photomorphogenesis (Timp. Kunkel, personal communication). The phytochrome apoprotein from *au*, after extraction and on *in vitro* incubation with phycocyanobilin, also failed to regain spectral activity (K.R. Reddy and R.P. Sharma, unpublished results). Recent investigations have indicated that, the *au* mutant is unable to convert heme or biliverdin IXα to phytochromobilin (Terry and Kendrick, 1996).

It is plausible that the low level of phytochrome apoprotein in *au* may result from its degradation in the absence of chromophore. The failure to reconstitute phytochrome extracted from *au* seedlings *in vitro* may result either from a lack of a component needed for phytochrome assembly, or the presence of an inhibitor in the extract which is specific for its assembly in tomato, or phytochrome present in etiolated seedlings of *au* is not capable of binding the chromophore. Although a role of a molecular chaperone has been indicated for assembly of denatured phytochrome (Mummert et al., 1993), whether the *au* mutant is deficient in such a molecule is yet to be investigated. The effect of the *au* mutation on the phytochrome level was more pronounced on PHYA which was reduced 5-fold compared to the wild type, whereas the phytochrome B apoprotein (PHYB) was present at a level equal to the wild type (Sharma et al., 1993). Mature *au* plants possessed a spectrally active phytochrome B (phyB)-like protein at a level equal to the wild type.

The tobacco *pew1* mutant also belongs to the category of chromophore-deficient mutants, and is similar to tomato *au* and *Arabidopsis hy1* and *hy2*. The *pew1* mutant possesses no spectrally active phyA, but retains PHYA at the wild-type level and can be rescued by application of the chromophore precursor, biliverdin. In comparison, the *pew2* mutant possesses a reduced level of PHYA like *au*, and is recalcitrant to spectral rescue by exogenous biliverdin (Kraepiel et al., 1994).

The chromophore-deficient mutants are highly valuable for deciphering the biosynthetic pathway of phytochromobilin. For example, in the *pcd1* mutant of pea (Weller et al., 1996) and the *yg2* mutant of tomato (Terry and Kendrick, 1996) the conversion of heme to biliverdin IXα is blocked, whereas in *au*, the block is between biliverdin IXα and 3Z-phytochromobilin (Terry and Kendrick, 1996). The chromophore deficiency is expected to cause an overall reduction in the functional level of all the phytochrome species, therefore, these mutants can not be used to investigate the effects of the phytochrome deficiency or a process regulated by a specific phytochrome species. For this purpose, specific phytochrome mutants are desirable.

Information about the effects of a reduction in phytochrome level on the physiological and morphogenic responses of plants can be illustrated in the case of *au* mutant. The *au* seedlings lack anthocyanin induction, possess longer hypocotyls under white, R and BL and FR, and show reduced germination in darkness compared to wild type (Kendrick et al., 1994). The *au* mutant also has a block in chloroplast development, and even after a prolonged white light exposure, only partial chloroplast development takes place (Koornneef et al., 1985; Neuhaus et al., 1993). It lacks R-mediated induction of the nuclear genes encoding plastidic proteins such as the small subunit of RUBISCO (*SSU*), chlorophyll *a/b* binding protein (*CAB*) and plastocyanin (Oelmüller et al., 1989; Oelmüller and Kendrick, 1991). Unlike the wild type which shows a significant induction of all the above responses, reduction in the spectrally functional phytochrome level retards the de-etiolation process of the *au* seedlings. While the *au* seedlings retain photomorphogenic responses to some degree under continuous R, no significant photoresponses are evoked by brief pulses of R (Kendrick et al., 1994).

Investigations into the photoregulation of enzymes under R revealed a dichotomy in the effect of the *au* mutation on induction of cytosolic and plastidic enzymes. Though the *au* mutant does not show a photo-induction of anthocyanin, brief pulses of R-induced phenylalanine ammonia lyase activity in cotyledons of the seedlings at a level equal to that of wild type, with similar profiles of induction (Goud et al., 1991). Similarly, in the hypocotyl, the time course of photo-induction of phenylalanine ammonia lyase in *au* was similar to the wild type. In both organs the photo-induction of phenylalanine ammonia lyase was accompanied by an increase in the level of immunodetectable phenylalanine ammonia lyase protein (K.V. Goud and R.P. Sharma, unpublished results). Further investigation revealed that *au* also retains a significant photo-induction of other cytosolic enzymes such as amylase and nitrate reductase with induction profiles essentially similar to the wild type (Goud and Sharma, 1994). Evidently, despite a severe reduction in the level of phytochrome, *au* seedlings retain photo-induction of cytosolic enzymes, indicating a functional operation of a residual pool of phytochrome.

The observed photo-induction of the enzyme levels in the *au* seedlings was however restricted only to the cytosolic enzymes. The plastidic enzymes such as nitrite reductase which show a normal photo-induction in wild type, show no significant induction in *au* even after prolonged exposure to R (Goud and Sharma, 1994). Though exposure of the *au* seedlings to R induces both the nitrate reductase and nitrite reductase transcripts (Becker et al., 1992), only activity of cytosolic nitrate reductase is induced (Goud and Sharma, 1994). Since *au* also lacks R-mediated photo-induction of other nuclear encoded plastidic proteins such as plastocyanin, *SSU* and *CAB*, its residual level of phytochrome would seem to be below the threshold required for photoregulation of plastidic enzymes. Alternatively, the *au* mutation has a pleiotropic effect on chloroplast development independent of the reduction in the phytochrome level. Such an effect may impair the capacity of the *au* plastids to import polypeptides from the cytosol, leading to a block in photo-induction of plastidic enzymes.

The fact that the *au* mutant, in spite of severe deficiency in phytochrome levels completes its life cycle normally, implies that it is a leaky mutant. The most severe effects are seen during de-etiolation of the *au* seedlings. Mature *au* plants show photoresponses similar to the wild type, such as the end-of the day FR (EODFR) response leading to an increase in plant height and decrease in anthocyanin content (López-Juez et al., 1990). Light-grown *au* plants also show spectrally active phytochrome, likely to be phyB at a level nearly equal to that of wild type (Sharma et al., 1993). However, *au* is less responsive in detecting small changes in the R:FR quantum ratio (Casal and Kendrick, 1993). The severe deficiency in the level of chlorophyll in *au* surprisingly does not result in a similar reduction in its photosynthetic efficiency, since it compensates for the lack of the chlorophyll by high photochemical activity (Becker et al., 1992).

The survival of the *au* mutant despite being deficient in phytochrome apparently depends on the coaction of other photoreceptors, particularly that absorbing in the BL. Although *au* shows no R induction of nuclear transcripts of plastidic proteins, a pre-irradiation with BL potentiates the induction of these transcripts. In *au* seedlings, the photo-induction of nitrate reductase is seen only when the seedlings were grown under white light, but not under BL or R alone. In conclusion, it appears that *au* is a leaky mutant that gradually accumulates phytochromes during development.

4.2 Phytochrome-Apoprotein Mutants

The realization that phytochrome is encoded by a multi-gene family led to a search for mutants lacking a particular phytochrome species. Known mutants of distinct phytochrome functions were considered potential candidates. The cucumber long hypocotyl (*lh*) mutant (Fig. 1) and the *hy3* mutant of *Arabidopsis* are defective in some of the phytochrome responses such as EODFR response and R-mediated inhibition of hypocotyl elongation, but at the same time possess normal spectrophotometrically detectable levels of phytochrome in etiolated seedlings (Lo′pez-Juez et al., 1992; Nagatani et al., 1991a; Somers et al., 1991). Immunochemical analysis of these mutants using an antibody raised against the gene product of PHYB showed that these mutants are severely depleted in PHYB (Nagatani et al., 1991a; Somers et al., 1991; Lo′pez-Juez et al., 1992). Similarly, the elongated *ein* mutant of *Brassica* also lacks PHYB (Devlin et al., 1992). While it was assumed that a white light elongated phenotype is associated with the deficiency of phyB, the tomato *tri* mutant, which also lacks PHYB, is only slightly elongated at the seedling stage (van Tuinen et al., 1995b). Similarly, a photoperiod insensitive mutant of *Sorghum* which is also deficient in light-stable phytochrome B does not have an elongated phenotype (Childs et al., 1992; 1997). The *lv* mutant of pea is also deficient in phyB (Weller et al., 1995).

Since the phyB-deficient mutants retained a normal response to FR mediated inhibition of hypocotyl elongation, it was evident that another phytochrome species participated in this response. A search for the FR-insensitive mutants in *Arabidopsis* by three groups resulted in the isolation and characterization of several mutants, many of which possessed a defective PHYA or lacked it all together (Parks and Quail, 1993; Nagatani et al., 1993; Whitelam et al.,

1993). The mutants *hy8-1* and *hy8-2*, *fhy2*, and *fre1* were deficient in PHYA. The molecular analysis of the *hy8-1* and *hy8-2* mutants showed that the *PHYA* gene sequences contained a single nucleotide change which inserted a translational stop codon in the protein coding sequence (Dehesh et al., 1993). These *phyA* mutants, when grown under white light, unlike *phyB* mutants, display a phenotype similar to the wild type and are able to complete their life cycle normally. These results suggest that phyA is dispensable, and plays only a minor role in plant photomorphogenesis under white light. In addition phyA-defective mutants have also been isolated in tomato (van Tuinen et al., 1995a).

Fig. 1. Phenotype of 2 week-old, white-light (16 h light/ 8 h dark) grown wild-type and *lh*-mutant seedlings of cucumber. Right, wild type; left, *lh* mutant. Note the extremely long hypocotyl of the *lh* mutant, which is deficient in phyB.

The molecular analysis of the FR insensitive *hy8-3* mutant demonstrated that it is defective in the regulatory domain of phyA action (Dehesh et al., 1993). This mutant lacks the FR-mediated inhibition of hypocotyl elongation but possesses a photoreversible phyA, which also shows the normal light-induced rapid turnover of the PHYA protein. The mutation which is localized at the residue 727 lies outside the dimerization range of the phytochrome protein and therefore represents the potential site involved in the regulatory domain of phyA, which may be involved in interaction with the next active component of the signal transduction chain.

The FR-insensitive mutants also provided evidence that the magnitude of some of the phytochrome-mediated responses is related to the absolute level of phytochrome in plants. The analysis of plants heterozygous for the *fhy2* allele showed that *fhy2/FHY2* plants, when grown in FR, have a hypocotyl length longer than that of homozygous wild-type plants (Whitelam et al., 1993). This indicates that the *fhy2* alleles are partially dominant, and it is possible that this exerts its function because two copies of the allele are needed to confer the wild-type phyenotype. It is therefore evident that a close relationship exists between phyA level and FR-mediated hypocotyl inhibition.

Physiological and biochemical analysis of these mutants, and double mutants derived from them, has yielded valuable information about the relative function and importance of these two phytochrome species. In general, phyB mutants flowered earlier than the wild type, and the mutant seeds, unlike wild type, showed a lower percentage of germination in darkness. Some information about the relative response of these mutants is summarized in Table 4 (Nagatani et al., 1993). The double mutants of both phytochrome survived, moreover retained many of the photoresponses, indicating the involvement of phytochrome species other than phyA and phyB. The phenotype of the mutants and the double mutants also indicated the redundancy in the action of these phytochrome species.

Table 3. Some photoreceptor mutants and their phenotypes

Species	Mutant	Phenotype	Lesion
Tomato	*au, yg-2*	Yellow green, elongated hypocotyl	Chromophore
	fri	Elongated hypocotyl	Phytochrome A
	tri	Elongated hypocotyl	Phytochrome B1
Arabidopsis	*hy1, hy2*	Yellow green, elongated hypocotyl	Chromophore
	hy3	Pale green, elongated hypocotyl in red light	Phytochrome B
	hy4 (cry1) (blu1-4)	Abnormally elongated hypocotyl in blue light	Blue light photoreceptor
	hy8, fre-1, fhy-2	Elongated hypocotyl in far-red light	Phytochrome A
	JK218, NPH	Loss of phototropism	not known
	uvr, uvh	hypersensitive to UV-B	not known
Tobacco	*pew1, pew2*	Yellow green	Chromophore
Sorghum	*ma_3R*	Early flowering	Phytochrome B
Brassica	*ein3*	Elongated plants	Phytochrome B
Pea	*lv*	Tall plants	Phytochrome B
	pcd1	Yellow-green	Chromophore
Cucumber	*lh*	Tall plants	Phytochrome B

Table 4. Photoresponses of *Arabidopsis phyA, phyB* and *phyA,phyB* double mutant under different light conditions (Adapted from Nagatani et al., 1993)

Response	Wild Type	*phyA*	*phyB*	*phyA,phyB*
Seed germination (%)				
Dark	97	98	83	98
White light	87	93	97	100
Red light	97	97	100	97
Far-red light	61	0	97	33
Hypocotyl length (mm)				
Dark	10.7	10.1	9.9	11.9
White light	1.0	1.3	4.3	6.7
Red light	4.0	5.5	10.8	15.2
Far-red light	1.6	5.8	1.5	6.6

4.3 Specific Blue-Light Mutants

The molecular identity of the BL photoreceptors has been more enigmatic. Although the action spectroscopy studies have indicated the putative chromophore molecules which could be associated with the BL responses, the molecular nature of the apoprotein moiety for these is still to be determined. In general, BL responses can be grouped into two categories; one leading to phototropic bending of organs towards unidirectional light and the other BL-mediated inhibition of hypocotyl elongation (Liscum and Hangarter, 1994). In addition, BL also stimulates the transcription of several genes and is involved in the photoregulation of opening and closing of stomata (Kaufman, 1993).

4.3.1 Phototropic mutants

The molecular identity of the photoreceptor and components involved in signal transduction for phototropism have lately been extensively investigated but with little success. Isolation of phototropic mutants provided, for the first time, some insight into the process of phototropism at the molecular level (for details see Chapter 26). One characteristic property of phototropic responses is a complicated fluence response curve, with first and second positive curvatures at low and high fluence radiation, interspersed with a negative response. A complicated model was invoked to explain the complex fluence response curve based on the operation of two photoreceptors (Zimmerman and Briggs, 1963). Among the first phototropic mutants isolated the JK 224 (*nph1-2*) provided genetic evidence for existence of the two photoreceptors, since the first positive curvature of this mutant is shifted 20–30 fold in BL, but is normal in green light (Khurana and Poff, 1989; Khurana et al., 1989). However, the isolation of null mutants of the phototropism *nph*1-1, JK218 (*nph* 3-3) provided evidence for a universal photoreceptor for phototropism, as these mutants lacked not only BL-mediated phototropism, but also lost response to UV-A and green light (Liscum and Briggs, 1995).

Since mutation at the same locus has a differential effect on the first and second positive curvatures, it has been suggested that this effect may be due to the existence of a dual or multi-chromophoric photoreceptor, which results in the complex fluence response curve observed for phototropism. Alternatively, the *NPH* alleles may encode for an essential signal transduction component through which photoreceptor regulation of the first and second positive curvature operates (Liscum and Briggs, 1995). It has often been suggested that the photoreceptor for phototropism may be localized in a membrane or operate via changes in membrane properties. In most species, the phototropic bending of organs can be induced by exposure to a brief pulse of BL. It is found that in many of these species such a BL pulse also causes rapid phosphorylation of a protein of ~120 kD molecular weight which is associated with the plasma membrane (Short and Briggs, 1994; Short et al., 1992; Raymond et al., 1992a). The likely participation of such a protein in the phototropic pathway is indicated by the observation that the phosphorylation of this protein is severely reduced in the *nph1-2* mutant and is totally absent in the *nph1-1* mutant (Raymond et al., 1992b; Liscum and Briggs, 1995). Moreover, the level of this protein in the *nph1-1* mutant is less than 10% of the wild-type level. It has been suggested that the 120-kD protein may be the putative photoreceptor for phototropism, which may be activated by autophosphorylation (Liscum and Briggs, 1995). In *Arabidopsis* seedlings the amplitude of phototropic curvature to blue light is enhanced by a prior exposure to red light acting via phytochrome. Comparison of fluence-response relationships of this response in wild type, *phyA*-, and *phyB*-deficient mutants, and transgenic plants overexpressing phyA or phyB, showed that phyA is necessary for the very-low to low-fluence response, but the high-fluence response is apparently controlled by a phytochrome other than phyA or phyB (Janoudi et al., 1997).

4.3.2 Hypocotyl elongation

The analysis of BL-insensitive mutants provided impeccable evidence for the existence of at least two genetically distinct pathways for the operation of phototropism and inhibition of hypocotyl elongation. The *hy4* mutant of *Arabidopsis* was the first mutant which distinctly showed that the BL-effect was different from that of phytochrome on hypocotyl elongation (Koornneef et al., 1980; Ahmad and Cashmore, 1993). This was followed by the isolation of other mutants such as *blu1-blu3* which were also defective in their BL response (Liscum and Hangarter, 1991), Specifically, these mutants failed to respond to high fluence BL but retained normal responses to low fluence BL and other wavelengths (Liscum and Hangarter, 1991). The phenotype of *blu* mutants is strongly similar to that of *hy4* and there is now evidence that *blu* mutants may represent weak alleles of *hy4*. (Liscum cited in Jenkins et al., 1995).

The usefulness of mutants to identify the components of photoperception and signal transmission is best illustrated by the case of the *hy4* mutant of *Arabidopsis.* (Ahmad and Cashmore, 1993). The deduced amino acid sequence for the protein encoded by the *HY4* locus (cry1 gene) showed homology to the DNA photolyases of the microbes in several respects. (The characterization of *cry1* mutant has been discussed in Chapter 26.)

The availability of phytochrome and BL mutants has helped to clarify the relative participation of these photoreceptors in controlling identical responses. The phytochrome chromophore-deficient mutants such as *hy2* and *hy6* lose the responsiveness to low fluence BL-mediated hypocotyl elongation inhibition, but respond normally to high fluence BL (Young et al., 1992; Goto et al., 1993). It is evident therefore that the low fluence BL response is mediated by the action of phytochrome, and the BL photoreceptor acts under high fluence BL. Additional support for this view is provided by the construction of the *blu(hy4), hy6* double mutant that displays an elongated hypocotyl under white, R, and FR, but is short under BL indicating that these two photoreceptors act independently of each other (Young et al., 1992). However, recent studies with *hy4* and specific phytochrome double mutants support the opposite conclusion (Ahmad and Cashmore, 1997) and results with *hy4, hy6* double mutants are explained by the leakiness of the *hy6* mutation (M. Ahmad and A.R. Cashmore, personal communication). Manipulations of the levels of active phytochrome (Pfr) with R or FR pulse subsequent to BL treatments altered the activity of CRY1. Moreover, in phytochrome-deficient mutants CRY1-mediated blue-light responses were considerably reduced, even though the level of CRY1 photoreceptor was unaffected. Apparently, CRY1 needs a coaction of active phytochrome for full expression (Ahmad and Cashmore, 1997).

The characterization of these mutants has similarly provided evidence for independent participation of phytochrome and the BL photoreceptor in the control of the apical hook opening in seedlings. The *blu(hy4)* mutant responded like wild type to low fluence BL induced hook opening, whereas mutants such as *hy1* and *hy2* had an altered response to apical hook opening in low fluence BL (Liscum and Hangarter, 1993b). The opposite action was observed under high fluence BL, with *blu* mutants being defective in hook opening, whereas the *hy1* and *hy2* mutants responded like wild type (Liscum and Hangarter, 1993b, c). It is therefore evident that low fluence BL effects are mediated by phytochrome and high fluence BL effects are mediated by the BL photoreceptor alone or in coaction with phytochrome. The phenomenon of the BL-induced gene expression in mutants has only been studied superficially. It was shown that the level of the CHS transcript is reduced to half that of the wild type under white light in *hy4* seedlings (Chory, 1992) and under BL is reduced 3-fold in *hy4* mutant seedlings (Liscum and Hangarter 1994). However, phyA and phyB are not required for blue/UV-A regulation of *CHS* expression and the light-responsive element of *CHS* mediates blue and UV-A induced transcription in the absence of the above phytochrome species (Batschauer et al., 1996).

In the literature, the photoresponses induced by BL and by UV-A have been traditionally grouped together. The differential responsiveness of mutants to BL and UV-A provided evidence for an independent operation of the UV-A and BL photoreceptor. The analysis of *blu(hy4)* and, *hy6* mutants under UV-A showed that these mutants, though deficient in the BL-induced responses, retained the wild-type responses to UV-A. The above observation therefore invokes a third photoreceptor in addition to phytochrome and the BL photoreceptor which absorbs in the UV-A (Young et al., 1992). In addition, analysis of mutants defective in root phototropism such as *rpt1* and *rpt2* show that these mutants retain normal phototropism of the hypocotyl indicating that the transduction pathway of these responses probably follow different routes (Okada and Shimura, 1992).

Comparison of BL and phototropic mutants showed the genetic independence of these two BL-mediated responses. The *blu(hy4)* mutants retained the phototropic responses at the wild-type level and the phototropic mutant *nph1-1* and *nph1-2* showed normal BL mediated inhibition of hypocotyl elongation. The *blu(hy4), nph1-1* double mutant lacked both the responses, and did not orient towards BL and showed no inhibition of hypocotyl elongation under high fluence BL (Liscum et al., 1992). The information obtained using mutants therefore clearly highlights that, similar to phytochrome, the BL mediated responses are also mediated by a battery of photoreceptors. It is expected that molecular analysis of the available mutants will help in the identification of the photoreceptors, and also the components encoding the signal transduction chain.

4.4 UV-B Mutants

Mutants which are specifically deficient in their response to UV-B have been isolated only recently. Since UV-B is deleterious to plant growth, *Arabidopsis* mutants which failed to grow after exposure to UV-B could be isolated as UV-B sensitive mutants. Using the sensitive reaction to UV-B, Harlow et al. (1994) isolated the *Arabidopsis uvh1* mutant which showed diminished growth in UV-B and also in UV-A. The mutant plants showed chlorosis, wilting and cell death on exposure to short intense exposure to UV-B or UV-C. However, such a sensitive reaction to UV light was not caused by the lack of flavonoids as reported for other UV sensitive *Arabidopsis* mutants (Li et al., 1993). The mutant possessed normal photoreactivation of DNA dimers, but probably was defective in some other aspects of the DNA repair, or was less tolerant to the DNA damage caused by UV-B.

One such mutant *uvr1* was found to be 6-fold more sensitive to UV-B in comparison to wild type for UV-B-inhibition of root elongation. The UV-B also affected the morphology of the aerial parts of the plants. The formation of the cyclobutane pyrimidine and pyrimidine-pyrimidinone (6–4) dimer in DNA is one of the major causes of the deleterious effects of UV-B on the growth of plants. While the mutant and wild type possessed a near identical capability to repair cyclobutane dimers, the *uvr1* mutant was found defective in the repair of 6–4 dimers. It is therefore likely that the *uvr1* is mutated in a gene involved in DNA repair (Britt et al., 1993). However, the above mutant does not provide a deep insight into UV-B perception and effect on plants other than to highlight that one of the primary targets of UV-B action is damage to DNA. It is not known if the damage to DNA is perceived as a signal for the presence of UV-B in the environment or alternate perception mechanisms also operate to induce UV-B-mediated photomorphogenesis. A recent study on the inhibition of hypocotyl elongation by UV-B using tomato *au* and wild-type tomato seedlings indicated that the photoreceptor chromophore involved in UV-B may be a flavin (Ballaré et al., 1995). The above study also indicated that DNA or aromatic residues in proteins are not the chromophore mediating the UV-B response.

5. Signal Transduction Chain Mutants

It is known that the photomorphogenic effect of light is caused by changes in the expression of the genes localized in the nucleus. Therefore after activation of the photoreceptor, a signal should be transmitted to the nuclei via a signal transduction chain which would ultimately regulate the transcription of the light-responsive genes. The components of the signal chain may be either specific to a given photoreceptor; or different photoreceptors may share a common pathway of the signal transmission for an identical photoresponse. Two strategies have been employed for the isolation of mutants deficient in the components of the signal transduction chain. The first strategy employed isolation of mutants which are normal with respect to the photoreceptor level and function, yet show altered photomorphogenic responses. The second approach employed isolation of mutants among dark-grown mutant seedlings with a phenotype similar to the light-grown seedlings. The analysis of these mutants indicated that the signal chain controlling photomorphogenic responses consists of novel signaling molecules, and these molecules may participate in the regulation of more than one developmental pathway. This analysis also showed that photomorphogenesis is the default pathway and a complement of genes is required to repress photomorphogenesis in darkness.

Among the first category of mutants are high pigment-1(*hp-1*) mutants of tomato and *hy5, fhy1-1* and *fhy3-1* mutants of *Arabidopsis.* The *hp-1* mutant of tomato showed exaggerated photoresponses such as super-induction of the photoregulated increase in anthocyanin and high chlorophyll level in mature fruits (Peters et al., 1989). The *hp-1* mutants do not require a BL pre-irradiation for the photo-induction of anthocyanin production in seedlings and show maximal anthocyanin induction in R alone. Since the *hp-1* mutation is recessive, it is proposed that the *HP-1* gene codes for a repressor of photomorphogenesis (Kendrick et al., 1994). In wild-type plants, phytochrome action appears to be under the constraint of the *HP-1* gene product, both BL and the *hp-1* mutation lower the level of the *HP-1* gene product leading to responsiveness amplification (Peters et al., 1992). Moreover, the phenotype of the *au, hp-1* double mutant resembles that of *au* indicating that the *au* mutation is epistatic to *hp-1* in the seedling stage. The *hp-1* hypocotyls are shorter than the wild type, and the mature plant is also a dwarf. It has been shown that the above phenotype can be partly overcome by application of gibberellins (van Wann, 1995).

Among the five long hypocotyl mutants initially isolated by M. Koornneef, the *hy5* mutant appears to be defective in a component of signal transduction (Chory, 1992). The mutant possesses an elongated hypocotyl in R and has normal levels of spectrally active phytochrome and the gene products of *PHYA, PHYB,* and *PHYC* (Somers et al., 1991). The *hy5* mutant also has reduced sensitivity to BL. Among the FR insensitive mutants isolated in *Arabidopsis,* the *fhy1-1* and *fhy3-1* mutants show spectrally normal levels of phyA, but reduced sensitivity to FR (Whitelam et al., 1993). These two mutants appear to be defective in a component in the signal transduction chain emanating from phyA. Since the phyB response is normal in these mutants this component is specific to the phyA-triggered signal chain.

The analysis of the light-triggered signal transduction chain was greatly aided by the isolation of mutants which grew in darkness with a phenotype similar to light-grown plants. The first one isolated by J. Chory in *Arabidopsis* was named as *det* (de-etiolated) mutant. The *det* mutant lacked the typical phenotype of wild type in darkness, had a short hypocotyl, expanded cotyledons and lacked a hypocotyl hook (Chory et al., 1989). In addition, the level of the transcripts of several light-induced genes in dark-grown *det* plants was higher than the wild type, indicating that these genes are constitutively expressed in the *det* mutant. The *DET1* gene appears to be required for the correct expression of the plastidic signal, since roots of the mutant seedlings developed chlorophyll and showed plastidic differentiation. The *DET1* gene product was cloned by chromosome walking and sequenced. It possesses amino acid sequences for the nuclear localization indicating that it is a nuclear protein (Pepper et al., 1994). Plants transformed with the *GUS-DET1* fusion gene demonstrated that the product is localized in the nuclei. The DET1 protein appears to be a negative regulator of photomorphogenesis, which though localized in the nuclei, does not show a specific domain which could classify it as one of the proteins or a cofactor which participates in nuclear transcription. It is assumed that the *DET1* product acts as a global regular of development since embryo development-defective *fusca* mutants are allelic to some *det* mutants (Castle and Meinke, 1994).

X.-W. Deng and his colleagues isolated another class of mutants which were phenotypically similar to *det* but were mutated at different loci. These mutants were named *cop* mutants to signify *constitutively induced photomorphogenesis* (Deng et al., 1991). A series of the *cop* mutants have been isolated and the genes encoding a few of the *COP* loci have also been cloned (Deng, 1994). Similar to the *DET1* the *COP1* gene also encodes with a nuclear localized protein, albeit with few differences (von Arnim and Deng, 1994). The *COP1* gene product is a hybrid molecule with dual regulatory features such as similarity of the N-terminus with an α-subunit of heterotrimeric G-proteins and similarity with the transcriptional activator complex of *Drosophila* (TFllb$_{80}$ complex) (Deng et al., 1992; Deng, 1994). Such features led to the conjecture that it is a signal transmitter protein, which may complex with G-proteins to perceive the light stimulus via phytochrome, and also migrate to the nuclei to initiate action. The studies conducted using a *GUS-COP1* fusion construct indicated that this fusion protein accumulates in the nuclei of dark-grown plants, whereas in light-grown plants, it is present in a diffused pattern in the cytosol of the cells.

X.-W. Deng isolated several *cop* mutants in *Arabidopsis*, and a few of them were found to be allelic to the previously described *fusca* mutants (Castle and Meinke, 1994; Misera et al., 1994) and also to hookless mutants of *Arabidopsis* (Ecker, 1995). The *cop3* mutant was found to be allelic to the *hl* mutant and the gene encoding this locus has also been cloned. Its predicted amino acid sequence shows similarity to a diverse group of N-acetyltransferases (Lehmann et al., 1996). This gene is involved in the apical hook opening which is mediated via ethylene. The *COP9* gene was also found to be allelic to the *FUS7* gene, and it encodes a small protein of 27 kD, which most likely functions as a multimeric complex associated with the *COP8* and *COP11* gene products (Wei et al., 1994; Castle and Meinke, 1994). Similarities in the phenotypes of *fus*, *hl,* and *cop3* mutants clearly highlight that the process of photomorphogenesis is routed through global regulators of plant development which participate in several morphogenic responses (Deng, 1994; Quail et al., 1995).

The notion that some of the *det/cop* mutations represents global regulators have been supported by the analysis of the *det2* mutant. The cloning and sequencing of the *DET2* gene showed strong homology of the gene product to a steroid 5-α-reductase enzyme (Li et al., 1997), which is involved in the biosynthetic pathway of the plant hormone, brassinosteroid. *DET2* gene product and mammalian steroid 5-α-reductase enzymes could substitute each other in the mammalian system and *det2* mutant, respectively (Li et al., 1997). These results strongly suggest a functional and evolutionary similarity of these gene products involved in the steroid hormone signaling system.

The inter-relationships and hierarchies in action of these signal transduction chain mutants and photoreceptors have been examined by making double mutants and observing the phenotypes of the hybrid plants. These experiments have shown that among the gene loci, *DET1* which acts on signal transmission lies upstream of other loci, followed by *COP1*, and other loci. The model depicting the inter-relationship between these loci are presented in Fig. 2, which has been derived by genetic analysis (Kwok et al., 1996). In *Arabidopsis,* 11 loci, 3 *DET* and 8 *COP*, have been identified. Of these, *DET2, DET3,*

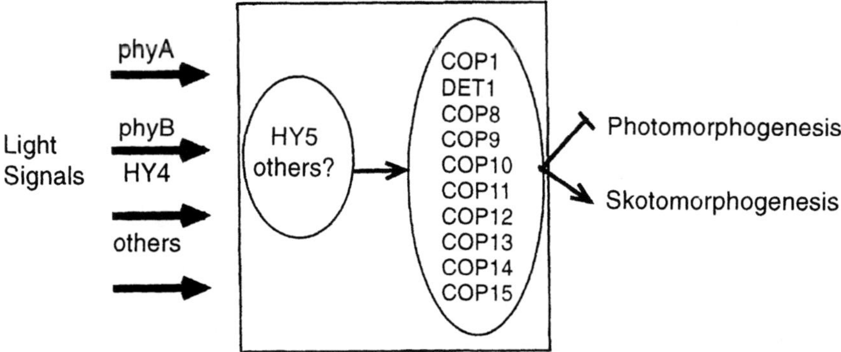

Fig. 2. A genetic model for the role of different pleiotropic *COP/DET* genes in light-regulated development of *Arabidopsis* seedlings (Adopted from Kwok et al., 1996). These genes encode for repressors that control the development of seedlings.

COP2, COP8, and *COP4* loci lead to partial photomorphogenesis in darkness, whereas *DET1, COP1, COP8* to *COP11* have a more pleiotropic effect in darkness. Kwok et al. (1996) identified four other loci *COP12* to *COP15,* in addition to the already identified six pleiotropic *Arabidopsis COP/DET* genes. They showed that photomorphogenesis in dark, is regulated by a complement of at least 10 pleiotropic *Arabidopsis* genes, that are necessary for the repression of photomorphogenic development. The pleiotropic nature of these loci also implies that these loci encode products that act early in the developmental pathway regulating cell differentiation, plastid development and gene expression. Since these loci are recessive, it is evident that light nullifies their suppressive action. These products also act to suppress chloroplast development in non-photosynthetic tissues. However, these loci are not involved in the regulation of all photomorphogenic responses as, phytochrome is still required for inducing germination of seeds in these mutants (Kwok et al., 1996).

Information about putative signal transduction mutants in plants other than *Arabidopsis* is limited. Although some observations have been made on the *lip* mutant of pea. The dark-grown *lip* seedlings partially resemble light-grown seedlings and possess partially developed agranal plastids. The dark-grown mutant plant has 10-fold less phytochrome and possesses higher levels of transcripts of *CAB* and *RBCS* (Frances et al., 1992).

Recently an alternative approach to decipher the signaling component has been attempted by isolating two suppressor mutants of the *hy2* mutation *Arabidopsis, shy1-1D* and *shy2-1D* located on chromosome 1 (Kim et al., 1996). Both mutations suppress the elongated phenotype of *hy2* by light-independent inhibition of hypocotyl growth as well as by increasing the effectiveness of light inhibition of hypocotyl elongation. It was found that the *shy1-1D* mutation suppressed hypocotyl elongation in R, but not in FR, and *shy 2-1D* suppressed hypocotyl elongation both in R and FR. It is likely that SHY1 and SHY2 represent a novel class of components involved in photomorphogenesis.

Another approach has been used by Pepper and Chory (1997) who used a morphological screen to isolate six extragenic *ted* (for reversal of the *det* phenotype) mutations, that partially or fully suppressed the seedling phenotype of *det1-1*. The analysis indicated that the *ted* mutations partially suppressed the expression of genes in darkness in *det1-1* and also partially suppressed the light-grown morphological phenotype of mature *det1-1* plants. The *ted1* and *ted2* mutations also suppressed the daylength insensitivity phenotype of *det1*. The *TED1, TED2* and *TED3* represent newly identified loci which apparently function in close association with *DET1*.

6. Transgenic Plants Bearing Modified Phytochrome or Over-Expressing Phytochrome

Since analysis of mutants provides only partial information about the role and function of the mutated gene loci, a direct approach has been to generate transgenics bearing native or modified gene products in plants. The likely functions of the introduced genes can be proposed by the analysis of gene products and observation of the mutant phenotypes. The initial studies on transgenic tomato, tobacco and *Arabidopsis* bearing heterologous phytochrome A showed a light-exaggerated phenotype with extremely dwarf dark-green plants (Boylan and Quail, 1989; Keller et al., 1989; Key et al., 1989). The basic conclusion derived from such studies was that the PhyA molecules can function across the monocot and dicot boundary and excess production of PhyA is detrimental to plant growth. The molecular cloning of the other phytochrome genes, i.e., of *PHYB* has also led to the generation of transgenics for *PHYB* (Wagner et al., 1991). Apparently, the phenotype of phyB and phyA over-expressers are quite similar, both showing the stunting of growth and intense accumulation of chlorophyll. Segregation of the transgenics in the next generation occurs in a Mendelian fashion (Boylan and Quail, 1991; Nagatani et al., 1991b).

Clues about the biological activity of photoreceptors have been obtained by site directed mutagenesis and expression of truncated phytochrome molecules in transgenic plants. The N-terminus of rice phytochrome is rich in alanine residues, and the conversion of these residues to serine made the phytochrome biologically more potent as assayed in transgenic tobacco carrying the rice *PHYA* gene (Stockhaus et al., 1992). In contrast, the deletion of as many as 6 kD of the N-terminal sequence from PHYA made it biologically inactive, though transgenics carrying this phytochrome showed normal photoreversibility of the PhyA molecules (Cherry et al., 1992). It was found that the N-terminus of PhyA contains two functional domains, one necessary for conformational stability and biological activity (residues 13–62), and the other for attenuating phytochrome responses (residues 6–12) (Jordan et al., 1996).

Using the short hypocotyl phenotype generated by over-expressing oat

PHYA in *Arabidopsis* Boylan et al. (1994) assayed the interaction between endogenous and transgene phyA by making an internal deletion, and C- and N-terminal deletions of oat PHYA molecule. These alterations of the oat phytochrome molecule caused dominant-negative interference and blocked the capacity of endogenous *Arabidopsis* phyA to inhibit growth under FR. This study highlighted that while the N-terminus is the site of perception of the light signal, it needs the C-terminus of the molecule to transmit the information to the next component of the signal transduction chain.

The dwarf phenotype caused by phyA over-expression in tobacco has been shown to be due to the attenuation of gibberellins levels and can be overcome by the exogenous application of gibberellic acid (Jordan et al., 1995). In transgenic seedlings, phyA is expressed preferentially in vascular tissues when driven by the CaMV promoter. The promoters belonging to that of *CAB* or ubiquitin which expressed phytochrome in tissues other than vascular tissue were less effective, indicating that accumulation of phytochrome in vascular tissue suppresses gibberellin production and generates the dwarf phenotype.

The deletion experiments carried out on phytochrome have been very useful in assigning sites of importance for biological action in the phytochrome as well as regions needed for chromophore attachment and photoreversion. These deletion experiments clearly show that the C-terminal part of the phytochrome molecule is equally important for biological activity, and transgenes carrying a deletion in the C-terminal-half of the molecule results in a product that is normally photoreversible but biologically inactive (Cherry et al., 1993; Wagner and Quail, 1995). Wagner et al. (1996) expressed a number of deletion and substitution derivatives of phyB in transgenic *Arabidopsis*. The expression of the N-terminal or C-terminal domains of phyB alone produced a fully photoactive monomeric molecule, and a photoinactive dimeric molecule respectively, both of which lacked regulatory activity, indicating that both domains are necessary for phyB activity. While the residues 6–57 contained the site for fluence rate of light perception, the residues 652–712 were necessary for maximal biological activity. Much information has been obtained regarding the functions of different domains on the phytochrome molecules by making chimeric phytochrome molecules bearing amino acid sequences transplanted from a related phytochrome molecule. These aspects are discussed in more detail in Chapter 25.

7. Restoration of Photomorphogenesis by Micro-Injection of Phytochrome and Other Molecules

Phytochrome-deficient mutants such as *tomato au* and *hy1* and *hy2* of *Arabidopsis* display a phenotype showing elongated pale-green seedlings. These seedlings also lack anthocyanin and have poorly developed chloroplasts. Apparently, these effects ensue as the level of spectrally active phytochrome in these mutants is not sufficient to initiate photomorphogenesis. It is therefore possible to restore photomorphogenesis in these seedlings by providing them with biologically active phytochrome. For example, the exogenous supply of biliverdin restores photomorphogenesis in *hy1* and *hy2* mutants of *Arabidopsis*. N.H. Chua and his colleagues successfully restored the photomorphogenic responses in tomato *au* seedlings by microinjection of oat phyA into single sub-epidermal hypocotyl cells (Neuhaus et al., 1993). They injected phytochrome into etiolated seedlings and after transfer to white light for 2 days, photoresponses such as anthocyanin synthesis and chloroplast development could be observed in injected cells. Therefore, the micro-injection of phytochrome in the tomato hypocotyl cell provided a convenient assay for dissecting the signal chain emanating from phytochrome. The action of injected phytochrome could however be blocked by inhibitors of G-grotein action, suggesting that the activation of a G-protein may be an early step in signal transmission. An observation in favor of such a view was that the phytochrome-mediated responses in *au* cells could be emulated by injecting GDP-γ-S which activates G-proteins. Injections of Ca^{2+} or activated calmodulin and cGMP revealed a split in the signal transmission pathway: (i) calcium/ calmodulin pathway controlling chloroplast development; (ii) cGMP controlling anthocyanin induction. These pathways also show a reciprocal control and have cross talk among them such as shut-off of the photosynthetic complex synthesis by high levels of cGMP and switch-off of the anthocyanin pathway by high levels of Ca^{2+} (Bowler et al., 1994). The details of the cellular basis of signal transduction is discussed in Chapter 30.

Kunkel et al. (1996) injected phytochrome-phycocyanobilin adducts (phyA*, phyB*) and oat phyA into etiolated *au* tomato seedlings and monitored the accumulation of anthocyanin and chlorophyll, and also the expression of light-regulated promoters (*CHS, LHCB1* and *FNR*). While phyA* micro-injection under white light caused anthocyanin and chlorophyll accumulation and mediated CHS-, LHCB1- and FNR-GUS expression, phyB* micro-injection induced only chlorophyll accumulation and mediated LHCB1-GUS and FNR-GUS expression, but chlorophyll accumulation and CHS-GUS expression was absent. These experiments indicated that phyB can not induce juvenile anthocyanin, and under R it needs interaction with other photoreceptors to mediate de-etiolation. In contrast, under FR phyA acts independently of other photoreceptors.

The gene involved in photomorphogenesis such as *COP* and *DET* are global genes for the regulation of plant development, whereas the biochemical effectors, such as calcium/calmodulin and cGMP are also universal secondary messengers. It is likely that some of these genes may be downstream of the primary action of phytochrome, which may involve activation of G-proteins. This may in turn regulate the levels of *COP* and *DET* gene products and initiate photocontrol of gene expression. There are additional steps in signal transmission such as components of the phosphoinositide cycles and the phosphorylation of proteins which are also involved in the photoregulation of gene expression (Chandok and Sopory, 1996; Raghuram and Sopory, 1995). Additionally, metabolite sucrose also appears to play an important role in regulating phytochrome-controlled responses (Dijkwel et al., 1997).

8. Mutants/Transgenic Plants and the Role of Phytochrome in Plant Development

In recent years, the analysis of the light-regulated genes and molecular mechanisms in phytochrome signal transduction have been analyzed using genetic screens which allow identification of mutations resulting in altered expression of the light-regulated genes. J. Chory's group have pioneered this work and have isolated three groups of mutants: *cue* (*CAB* underexpresser), *doc* (dark over-expression of *CAB*) and *gun* (gene uncoupled). The strategy to isolate these mutants is simple and relies on the screening of transgenic *Arabidopsis* which are carrying a specific promoter hooked to a reporter gene. The mutagenization of these transgenics and the screening of the mutants with altered GUS expression allows identification of mutants that have lost one or more components of the signal transduction pathway. This approach has produced several mutants, whereas conventional screens have yielded only one authentic phytochrome signal transduction chain mutant (*hy5* of *Arabidopsis*). Using this technique Li et al. (1995) reported that the mutation at the *cue1* locus results in defective regulation of light-regulated genes in mesophyll but not in the bundle-sheath cells. Moreover, the mutation also affected chloroplast development in these cells, but, at the same time, another light-regulated gene *CHS* expressed normally in the same cells. Since *cue1* mutants show reduced response to both R and BL, it is likely that the signal transduction chains emanating from more than one photoreceptor share a common pathway. It is also likely that *CUE* genes act downstream of these photoreceptors and as a cell specific positive regulator, linking light and the developmental program in mesophyll cells.

The above strategy of isolating mutants was also used in a reverse fashion which permitted the isolation of the *doc* mutants, where the *CAB* gene was over-expressed in dark-grown plants. Three loci were identified and the mutation in these loci did not affect the morphological development in dark-grown seedlings, as in the case of *det* and *cop* mutants. At the same time, the *CAB* gene showed enhanced expression without light. This study also highlighted that *CAB* expression can be separated from that of *RBCS* and the mutants also had an impaired growth response under short day conditions (Li et al., 1994). Similarly, the mutations at the *GUN* locus show that light-triggered gene expression of *RBCS* and *CAB* promoters can be uncoupled from the presence of chloroplasts (Susek et al., 1993).

In essence these studies along with the others where the *CAB* promoter has been used to identify the loci regulating the circadian expression of genes (Miller et al., 1995), emphasize the plurality in the components of the signal transduction pathways that regulate the expression of *CAB*. It is essential that the gene products encoded by these loci are identified to determine their specific functions.

Transgenic plants bearing phytochrome genes have been largely constructed using monocot PHYA genes (Kay et al., 1989; Boylan and Quail, 1989, 1991; Keller et al., 1989). In all cases, the transgenic seedlings display an enhanced sensitivity to FR for inhibition of hypocotyl elongation (McCormac et al., 1992). In green plants, over-expression of phyA causes growth inhibition under low R/FR ratio, probably as a result of the retention of the FR-mediated high irradiance reaction (HIR) (McCormac et al., 1992).

Transgenic plants constructed using *PHYB* genes also display a short hypocotyl in *Arabidopsis* seedlings indicating that both phyA and phyB can influence the elongation process (Wagner et al., 1991). However, the over-expression of *PHYB* in transgenic *Arabidopsis* seeds caused very high dark germination levels, whereas phyB-deficient *hy3* seeds germinate poorly in darkness (McCormac et al., 1993).

Heyer et al. (1995) generated transgenic potato plants containing the homologous *PHYA* gene in both a sense and an antisense direction under a strong promoter. In both cases, the increase or decrease of phytochrome in plants did not display a strong effect on the growth and development of sprouts under white light. The phyA over-expressers showed enhanced leaf expansion and hook opening under, FR, which was delayed in phyA under-expresser plants. Interestingly, the sprouts with reduced phyA levels were also less sensitive to R with respect to stem extension and *SSU* expression. The increased level of phyA in transgenic plants reduced the shade-avoidance response in over-expressers.

The availability of genomic clones of *COP1* and *DET1* has permitted the construction of over-expressers for these genes. The constitutive expression of the *COP1* gene product results in a phenotype typical of shade avoiding plants. The shade avoidance phenotype observed in nature may, perhaps result from the lack of down-regulation of a gene product similar to *COP1* in plants grown under low R/FR ratio (McNellis et al., 1994). Some information about the role of *COP1* in light control of seedling development has been obtained by expressing mutant cDNA encoding the N-terminal 282 amino acids (N 282) of *COP1* in transgenic plants (McNellis et al., 1996). High-level expression of this fragment caused a phenotype similar to *cop1* mutants with hypersensitivity of hypocotyl elongation inhibition by white, BL, R, and FR and in dark N282 expression led to pleiotropic photomorphogenesis such as cotyledon development and gene expression.

Comparison of photoresponses in phyA and phyB mutants has yielded information about the role of these two phytochrome species in plant development. One of the obvious observations is that phyB functions to perceive R and phyA functions to detect FR in the process of photo-inhibition of hypocotyl elongation (Reed et al., 1994). However, their actions are not discrete and mutant plants missing either of these phytochromes survive and complete their life cycle.

The photoperiodic control of flowering is a response where phytochrome acts as a photoreceptor. The flowering response in mutants and transgenic plants of *Arabidopsis* was compared with wild type to investigate the role of different phytochrome species in the process. *Arabidopsis* is a long-day plant and lack of phyA delays flowering (Johnson et al., 1994). Apparently, phyA functions to detect the daylength in *Arabidopsis* and mutations in *PHYA* lead to a delay in flowering. In contrast, phyB-deficient mutants flower somewhat earlier (Reed et al., 1993). Since *PHY* mutations also affect the overall growth and morphological responses of plants, the altered flowering behavior may be a consequence of these changes. A critical evaluation of flowering behavior of mutants and transgenic seedlings showed that the EODFR response is similar in both *phyA* and *phyB* mutants and wild type (Bagnall et al., 1995). The FR exposure promoted flowering and R exposure nullified the FR effect. In transgenic over-expressers of phyA the response was similar to the mutant, but in phyB over-expressers EODR promoted flowering and EODFR inhibited flowering. In essence, these results indicate that neither phyA nor phyB alone is sufficient to regulate flowering, in fact the *phyA,phyB* double null mutants also respond to EODFR treatments by flowering early, suggesting that this response is also controlled by other phytochrome species (Devlin et al., 1996). Ahmad and Cashmore (1996) have isolated several early-flowering mutants of *Arabidopsis*, one such mutant *pef1* was insensitive to both R and FR in inhibition of hypocotyl elongation response. The *pef1* mutant seedlings possessed normal levels of phyA, but were defective in both phyA and phyB-mediated signaling pathways.

The *phyB* mutants show a phenotype reminiscent of that due to the shade-avoidance reaction indicating that phyB serves as a mechanism for detecting vegetational shade (López-Juez et al., 1992). Using mutants it has also been shown that phyB regulates germination of *Arabidopsis* seeds in darkness and lack of it abolishes dark germination in *hy3* mutants (Shinomura et al., 1994). Light-mediated cotyledon expansion is also inhibited in *phyB*-deficient *Arabidopsis* seedlings indicating a role of phyB in this process (Neff and van Volkenburg, 1994).

Studies on "twilight-inducible" expression of a plant homeobox gene (*ATHB-2*) revealed that a phytochrome species other than phyA and phyB regulates this phenomena. Photoregulation of the *ATHB-2* gene is uniquely mediated by changes in the R/FR ratio, which normally occur at dawn and dusk or under canopy. *ATHB-2* plays a role in mediating cell elongation, since this gene is expressed at high levels in etiolated seedlings and R or FR inhibition of seedling elongation also down-regulates the expression of *ATHB-2* mRNA (Carabelli et al., 1996).

In general, the information obtained by the use of mutants indicates the plurality in action of several photoreceptors with a certain amount of built-in redundancy. The fact that most photoreceptor mutants are able to survive and complete their life cycle indicates that there is a great degree of overlap between the functions of several species. However, this redundancy in the laboratory does not mean they are redundant in the natural environment. To date, few mutants defective in components of the signal transduction chain are available.

9. Future Direction

The use of mutants and transgenic plants has provided valuable information about functional sites of photoreceptor molecules and has led to the identification of likely components of their signal transduction chains. The identity of three important groups of photoreceptors: UV-B, UV-A and BL photoreceptors has to be conclusively established. For the first two, very little is known about their molecular identity, but one of the BL photoreceptor genes has been cloned. The use of appropriately designed transgenic mutants can enable screens to be designed for putative photoreceptor mutants. For example, due to the deleterious nature of UV-B it is not possible to isolated mutants which have a lesion in the UV-B-inducible genes. Nevertheless, using a UV-B sensitive promoter hooked with an appropriate reporter gene, it should be possible to screen for transgenic mutants plants which are deficient in perception and transmission of UV-B. A similar approach can also be used for screening for BL receptor mutants using a BL-sensitive promoter sequence. Isolation of mutants defective in other phytochrome species is also needed to analyze the functions of these phytochromes. Since the gene sequences for these phytochromes are known, it should be possible to make transgenic plants bearing *PHYC, D* and *E* genes in the antisense direction.

An important question is; how the information gathered using mutants and transgenics can be exploited to improve the yield of selected crop plants? The phenomenon of proximity perception to detect neighboring plants, and of shade avoidance in dense canopy reduces crop yield. Recently it has been demonstrated that plants over-expressing phyA can inhibit proximity-conditional dwarfing in the field situation, providing direct evidence for improved crop performance at high crop densities (Robson et al., 1996). Environmental regulation of flowering is a potential area where a great amount of information can still be obtained using the molecular-genetic approach. Understanding of these mechanisms may enable us to tailor crops that can be grown more densely without reducing yield. Regulation of flowering is also of great importance in the floriculture industry, and a clear understanding of the role of phytochrome would allow precise regulation of flowering in ornamental species. The imminent danger of an increase in UV-B radiation on loss of crop productivity is also recognized, and an understanding of its action and the molecular nature of its photoreceptor, would be highly useful to devise means to protect plants against its damaging effect.

References

Adamse, P., Jaspers, P.A.M.P., Bakker, J.A., Wesseius, J.C., Herringa, G.H., Kendrick, R.E. and Koornneef, M. 1988. Photophysiology of tomato mutant deficient in labile phytochrome. *J. Plant Physiol.* **133**: 436–440.

Ahmad, M. and Cashmore, A.R. 1993. *HY4* gene of *Arabidopsis thaliana* encodes a protein with characteristics of a blue-light photoreceptor. *Nature* **366**: 162–166.

Ahmad, M. and Cashmore, A.R. 1996. The pef mutants of *Arabidopsis thaliana* define lesions early in the phytochrome signaling pathway. *Plant J.* **10:** 1103–1110.

Ahmad, M. and Cashmore, A.R. 1997. The blue-light receptor cryptochrome 1 shows functional dependence on phytochrome A or phytochrome B in *Arabidopsis thaliana. Plant J.* **11**: 421–427.

Bagnall, D.I., King, R.W., Whitelam, G.C., Boylan, M.T., Wanger, D. and Quail, P.H. 1995. Flowering responses to altered expression of phytochrome in mutants and transgenic lines of *Arabidopsis thaliana* (L.) Heynh. *Plant Physiol.* **108:** 1495–1503.

Ballaré, C.L., Barnes, P.W. and Flint, S.D. 1995. Inhibition of hypocotyl elongation by ultraviolet-B radiation in de-etiolating tomato seedlings I. The photoreceptor. *Physiol. Plant.* **93:** 584–592.

Batschauer, A., Rocholl, M., Kaiser, T., Nagatani, A., Furuya, M. and Schäfer, E. 1996. Blue and UV-A light-regulated CHS Expression in *Arabidopsis* independent of phytochrome A and phytochrome B. *Plant J.* **9:** 63–69.

Backer, T.W., Foyer, C. and Caboche, M. 1992. Light-regulated expression of the nitrate-reductase and nitrite-reductase genes in tomato and in the phytochrome-deficient *aurea* mutant of tomato. *Planta* **188:** 39–47.

Bowler, C., Neuhaus, G., Yamagata, H. and Chua, N.H. 1994. Cyclic GMP and calcium mediated phytochrome phototransduction. *Cell* **77:** 73–81.

Boylan, M., Douglas, N. and Quail, P.H. 1994. Dominant negative suppression of *Arabidopsis* photoresponses by mutant phytochrome A sequences identifies spatially discrete regulatory domains in the photoreceptor. *Plant Cell* **6:** 449-460.

Boylan, M.T. and Quail, P.H. 1989. Oat phytochrome is biologically active in transgenic tomatoes. *Plant Cell* **1:** 765–773.

Boylan, M.T. and Quail, P.H. 1991. Phytochrome A over-expression inhibits hypocotyl elongation in transgenic *Arabidopsis. Proc. Nat. Acad. Sci. USA* **88:** 10806–10810.

Britt, A.B., Chen, J.J., Wykoff, D. and Mitchell, D. 1993. A UV-sensitive mutant of *Arabidopsis thaliana* defective in repair of pyrimidine (6–4) pyrimidinone dimers. Science **261:** 1571–1574.

Carabelli, M., Morelli, G., Whitelam, G. and Ruberti, I. 1996. Twilight-zone and canopy shade induction of *Athb-2* homeobox gene in green plants. *Proc. Natl. Acad. Sci. USA* **93:** 3530–3535.

Casal, J.J. and Kendrick, R.E. 1993. Impaired phytochrome-mediated shade-avoidance responses in the *aurea* mutant of tomato. *Plant Cell Environ.* **16:** 703–710.

Castle, L.A. and Meinke, D.W. 1994. A *FUSCA* gene of *Arabidopsis* encodes a novel protein essential for plant development. *Plant Cell* **6:** 25–41.

Chandok, M.R. and Sopory, S.K. 1996. Phosphorylation/dephosphorylation steps are key events in the phytochrome mediated enhancement of nitrate reductase mRNA levels and enzyme activity in maize. *Mol. Gen. Genet.* **25:** 599–608.

Cherry, J.R., Hondred, D., Walker, J.M., Keller, J.M., Hershey, H.P. and Vierstra, R.D. 1993. Carboxy-terminal deletion analysis of oat phytochrome A reveals the presence of separate domains required for structure and biological activity. *Plant Cell* **5:** 565–575.

Cherry, J.R., Hondred, D., Walker, J.M. and Vierstra, R.D. 1992. Phytochrome requires the 6-kD N-terminal domain for full biological activity. *Proc. Natl. Acad. Sci. USA* **89:** 5039–5043.

Childs, K.L., Miller, F.R., Cordonnier-Pratt, M.M., Pratt, L.H., Morgan, P.W. and Mullet, J.E. 1997. The sorghum photoperiod sensitivity gene, *ma3*, encodes a phytochrome B. *Plant Physiol.* **113:** 611–619.

Childs, K.L., Cordonnier-Pratt, M.M., Pratt, L.H. and Morgan, P.W. 1992. Genetic regulation of *Sorghum bicolor.* V II *ma3R* flowering mutant lacks a phytochrome that predominates in green tissue. *Plant Physiol.* **99:** 765–770.

Chory, J. 1992. A genetic model for light-regulated seedling development in *Arabidopsis. Development* **115:** 337–357.

Chory, J., Peto, C., Feinbaum, R., Pratt, L.H. and Ausubel, F. 1989. *Arabidopsis thaliana* mutant that develops as a light-grown plant in the absence of light. *Cell* **58:** 991–999.

Clack, T., Mathews, S. and Sharrock, R.A. 1994. The phytochrome apoprotein family in *Arabidopsis* in encoded by five genes: The sequences and expression of PHYD and PHYE. *Plant Mol. Biol.* **25:** 413–427.

Dehesh, K., Franci, C., Parks, B.M., Seely, K.A., Short, T.W., Tepperman, J.M. and Quail, P.H. 1993. *Arabidopsis* HY8 locus encodes phytochrome A. *Plant Cell* **5:** 1081–1088.

Deng, W. 1994. Fresh view of light signal transduction in plants. *Cell* **76:** 423–426.

Deng, X.-W., Caspar, T. and Quail. P.H. 1991. COP1: a regulatory locus involved in light-controlled development and gene expression in *Arabidopsis. Genes Dev.* **5:** 1172–1182.

Deng, X.-W., Matsui, M., Wei, N., Wagner, D., Chu, A.M., Feldmann, K.A. and Quail, P.H. 1992. COP1 an *Arabidopsis* regulatory gene encodes a protein with both a zinc-binding motif and a Gβ homologous domain. *Cell* **71:** 791–801.

Devlin, P.F., Halliday, K.J., Harberd, N.P. and Whitelam, G.C. 1996. The rosette habit of *Arabidopsis thaliana* is dependent upon phytochrome action-novel phytochromes control internode elongation and flowering time. *Plant J.* **10:** 1127–1134.

Devlin, P.F., Rood, S.B., Somers, D.E., Quail, P.H. and Whitelam, G.C. 1992. Photophysiology of the elongated internode (*ein*) mutant of *Brassica napus.Plant Physiol.* **100**: 1442–1447.

Dijkwel, P.P., Hüijser, C., Weisbeek, P.J., Chua, N.H. and Smeekens, S.C.M. 1997. Sucrose control of phytochrome A signaling in *Arabidopsis. Plant Cell* **9:** 583–595.

Ecker, J.R. 1995. The ethylene signal transduction pathway in plants. *Science* **268:** 667–675.

Elich, T.D., McDonagh, A.F., Palma, L.A. and Lagarias, J.C. 1989. Phytochrome chromophore biosynthesis. Treatment of tetrapyrrole-deficient *Avena* explant with natural and non-natural bilitrienes leads to formation of spectrally active holoproteins. *J. Biol. Chem.* **264:** 183–189.

Elich, T.D. and Lagarias, J.C. 1988. 4-Amino-5-hexynoic acid a potent inhibitor of tetrapyrrole biosynthesis in plants. *Plant Physiol.* **88:** 747–751.

Frances, S., White, M.J., Edgerton, M.D., Jones, A.M., Elliot, R.C. and Thompson, W.F. 1992. Initial characterization of a pea mutant with light-independent photomorphogenesis. *Plant Cell* **4:** 1519–1530.

Furuya, M. 1993. Phytochromes: their molecular species, gene families and functions. *Annu. Rev. Plant Physiol. Plant Mol. Biol.* **44:** 617–645.

Gardner, G. and Gorton, H.L. 1985. Inhibition of phytochrome synthesis by gabaculine. *Plant Physiol.* **77:** 540–543.

Goto, N., Yamamoto, K.T. and Watanabe, M. 1993. Action spectra for inhibition of hypocotyl growth of wild-type plant and of the *hy2* long hypocotyl muntants of *Arabidopsis thaliana* L. *Photochem. Photobiol.* **57:** 867–871.

Goud, K.V. and Sharma, R. 1994. Retention of photoinduction of cytosolic enzymes in *aurea* mutant of tomato. *Plant Physiol.* **105**: 643–650.

Goud, K.V., Sharma, R., Kendrick, R.E. and Furuya, M. 1991. Photoregulation of phenylalanine ammonia lyase is not correlated with anthocyanin induction in photomorphogenetic mutants of tomato (*Lycopersicon esculentum*). *Plant Cell Physiol.* **32:** 1251–1258.

Harlow, G.R., Jenkins, M.E., Pittalwala, T.S. and Mont, D.W. 1994. Isolation of *uvh1* an *Arabidopsis* mutant hypersensitive to ultraviolet light and ionizing radiation. *Plant Cell* **6:** 227–235.

Heyer, A.G., Mozley, D., Landscütze, V., Thomas, B. and Gatz, C. 1995. Function of phytochrome A in potato plants as revealed through the study of transgenic plants. *Plant Physiol.* **109:** 53–61.

Janoudi, A.K., Gordon, W.R., Wagner, D., Quail, P.H. and Poff, K.L. 1997. Multiple phytochromes are involved in red-light-induced enhancement of first-positive phototropism in *Arabidopsis thaliana. Plant Physiol.* **113:** 975–979.

Jenkins, G.I., Chrisite, J.M., Fuglevand, G., Long, J.C. and Jackson, J.A. 1995. Plant responses to UV and blue light. *Plant Science* **112:** 117–138.

Johnson, E., Harberd, N.P. and Whitelam, G.C. 1994. Photoresponses of light-grown phyA mutants of *Arabidopsis.* Phytochrome A is required for the perception of day length extensions. *Plant Physiol.* **105:** 141–149.

Jordan, E.T., Cherry, J.R., Walker, J.M. and Vierstra, R.D. 1996. The amino terminus of phytochrome A contains two distinct functional domain. *Plant Journal* **9:** 243–257.

Jordon, E.T., Hatfield, P.M., Hondred, D., Talon, M., Zeevart, J.A.D. and Vierstra, R.D. 1995. Phytochrome A over-expression in transgenic tobacco. Correlation of dwarf phenotype with high concentration of phytochrome in vascular tissues and attenuated gibberellin levels. *Plant Physiol.* **107**: 797–805.

Kaufmann, L.S. 1993. Transduction of blue-light signals. *Plant Physiol.* **102:** 333–337.

Kay, S.A., Nagatani, A., Keith, B., Deak, M., Furuya, M. and Chua, N.H. 1989. Rice phytochrome is biologically active in transgenic tobacco. *Plant Cell* **1:** 775–782.

Keller, J., Shanklin, J., Vierstra, R.D. and Hershey, H.P. 1989. Expression of a functional monocotyledonous phytochrome in transgenic tobacco. *EMBO J.* **8:** 1005–1012.

Kendrick, R.E. and Kronenberg, G.H.M. 1994. Photomorphogenesis in plants. Kluwer Academic publishers, Dordrecht, The Netherlands.

Kendrick, R.E., Kerckhoffs, L.H.J., Pundsnes, S., Van Tuinen, A., Koornneef, M., Nagatani, A., Terry, M.J., Tretyn, A., Cordonnier-Pratt, M.M., Hauser, B. and Pratt, L.H. 1994. Photomorphogenic mutants of tomato. *Euphytica* **79**: 227–234.

Khurana, J.P. and Poff, K.L. 1989. Mutants of *Arabidopsis thaliana* with altered phototropism. *Planta* **178:** 400–406.

Khurana, J.P., Ren, Z., Steinitz, B., Parks, B., Best, T.R. and Poff, K.L. 1989. Mutants of *Arabidopsis thaliana. Mol. Gen. Genet.* **225:** 468–473.

Kim, B.C., Soh, M.S., Kang, B.J., Furuya, M. and Nam, H.G. 1996. Two dominant photomorphogenic mutations of *Arabidopsis thaliana* identified as supressor mutation of *hy2. Plant Journal* **9:** 441–456.

Koornneef, M., Rolff, E. and Spruit, C.J.P. 1980. Genetic control of light-inhibited hypocotyl elongation in *Arabidopsis thaliana* L. Heynh *Z Pflanzenphysiol.* **100:** 147–160.

Koornneef, M., Cone, J.W., Dekens, R.G., Herne-Robers, E.G., Spruit, C.J.P. and Kendrick, R.E. 1985. Photomorphogenic responses of long hypocotyl mutants of tomato. *J. Plant Physiol.* **120:** 153–165.

Kraepiel Y., Julien, M., Cordonnier-Pratt, M.M., and Pratt, L.H. 1994. Identification of two loci involved in phytochrome expression in *Nicotiana plumboginifolia* and lethality of the corresponding double mutant. *Mol. Gen. Genet.* **242:** 559–565.

Kunkel, T., Neuthaus, G., Batschaeur, A., Chua, N.H. and Schäfer, E. 1996. Functional analysis of yeast-derived phytochrome A and B phycocyanobilin *adducts. Plant Journal* **10:** 625–636.

Kwok, S.F., Piekos, B., Misera, S. and Deng, X.-W. 1996. A complement of ten essential and pleiotropic *Arabidopsis COPT/DET/FUS* genes is necessary for repression of photomorphogenesis in darkness. *Plant Physiol.* **110**: 731–742.

Lehman, A., Black, R. and Ecker, J. 1996. *HOOKLESS1*, an ethylene response gene is required for differential cell elongation in *Arabidopsis* hypocotyl. *Cell* **85:** 183–194.

Li, H.M., Culligan, K., Dixon, R.A. and Chory, J. 1995. *CUE1*: A mesophyll cell-specific positive regulator of light-controlled gene expression in *Arabidopsis. Plant Cell* **7:** 1599–1610.

Li, H.M., Altschmied, L. and Chory, J. 1994. *Arabidopsis* mutants define downstream branches in the phototransduction pathway. *Genes Dev.* **8:** 339–349.

Li, J., Ou-Lee, T.M., Raba, R., Amundson, R.G. and Last, R.L. 1993. *Arabidopsis* flavonoid mutants are hypersensitive to UV radiation. *Plant Cell* **5:** 171–179.

Li, J.M., Biswas, M.G., Chao, A., Russell, D.W. and Chory, J. 1997. Conservation of function between mammalian and plant steroid 5-alpha-reductases *Proc. Nat. Acad. Sci. USA* **94:** 3554–3559.

Lin, C., Ahmed, M., Gordon, D. and Cashmore, A. 1995. Expression of an *Arabidopsis* cryptochrome gene in transgenic tobacco results in hypersensitivity to blue, UV-A and green light. *Proc. Nat. Acad. Sci. USA* **92:** 8423–8427.

Liscum, E., Young, J.C., Poff, K.L. and Hangarter, R.P. 1992. Genetic separation of phototropism and blue light inhibition of stem elongation. *Plant Physiol.* **100:** 267–271.

Liscum, E. and Hangarter, R.P. 1991. *Arabidopsis* mutants lacking blue light-dependent inhibition of hypocotyl elongation. *Plant Cell* **3:** 685–694.

Liscum, E. and Hangarter, R.P. 1993a. Genetic evidence that the red-absorbing form of phytochrome B modulates gravitropism in *Arabidopsis thaliana. Plant Physiol.* **103:** 15–19.

Liscum, E. and Hangarter, R.P. 1993b. Light-stimulated apical hook opening in wild-type *Arabidopsis thaliana* seedlings. *Plant Physiol.* **101**: 567–572.

Liscum, E. and Hangarter, R.P. 1993c. Photomorphogenic mutants of *Arabidopsis thaliana* reveal activities of multiple photosensory system during light-stimulated apical hook opening. *Planta* **191**: 214–221.

Liscum, E. and Hangarter, R.P. 1994. Mutational analysis of blue-light sensing in *Arabidopsis. Plant Cell Environ.* **17**: 639–648.

Liscum, E. and Briggs, W.R. 1995. Mutational in the *NPH1* locus of *Arabidopsis* disrupt the perception of phototropic stimuli. *Plant Cell* **7:** 473–485.

López-Juez, E., Nagatani, A., Tomizawa, K.I., Deak, M., Kern, R., Kendrick, R.E. and Furuya, M. 1992. The cucumber long hypocotyl mutant lacks a light stable PHYB like protein. *Plant Cell* **4:** 241–251.

López-Juez, E., Nagatani, A., Buurmeier, W.F., Peters, J.L., Furuya, M., Kendrick, R.E. and Wesselius, J.C. 1990. Responses of light-grown wild type and *aurea* mutant tomato plants to end of day far-red light. *J. Photochem. Photobiol.* **4:** 391–405.

McCormac, A.C., Whitelam, G.C. and Smith, H. 1992. Light-grown plants of transgenic tobacco expressing an introduced oat phytochrome a gene under the control of constitutive viral promoter exhibit persistent growth inhibition by far-red light. *Planta* **188:** 173–181.

McCormac, A.C., Smith, H. and Whitelam, G.C. 1993. Photoregulation of germination in seed of transgenic lines of tobacco and *Arabidopsis* which express an introduced cDNA encoding phytochrome A or phytochrome B. *Planta* **191:** 386–393.

McNellis, T.W., Torii, K.U. and Deng, X.-W. 1996. Expression an N-terminal fragment of cop1 confers a dominant-negative effect on light-regulated seedling development in *Arabidopsis. Plant Cell* **8:** 1491–1503.

McNellis, T.W., Von Arnim, A.G. and Deng, X.-W. 1994. Over-expression of *Arabidopsis* COP1 results in partial suppression of light-mediated development: Evidence for a light-inactivable repressor of photomorphogenesis. *Plant Cell* **6:** 1391–1400.

Miller, A.J., Straume, M., Chory, J., Chua, N.H. and Kay, S.A. 1995. The regulation of circadian period by phototransduction pathways in *Arabidopsis*. *Science* **267:** 1163–1166.

Misera, S., Miller, A.J., Weiland-Heidecker, U. and Jurgens, G. 1994. The *FUSCA* genes of *Arabidopsis*: Negative regulation of light responses. *Mol. Gen. Genet.* **244:** 242–252.

Mummert, E., Grimm, R., Speth, V., Eckerson, C., Schitz, E., Gatenby, A.A. and Schäfer, E. 1993. A TCP1-related chaperone from plants refolds phytochrome to its photoreversible form. *Nature* **363**: 644–648.

Nagatani, A., Chory, J. and Furuya, M. 1991a. Phytochrome B is not detectable in the *hy3* mutant of *Arabidopsis* which is deficient in responding to end of day far-red light treatments. *Plant Cell Physiol.* **32:** 1119–1122.

Nagatani, A., Reed, J.W. and Chory, J. 1993. Isolation and initial characterization of *Arabidopsis* mutants that are deficient in phytochrome *A*. *Plant Physiol.* **102:** 269–277.

Nagatani, A., Kay, S.A., Deak, M., Chua, N.-H. and Furuya, M. 1991b. Rice type 1 phytochrome regulates hypocotyl elongation in transgenic tobacco seedlings. *Proc. Nat. Acad. Sci. USA* **88:** 5207–5211.

Neff, M.M. and van Volkenburgh, E. 1994. Light-stimulated cotyledon expansion in *Arabidopsis* seedlings. The role of phytochrome B. *Plant Physiol.* **104:** 1027–1032.

Neuhaus, G., Bowler, C., Kern, R. and Chua, N.-H. 1993. Calcium/calmodulin dependent and independent phytochrome signal transduction pathways. *Cell* **73:** 937–952.

Oelmüller, R. and Kendrick, R.E. 1991. Blue light is required for survival of the tomato phytochrome deficient *aurea* mutant and the expression of four nuclear genes coding for plastidic proteins. *Plant Mol. Biol.* **16:** 293–299.

Oelmüller, R., Kendrick, R.E. and Briggs, W.R. 1989. Blue light mediated accumulation of nuclear encoded transcripts coding for proteins of the thylakoid membrane is absent in the phytochrome-deficient *aurea* mutant of tomato. *Plant Mol. Biol.* **13:** 223–232.

Okada, K. and Shimura, Y. 1992. Mutational analysis of root gravitropism and phototropism of *Arabidopsis thaliana* seedlings. *Aust. J. Plant Physiol.* **19:** 439–448.

Parks, B.M., Jones, A.M., Adamse, P., Koornneef, M., Kendrick, R.E. and Quail, P.H. 1987. The *aurea* mutant of tomato is deficient in spectrophotometrically and immunochemically detectable phytochrome. *Plant Mol. Biol.* **9:** 97–107.

Parks, B.M., Shanhklin, J., Koornneef, M., Kendrick, R.E. and Quail, P.H. 1989. Immunochemically detectable phytochrome is present at normal levels but is photochemically non-functional in the *hy1* and *hy2* long hypocotyl mutants of *Arabidopsis*. *Plant Mol. Biol.* **12:** 425–437.

Parks, B.M. and Quail, P.H. 1991. Phytochrome-deficient *hy1* and *hy2* long hypocotyl mutants of *Arabidopsis* are defective in phytochrome chromophore biosynthesis. *Plant Cell* **3:** 1177–1186.

Parks, B.M. and Quail, P.H. 1993. *hy8* a new class of *Arabidopsis* long hypocotyl mutants deficient in functional phytochrome A. *Plant Cell* **5:** 39–48.

Pepper, A., Delancy, T., Washburn, T., Poole, D. and Chory, J. 1994. DET1 a negative regulator of light-mediated development and gene expression in *Arabidopsis* encodes a novel nuclear-localized protein. *Cell* **78:** 109–116.

Pepper, A.E. and Chory, J. 1997. Extragenic suppressors of the *Arabidopsis det1* mutant identify elements of flowering-time and light-response regulatory pathways. *Genetics* **145:** 1125–1137.

Peters, J.L., Van Tuinen, A., Adamse, P., Kendrick, R.E. and Koornneef, M. 1989. High pigment mutants of tomato exhibit high sensitivity for phytochrome action. *J Plant Physiol.* **134:** 661–666.

Peters, J.L., Schreuder, M.E.L., Verduin, S.J.W. and Kendrick, R.E. 1992. Physiological characterization of a high-pigment mutant of tomato. *Photochem Photobiol.* **56:** 75–82.

Pratt, L.H. 1995. Phytochromes: Differential properties expression patterns and molecular evolution. *Photochem Photobiol.* **61:** 10–21.

Quail, P.H. 1991. Phytochrome: A light activated molecular switch that regulates plant gene expression. *Ann. Rev. Genet.* **25:** 389–409.

Quail, P.H., Boylan, M.T., Parks B.M., Short, T.W., Xu, Y. and Wagner, D. 1995. Phytochromes: Photosensory perception and signal transduction. *Science* **268**: 675–680.

Raghuram, N. and Sopory, S.K. 1995. Evidence for some common signal transduction events for opposite regulation on nitrate reductase and phytochrome-1 gene expression by light. *Plant Mol. Biol.* **29:** 25–35.

Reed, J.W., Nagatani, A., Elich, T.D., Fagan, M. and Chory, J. 1994. Phytochrome A and Phytochrome B have overlapping but distinct functions in *Arabidopsis* development. *Plant Physiol.* **104:** 1139–1149.

Reed, J.W., Nagpal, P., Poole, D.S., Furuya, M. and Chory, J. 1993. Mutations in the gene for the red/far-red light receptor phytochrome B alter cell elongation and physiological responses throughout *Arabidopsis* development. *Plant Cell* **5:** 147–157.

Reymond, P., Short, T.W. and Briggs, W.R. 1992a. Blue light activates a specific protein kinase in higher plants. *Plant Physiol.* **100:** 655–661.

Reymond, P., Short, T.W., Briggs, W.R. and Poff, K.L. 1992b. Light induced phosphorylation of a membrane protein plays an early role in signal transduction for phototropism in *Arabidopsis thaliana. Proc. Natl. Acad. Sci. USA* **89**: 4718–4721.

Robson, P.R.H., McCromac, A.C., Irvine, A.S. and Smith, H. 1996. Genetic engineering of harvest index in tobacco through overexpression of phytochrome gene. *Nature Biotechnology* **14:** 995–998.

Sharma, R., López-Juez, E., Nagatani, A. and Furuya, M. 1993. Identification of photo-active phytochrome A in etiolated seedlings and photo-active phytochrome B in green leaves of *aurea* mutant of tomato. *Plant Journal* **4:** 1035–1042.

Shinomura, T., Nagatani, A., Chory, J. and Furuya, M. 1994. The induction of seed germination in *Arabidopsis thaliana* is regulated by phytochrome B and secondarily by Phytochrome A. *Plant Physiol.* **104:** 363–371.

Short, T.W., Porst, M. and Briggs, W.R. 1992. A photoreceptor system regulating *in vivo* and *in vitro* phosphorylation of a pea plasma membrane protein. *Photochem. Photobiol.* **55:** 773–781.

Short, T.W. and Briggs, W.R. 1994. The transduction of blue light signals in higher plants. *Ann. Rev. Plant Physiol. Plant Mol. Biol.* **45:** 143–171.

Somers, D.E., Sharrock, R.A., Tepperman, J.M. and Quail, P.H. 1991. The *hy3* long hypocotyl mutant of *Arabidopsis* deficient in phytochrome B. *Plant Cell* **3:** 1263–1274.

Stockhaus, J., Nagatani, A., Halfter, U., Kay, S., Furuya, M. and Chua, N.-H. 1992. Serine-to-Alanine substitutions at the amino terminal region of phytochrome A result in an increase in biological activity. *Genes Dev.* **6:** 2364–2372.

Susek, R.E., Ausubel, F.M. and Chory, J. 1993. Signal transduction mutants of *Arabidopsis* uncouple nuclear *CAB* and *RBCS gene* expression from chloroplast development. *Cell* **74:** 787–799.

Terry, M.J., Waleithner, J.A. and Lagarias, J.C. 1993. Biosynthesis of the plant photoreceptor phytochrome. *Arch. Biochem. Biophys.* **306:** 1–15.

Terry, M.J. and Kendrick, R.E. 1996. The *aurea* and *yellow-green mutants* of tomato are deficient in phytochrome chromophore biosynthesis. *J. Biol. Chem.* **271:** 21681–2168.

Van Tuinen, A., Hanhart, C.J., Kerckhoffs, L.H.J., Nagatani, A., Boylan, M.T., Quail, P.H., Kendrick, R.E. and Koornneef, M. 1996. Analysis of phytochrome-deficient *yellow-green* and *aurea* mutants of tomato. *Plant Journal* **9:** 173–182.

Van Tuinen, A., Kerckhoffs, L.H.J., Nagatani, A., Kendrick, R.E. and Koornneef, M. 1995a. Far-red light-insensitive mutants of tomato. *Mol. Gen. Genet.* **246:** 133–141.

Van Tuninen, A., Kerckhoffs, L.H.J., Nagatani, A., Kendrick, R.E. and Koornneef, M. 1995b. A temporarily red light-insensitive mutant of tomato lacks a light stable, B-like phytochrome. *Plant Physiol,* **108:** 939–947.

Van Wann, E. 1995. Reduced plant growth in tomato mutants *high pigment* and *dark green* partially overcome by gibberellin. *Hort. Science* **30:** 379.

Von Arnim, A.G. and Deng, X.-W. 1994. Light inactivation of *Arabidopsis* photomorphogenic repressor COP1 involves a cell-specific regulation of its nucleo-cytoplasmic partitioning. *Cell* **79:** 1035–1045.

Wagner, D., Fairchild, C.D., Kuhn, R.M. and Quail, P.H. 1996. Chromophore-bearing NH_2-terminal domains of phytochrome A and B determines their photosensory specificity and differential light-lability. *Proc. Nat. Acad. Sci. USA* **93:** 4011–4015.

Wagner, D., Tepperman, J.M. and Quail, P.H. 1991. Over-expression of phytochrome B induces a short hypocotyl phenotype in transgenic "*Arabidopsis*". *Plant Cell* **3:** 1275–1288.

Wagner, D., Koloszvari, M. and Quail, P.H. 1996. Two small spatially distinct regions of phytochrome B are required for efficient signaling rates. *Plant Cell* **8:** 859–871.

Wagner, D. and Quail, P.H. 1995. Mutational analysis of phytochrome B identifies a small COOH-terminal domain region critical for regulatory activity. *Proc. Nat. Acad. Sci. USA* **93:** 8590–8600.

Wei, N., Chamovitz, D.A. and Deng, X.-W. 1994. *Arabidopsis* COP9 is a component of a nopvel signaling complex mediating light control development. *Cell* **78:** 117–124.

Weller, J.L., Nagatani, A., Kendrick, R.E., Murfeit, I.E. and Reid, J.B. 1995. New lv mutants of pea are deficient in phytochrome B. *Plant Physiol.* **108:** 525–532.

Weller, J.L., Terry, M.J., Rammeau Reid, J.B. and Kendrick, R.E. 1996. The phytochrome deficient *pcd1* mutant of pea is unable to covert heme to biliverdin IXa. *Plant Cell* **8:** 55 67.

Whitelam, G.C., Johnson, E., Peng, J., Carol, P., Anderson, M.L., Cowl, J.S. and Harberd, N.P. 1993. Phytochrome A null mutants of *Arabidopsis* display a wild type phenotype in white light. *Plant Cell* **5:** 757–768.

Young, J.C., Liscum, E. and Hangarter, R.P. 1992. Spectral dependence of light-inhibited hypocotyl elongation in photomorphogenic mutants of *Arabidopsis*: Evidence for a UV-A photosensor. *Planta* **188:** 106–114.

Zimmerman, B.K. and Briggs, W.R. 1963. A kinetic model for phototropic responses of oat coleoptiles. *Plant Physiol.* **38:** 253–61.

Index

MIX
Papier aus verantwortungsvollen Quellen
Paper from responsible sources
FSC® C105338

Printed by Books on Demand, Germany